Optical Systems Engineering

About the Author

Keith J. Kasunic has more than 20 years of experience developing optical, electro-optical, and infrared systems. He holds a Ph.D. in Optical Sciences from the University of Arizona, an MS in Mechanical Engineering from Stanford University, and a BS in Mechanical Engineering from MIT. He has worked for or been a consultant to a number of organizations, including Lockheed Martin, Ball Aerospace, Sandia National Labs, Nortel Networks, and Bookham. He is also an Instructor and Graduate Faculty Scholar at Univ. of Central Florida's CREOL—The College of Optics and Photonics, as well as an Affiliate Instructor with Georgia Tech's Distance Learning and Professional Education.

Optical Systems Engineering

Keith J. Kasunic

Univ. of Arizona, College of Optical Sciences
Univ. of Central Florida, The College of Optics and Photonics
Georgia Tech, Distance Learning and Professional Education
Lockheed Martin Corp.

New York   Chicago   San Francisco
Lisbon   London   Madrid   Mexico City
Milan   New Delhi   San Juan
Seoul   Singapore   Sydney   Toronto

The **McGraw·Hill** Companies

Cataloging-in-Publication Data is on file with the Library of Congress

McGraw-Hill books are available at special quantity discounts to use as premiums and sales promotions or for use in corporate training programs. To contact a representative, please e-mail us at bulksales@mcgraw-hill.com.

Optical Systems Engineering

Copyright © 2011 by The McGraw-Hill Companies, Inc. All rights reserved. Printed in the United States of America. Except as permitted under the United States Copyright Act of 1976, no part of this publication may be reproduced or distributed in any form or by any means, or stored in a database or information retrieval system, without the prior written permission of the publisher.

1 2 3 4 5 6 7 8 9 0   DOC/DOC   1 9 8 7 6 5 4 3 2 1

ISBN 978-0-07-175440-8
MHID 0-07-175440-7

The pages within this book were printed on acid-free paper.

Sponsoring Editor	**Project Manager**	**Production Supervisor**
Michael Penn	Aloysius Raj	Richard C. Ruzycka
	Newgen Publishing	
Acquisitions Coordinator	and Data Services	**Composition**
Michael Mulcahy		Newgen Publishing
	Copy Editor	and Data Services
	Matt Darnell	
Editorial Supervisor		**Art Director, Cover**
David E. Fogarty	**Proofreader**	Jeff Weeks
	Kathrin Immanuel	
	Newgen Publishing	
	and Data Services	

Information contained in this work has been obtained by The McGraw-Hill Companies, Inc. ("McGraw-Hill") from sources believed to be reliable. However, neither McGraw-Hill nor its authors guarantee the accuracy or completeness of any information published herein, and neither McGraw-Hill nor its authors shall be responsible for any errors, omissions, or damages arising from the use of this information. This work is published with the understanding that McGraw-Hill and its authors are supplying information but are not attempting to render engineering or other professional services. If such services are required, the assistance of an appropriate professional should be sought.

For anyone who finds this book useful

Contents

Preface ix
Acknowledgments xiii

1. **Introduction** 1
   1.1 Optical Systems 4
   1.2 Optical Engineering 7
   1.3 Optical Systems Engineering 13
       Problems 20
       Notes and References 21

2. **Geometrical Optics** 23
   2.1 Imaging 25
   2.2 Field of View 29
   2.3 Relative Aperture 34
   2.4 Finite Conjugates 37
   2.5 Combinations of Lenses 38
   2.6 Ray Tracing 40
   2.7 Thick Lenses 44
   2.8 Stops, Pupils, and Windows 50
   2.9 Afocal Telescopes 56
       Problems 64
       Notes and References 66

3. **Aberrations and Image Quality** 67
   3.1 Aberrations 69
   3.2 Diffraction 113
   3.3 Image Quality 117
       Problems 129
       Notes and References 131

4. **Radiometry** 133
   4.1 Optical Transmission 135
   4.2 Irradiance 148
   4.3 Etendue, Radiance, and Intensity ... 151
   4.4 Conservation Laws 162
   4.5 Stray Light 167

Contents

		Problems	179
		Notes and References	181
5.	**Optical Sources**		**183**
	5.1	Source Types	186
	5.2	Systems Design	210
	5.3	Source Specifications	216
	5.4	Source Selection	227
		Problems	232
		Notes and References	234
6.	**Detectors and Focal Plane Arrays**		**237**
	6.1	Detector Types	241
	6.2	Focal Plane Arrays	259
	6.3	Signals, Noise, and Sensitivity	275
	6.4	Detector Specifications	306
	6.5	Detector Selection	327
		Problems	336
		Notes and References	338
7.	**Optomechanical Design**		**341**
	7.1	Fabrication	344
	7.2	Alignment	364
	7.3	Thermal Design	370
	7.4	Structural Design	387
	7.5	Component Specifications	403
		Problems	408
		Notes and References	410
	Index		**413**

Preface

> I want to solve a problem.... I want to see how I can clarify the issue in order to reach the solution. I try to see how the area is determined, how it is built up in this figure.... Instead, someone comes and tells me to do this or that; viz., something like $1/a$ or $1/b$ or $(a - b)$ or $(a - b)^2$, things which clearly have no inner relation to the issue.... Why do just this? I am told "Just do it"; and then another step is added, again ununderstandable in its direction. The steps drop from the blue; their content, their direction, the whole process...appears arbitrary, blind to the issue of how the area is built up.... In the end, the steps do lead to a correct, or even proved answer. But the very result is seen in a way that gives no insight, no clarification.
>
> Max Wertheimer, *Productive Thinking*

> The result is that people who have understood even the simplest, most trivial-sounding economic models are often far more sophisticated than people who know thousands of facts and hundreds of anecdotes, who can use plenty of big words, but have no coherent framework to organize their thoughts.
>
> Paul Krugman, *The Accidental Theorist*

This book is for anyone developing optical hardware. With many optical engineering texts already on the market with a similar theme, what is there about this book that distinguishes it from the others?

On the simplest level, there are few books that approach optical systems engineering as a unique field of knowledge. For example, the range of system engineering skills useful in industry—system architecture trades, feasibility studies, performance modeling, requirements analysis and flow-down, allocation of error budgets, subsystem and component specifications, tying together the interfaces between subsystems, and evaluating vendor progress to ensure performance of critical hardware—are rarely addressed. These skills all build on a strong understanding of optical engineering fundamentals but are quite different from the traditional testing and lens design aspects of optical engineering found in other books.

At the same time, *no* book can provide the intuitive eye-hand-mind connections required for deep learning and the "aha!" moments that come from the first light of understanding. This intuition requires a kinesthetically enhanced, learning-by-doing-and-thinking (*mens et manus*) approach. However, the tone, language, and style of a book strongly affect the "mind" part of this balance, and the ability to imagine new designs is often limited not by a lack of inherent creative ability but rather by *how* the material was learned and the resulting associations formed.

As a result, and as implied by the opening quotations from Wertheimer and Krugman, the pedagogical goal is a book that encourages independent thinking—not memorization, anecdotes, or algebraic "flute music." This book is by no means perfect in that regard, but considerable effort has been made to emphasize physical understanding more so than algebra. Complex algebraic derivations are *not* a useful engineering skill, yet they seem to have become the norm for many university lectures. This development does not stem from any profundity in their content; instead, it has arisen because the academic system generally doesn't reward time put into teaching. Status and money are usually awarded to those professors who excel at winning research grants, with the result that students—and the faculty who enjoy working with them—are often left behind.

In contrast to these trends, the focus of this book is on the practical aspects of optical systems engineering that are useful in industry. The most important of these is the use of physical reasoning to understand design trends. For example, is a bigger or smaller engine needed to pull a heavy load up a steep mountain? It's not necessary to use Newton's laws to answer this question; with a few assumptions about what is meant by "heavy" and "steep," our experience with cars and mountains allows us to answer it immediately. Similar questions can be asked of optical systems: Is a bigger or smaller aperture needed for better image quality? Does the answer depend on pixel size? What about the amount of light the system collects—does it depend on the size of the aperture, the size of the pixels, or both? Unfortunately, much engineering instruction abounds with examples of what is *not* physical thinking. Geometrical optics, for example, is sometimes taught as if it were nothing more than a mathematical game of manipulating chief and marginal rays. Such an approach is sure to turn away students who are new to the field and would interest only those of an accounting mindset—who are not likely to be in an engineering classroom in the first place.

All this is not to say that mathematics has no value in hardware development. On the contrary, if a concept hasn't been quantified then it is equivalent to "viewgraph engineering," from which no useful hardware has ever been built. The correct strategy is not to ignore the mathematics but rather to ensure that comprehension and physical reasoning *precede* analysis, after which rules of thumb and back-of-the-envelope calculations—which are correct to an order of magnitude—can be applied. Such systems-level feasibility analysis can then be followed up with the use of specialized, "back-of-the-elephant" design software for lens design, optical filters, stray light modeling, STOP analysis, and so forth.

For additional background, the field of optical engineering has been blessed with a number of excellent books. These include Jenkins and White's *Fundamentals of Optics*, Hecht's *Optics*, Smith's *Modern Optical Engineering*, Fischer's *Optical System Design*, Friedman and Miller's *Photonics Rules of Thumb*, and Hobbs's *Building Electro-Optical Systems*. Although not strictly a text on optics, Frank Crawford's *Waves* is a work of teaching genius. The only prerequisites for this book are either Hecht or Jenkins and White along with familiarity with Snell's law, the lens equation for simple imaging, and the concepts of wavelength and wavefronts.

Keith J. Kasunic
Boulder, Colorado

Acknowledgments

For the many colleagues I have had the opportunity to learn from over the years, including: Earl Aamodt, Dave Adams, Marc Adams, Jasenko Alagic, Tom Alley, Karamjeet Arya, Jason Auxier, Theresa Axenson, Ramen Bahuguna, James Battiato, Dave Begley, Jim Bergstrom, Ian Betty, Charles Bjork, Pete Black, Don Bolling, Glenn Boreman, Rob Boye, Bob Breault, Gene Campbell, Scott Campbell, Andrew Cheng, Ed Cheng, Art Cockrum, Jasmin Cote, Marie Cote, Tiffanie D'Alberto, Tom Davenport, Michael Dehring, Jack Doolittle, David Down, Ken Drake, Patrick Dumais, Bente Eegholm, Kyle Ferrio, Paul Forney, Fred Frohlich, John Futch, Phil Gatt, Gary Gerlach, Dave Giltner, Mark Goodnough, Mike Griffin, Ron Hadley, Charley Haggans, Pat Hamill, Rob Hearst, Jon Herlocker, Karin Hinzer, Terry Hoehn, Thomas Hoft, Mike Horn, Joan Howard, Ike Hsu, Bruce Jurcevich, Bob Kaliski, Josh Kann, Shanalyn Kemme, Peter Kondratko, Tim Koogle, Sarma Lakkaraju, Igor Landau, Patrick Langlois, Leo Laux, Arno Ledebuhr, Rob Ligon, Aseneth Lopez, Romain Majieko, Billy Maloof, Masud Mansuripur, Bob Manthy, Brian Marotta, Bob Marshalek, Mark McCall, Steve McClain, Steve Mechels, Bill Meersman, Tom Mirek, Rud Moe, Jerry Moloney, Phil Morris, Howard Morrow, Drew Nelson, Terry Nichols, Eric Novak, Matt Novak, Dan O'Connor, Mike O'Meara, Toby Orloff, Jim Palmer, George Paloczi, Michael Parker, Scott Penner, Kevin Peters, Nasser Peyghambarian, Jim Pinyan, Ron Plummer, Sierk Potting, Eric Ramberg, Andy Reddig, Brian Redman, Bruce Reed, Benoit Reid, Steve Rentz, Mike Rivera, Paul Robb, Tom Roberts, Clark Robinson, Roger Rose, Jeff Ruddock, Rich Russell, Sandalphon, Michael Scheer, Jeff Scott, Darwin Serkland, Alex Shepherd, Joe Shiefman, Russ Sibell, Yakov Sidorin, Dan Simon, Dave Stubbs, Charles Sullivan, Gene Tener, Jennifer Turner, Radek Uberna, Jim Valerio, Tim Valle, Allen Vawter, Bill Vermeer, Mark Von Bokern, Howard Waldman, Chanda Walker, Jon Walker, Steve Wallace, William Wan, Dan Welsh, Kenton White, Kevin Whiteaker, Dave Wick, Carl Wiemer, Stu Wiens, Phil Wilhelm, James Wong, Ian Woods, Ewan Wright, and Jim Wyant.

CHAPTER 1
Introduction

The telescope was almost as big as a school bus, the largest ever launched into space. It was also one of the most sophisticated optical systems ever built, capable of imaging galaxies near the beginning of the universe with a quality and sensitivity designed to exceed the best cameras used by professional photographers.

The engineers, scientists, technicians, and managers who worked on its development waited anxiously as results from its collection of "first light" started to come in. In the month since its release from the Shuttle bay into an orbit around the Earth (Fig. 1.1), the Hubble Space Telescope (HST) had gone through a number of performance checks and was now ready to collect images.

Elation greeted those first images to come in, but it was soon followed by skepticism from one of the astronomers evaluating the results. Instead of crisp, high-quality images of far-away galaxies, the pictures were fuzzy—blurred to the point where it was soon clear to even the untrained eye that something had gone wrong. Rather than the vision of a young fighter pilot, NASA's new telescope was acting middle-aged and needed glasses.

As the root cause of the problem was slowly uncovered, it became clear that a simple error in measuring the shape of the largest mirror had turned a complex, $1.5 billion project into one of the most embarrassing mistakes in the history of optical systems development. At great cost, corrective "glasses" were eventually installed on the Hubble, which enabled images of the type shown in Fig. 1.2. Unfortunately, the company responsible for the error could not also repair its reputation and eventually sold off its once-profitable optics division, the price to be paid for engineering mistakes on this scale.

There were numerous opportunities to uncover the error before the telescope was launched. The reason it was not discovered evidently stems from an incorrect view of systems engineering that continues to this day—namely, that working hardware can be produced simply by "checking the boxes" for completed tasks. Unfortunately, this perspective misses some critical questions: Have the tasks been correctly defined, and have they been *successfully*

FIGURE 1.1 The Hubble Space Telescope in orbit around the Earth. This photo was taken from the Space Shuttle. (*Photo credit: NASA, www.nasa.gov.*)

FIGURE 1.2 Hubble image of the M104 (Sombrero) Galaxy, located 30 million light years from Earth. The galaxy contains several hundred billion stars in a region 60,000 light years in diameter. (*Photo credit: NASA, www.nasa.gov.*)

completed? More importantly: Who is in a position to know? The NASA *Failure Report* described the cause in a similar manner:[9]

- "Although the telescope was recognized as a particular challenge, with a primary mirror requiring unprecedented performance, there was a surprising lack of optical experts with experience in the manufacture of large telescopes during the fabrication phase."
- "Fabrication of the HST mirror was the responsibility of the Optical Operations Division of Perkin-Elmer, which did not include optical design scientists and which did not use the skills external to the division which were available to Perkin-Elmer."
- "Perkin-Elmer line management did not review or supervise their Optical Operations Division adequately. In fact, the management structure provided a strong block against communication between the people actually doing the job and higher-level experts both within and outside Perkin-Elmer."
- "The Optical Operations Division at Perkin-Elmer operated in a 'closed door' environment which permitted discrepant data to be discounted without review."
- "The Perkin-Elmer Technical Advisory Group did not probe at all deeply into the optical manufacturing process . . . . This is particularly surprising since the members were aware of the history . . . where spherical aberration was known to be a common problem."
- "The quality assurance people at Perkin-Elmer . . . were not optical experts and, therefore, were not able to distinguish the presence of inconsistent data results from the optical tests."

While it's also easy to blame a "not invented here" (NIH) attitude for the lack of communication and cooperation, a more likely cause is short schedules and tight budgets. For example, the phrase "This is particularly surprising" that was used to describe the Perkin-Elmer Technical Advisory Group is a clue that project schedules took precedence over technical expertise. It's not difficult to imagine program managers, faced with a tight schedule and slowly realizing the complexity of the task, saying: "We don't have time for all this 'review' and 'oversight' stuff!"

The pressures of tight schedules can be overwhelming, but familiarity and experience with the tasks reduces the stress considerably. It seems that the "high-level experts" on the HST were so trained but that "the people actually doing the job" were not. Quality assurance people were singled out in the Hubble *Failure Report* as not being optical experts, but the same could probably be said of the technicians, assemblers, and machinists as well as a large fraction of the engineers.

Although not everyone can be an expert, the type of errors that occurred with the Hubble might have been avoided with appropriate education and training. Given such training, those who are doing the actual work form an effective backup system to counter weaknesses in the management structure. Such a system recognizes that mistakes can and will be made even by the most experienced and talented; this means that critical results should be verified and validated with a "second set of eyes."[10] When failures do occur, this backup system (the "second set of eyes") becomes the primary one. Although such bottom-up management has its own risks, at least there is a backup in place.

Looking beyond its initial failures, many other things on Hubble were done right. Thousands of images of the universe's incomprehensible complexity have been captured since corrective optics were installed during a dramatic in-orbit repair in 1993. With periodic hardware upgrades, Hubble is still operating more than 15 years later, serving as a platform for new instruments and collecting images more sophisticated than were ever thought possible at the beginning of its life.[8]

Despite the lessons learned from Hubble, decisions made during the development of new optical systems continue to be misguided. A more recent example was covered in depth by the *New York Times* ("Death of a Spy Satellite Program," 11 November 2007), where it was disclosed that a major spy satellite program was unable to deliver the performance promised by the aerospace contractor.[11] As shown in Chap. 7, one cause was—even more directly in this case—inadequate education, training, and experience in the development of optical systems.

1.1 Optical Systems

Optical systems can usually be classified as being one of five different types: imagers, radiometers, interferometers, spectrometers, and polarimeters. Without even knowing the meaning of these words, we can also combine them into "imaging radiometers," "imaging spectrometers," "spectropolarimeters," and so on, but the basic concepts remain the same. Complete mastery of optical systems engineering requires knowledge of all these types; however, the focus of this introductory book is on imagers and radiometers, which are prerequisites for understanding more advanced systems.

Imagers such as cameras capture light from a scene, as with the view provided by Google Earth from 500 km above the planet. The lenses in an imager focus light onto an array of detectors known as a focal plane array (FPA) to recreate the scene (or create an image). While conceptually simple, professional-level imaging lenses often have 15 to 20 individual lens elements to improve image quality far beyond that achievable with inexpensive cell-phone cameras. The concept of image quality reflects our intuitive sense of what makes for a good or bad photograph; for example, a good photo is "crisp" and "cheery" whereas a bad photo is fuzzy and dark. Cinematographers may turn these concepts upside down to communicate a message, but they still rely on a high level of performance from the optical system to make their statement.[1,2]

Radiometers are not designed for image quality; instead, they are used to accurately measure how *much* light is coming from a scene. Smokestack gases, for example, emit more light as they get hotter, possibly indicating an underlying problem with an industrial process that creates excess pollutants. Radiometers can be used to measure the difference in average gas temperature (and therefore light emitted) as a metric of process quality. Industrial monitoring of this sort does not typically require high-fidelity images of the scene, but the measuring equipment must be sensitive enough to detect small differences in temperature. Optical systems have also been designed to measure small differences in emitted or reflected light at every point in an image; such systems are known as *imaging radiometers* (or radiometric imagers).

What makes an imager or radiometer "optical" is the wavelength of light, or the distance between peaks of the waves carrying electromagnetic energy.[1,2] Wavelengths that, in a vacuum, range from about 0.1 to 30 micrometers (or microns, symbol μm) are usually classified as optical. The broad categories are ultraviolet (UV), visible (VIS), and infrared (IR); Table 1.1 shows that the associated range (or "band") of wavelengths are 0.1 to 0.4 μm, 0.4 to 0.7 μm, and 0.7 to 30 μm, respectively.

Wavelength Band	Abbreviation	Wavelength
Vacuum ultraviolet	VUV	0.10–0.18 μm
Deep ultraviolet	DUV	0.18–0.32 μm
Near ultraviolet	NUV	0.32–0.40 μm
Visible	VIS	0.4–0.7 μm
Near infrared	NIR	0.7–1 μm
Shortwave infrared	SWIR	1–3 μm
Midwave infrared	MWIR	3–5 μm
Longwave infrared	LWIR	8–12 μm
Very longwave infrared	VLWIR	12–30 μm

TABLE 1.1 Wavelengths corresponding to the associated bands used in optical systems. The wavelengths from 5 to 8 μm, which are strongly absorbed by the Earth's atmosphere, are sometimes included as part of the MWIR band.

The UV band is subdivided into vacuum UV (0.10–0.18 µm, wavelengths that are strongly absorbed by air and so require a vacuum), deep UV (0.18–0.32 µm), and near UV (0.32–0.40 µm). The IR band is subdivided into near IR (0.7–1.0 µm), shortwave IR (1–3 µm), midwave IR (3–5 µm, an atmospheric transmission band), longwave IR (8–12 µm, another atmospheric transmission band), and very longwave IR (12–30 µm). The basic unit of microns is very small compared with typical mechanical dimensions; this fact is a major contributor to the difficulty of building optical systems.

A wide range of optical systems have been built around these wavelengths. The components of a generic optical system are illustrated in Fig. 1.3. These components include: optical sources emitting energy; objects reflecting that energy; an atmosphere (or vacuum) through which the energy propagates on its way to the optics; lenses, mirrors, and other optical components used to collect this energy; detectors that capture an image of the source (an imager), measure its energy (a radiometer), or both (an imaging radiometer); electronics to convert the electrons from the detector into usable signals; and software and displays (such as high-definition TVs) to help interpret the results.

Figure 1.4 offers a schematic view of the Kepler Space Telescope, a not-so-typical optical system designed to search the skies for Earthlike planets that could support life. The source in this case is the universe of stars and extrasolar planets in the telescope's field of view (FOV) as well as some light from outside the FOV that makes its way onto the detector. This light-collecting telescope is relatively simple and consists of a primary mirror and corrector plate that brings the scene into slightly blurred focus on the detector. The detector comprises 42 rectangular arrays consisting of many individual detectors, called picture elements (or pixels), that create the instrument's focal plane array.

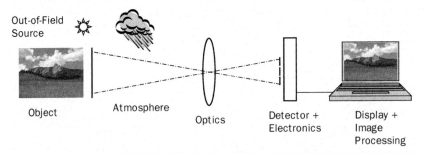

Figure 1.3 Conceptual diagram of the components of an optical system, which include an object illuminated by a source (such as the Sun), the atmosphere (for terrestrial systems), optics, detector, electronics, display, and image processing software. (*Photo credit: Mr. Brian Marotta, Louisville, Colorado.*)

Introduction 7

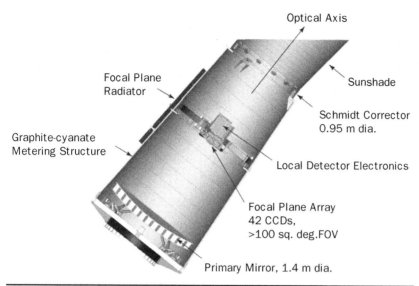

FIGURE 1.4 Schematic of the Kepler Space Telescope, a radiometer designed to search for Earthlike planets that could support life. (*Credit: NASA, www.kepler.nasa.gov.*)

Different types of space telescopes that point back toward Earth are used for remote sensing and environmental monitoring, revealing areas where pollution or degradation are prevalent. These source-plus-optics-plus-detector systems are the emphasis of this book and are found in a variety of applications, including cell-phone cameras, high-power microscopes, CD and DVD players, laser radar systems, fiber-optic communication networks that are the backbone of the Internet, and biomedical products such as confocal fluorescence microscopes for three-dimensional imaging of tumors.

Components such as high-efficiency solar cells and light-emitting diodes (LEDs), both of which play a key role in reducing greenhouse gases, also belong to the world of optical systems. Industrial applications such as fish-eye lenses for full-hemisphere imaging (Fig. 1.5), highly specialized lenses for the semiconductor lithography process used to manufacture integrated circuits, machine vision for automated inspection of food quality, and real-time inspection of heat loss (MWIR and LWIR radiation) from buildings are all common applications of optical systems within larger systems.

1.2 Optical Engineering

Designing and building optical systems requires a specific set of skills, typically classified as optical engineering, that also include aspects of mechanical, software, and electrical engineering.[1-7] The

FIGURE 1.5 Highly specialized multielement lens used to obtain fish-eye images over a 180-degree field of view. (*Credit: Warren J. Smith, Modern Lens Design, McGraw-Hill, www.mcgraw-hill.com.*)

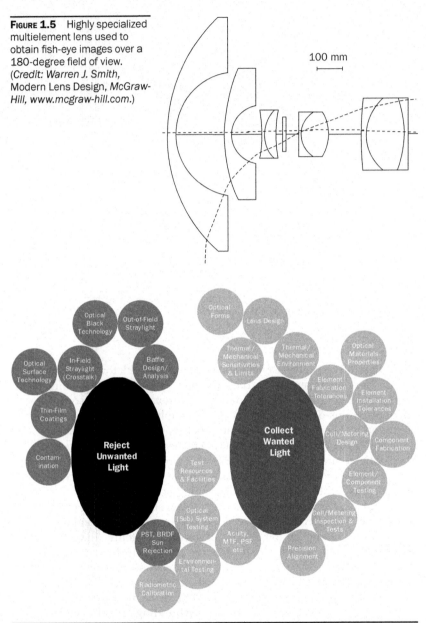

FIGURE 1.6 Summary of optical engineering components organized into two categories: collecting the light that's important and rejecting the light that's not. (*Courtesy: Howard Morrow and Paul Forney of Lockheed Martin Corp.*)

many components of optical engineering are summarized in Fig. 1.6, which shows how the field can be organized into two categories: collecting the light that's important and rejecting the light that's not. The figure displays many aspects of each category, but we can identify six common themes: geometrical optics, aberrations and image quality, radiometry, sources, detectors, and optomechanical design.

Geometrical optics (with Snell's law approximated by $n \sin \theta \approx n\theta$) is the basis of any optical system where "wanted light" is controlled to create an image. This includes the classical examples that most of us are familiar with: eyeglasses, magnifying glasses, cameras, binoculars, microscopes, and telescopes. As shown in Chap. 2, the lenses and mirrors in these systems create images by reshaping wavefronts emitted by sources. The usual example (illustrated in Fig. 1.7) is an optical source or "object"—at a large distance from a simple ("thin") lens—and brought to an image at the lens's focal length. The result is an image in a particular location and of a certain size and orientation.

Knowledge of image size and location is not the complete story, however, because image quality is not determined solely by geometrical optics. Design decisions as simple as which side of a lens faces the object, for example, dramatically affect the image quality (see Fig. 1.8). The resulting image errors ("aberrations") are not predicted by geometrical optics, but they have a significant effect on

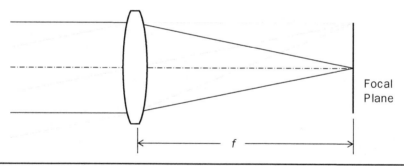

Figure 1.7 Objects a large distance from a lens are brought to focus at the focal length f of a positive lens. The center of the field is brought into focus at the center of the image plane (the focal plane) while the bottom of the field is brought into focus at the top of the image plane.

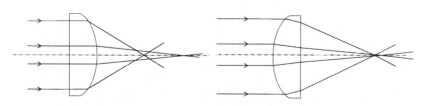

Figure 1.8 Illustration of the (nonideal) lens property known as *spherical aberration*. The figure shows how image quality is affected by which side of the lens faces the source; the lens on the right produces a crisper image of higher quality.

the choice of design and number of elements needed to satisfy image quality requirements. As shown in Chap. 3 (and in Fig. 1.5), this leads to more sophisticated designs than the simple lens shown in Fig. 1.7. This topic is covered in great detail in the branch of optical engineering known as lens design.[13,14]

Even with crisp, high-quality images, it is the rare optical system that captures enough light to produce bright images with high contrast. More typically, the amount of light is at a premium, and specific measures must be taken to collect as much light as possible and to transmit that light through the system with minimal loss. A simple example is that of attempting to start a campfire with a magnifying glass and the sunlight reflected from the Moon. We know from experience that a standard-size magnifier cannot get the fire going, but this result can also be predicted using simple analysis. The methods for analyzing the amount of light collected by and transmitted through an optical system are collectively known as *radiometry* and are reviewed in Chap. 4.

Geometrical optics, aberrations, and radiometry constitute our fundamentals, and our hardware building blocks are sources, detectors, optical, and optomechanical components. Most optical sources occur naturally, either as reflected sunlight or naturally emitted IR radiation. Artificial sources such as tungsten lamps and light-emitting diodes allow light to be added to the system (Fig. 1.9). When this is not feasible, nighttime imaging and radiometry are still possible using the light emitted by the objects themselves, rather than the reflected light from an external source such as the Sun or Moon. As we will see in Chap. 5, such IR systems commonly use the MWIR and LWIR bands because of the strong emission of these wavelengths from objects at room temperature.

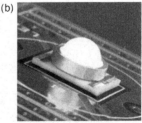

FIGURE 1.9 Optical sources such as (a) tungsten lamps and (b) high-brightness light-emitting diodes add light to the system. *(Photo credits: (a) Wikipedia author Stefan Wernli; (b) Cree, Inc., used by permission.) (See also color insert.)*

Detectors and focal plane arrays are equally critical components of an optical system. Without FPAs, for example, the HST would require a complex scanning mechanism to collect images, while photographic film does not allow for images to be converted to electrons and sent via radio waves back to Earth. Given their success on Hubble, FPAs in one form or another have since replaced photographic film in just about every area of optical engineering (Fig. 1.10). As with any new technology, new challenges and architectures have also arisen, many of which are reviewed in Chap. 6.

In addition to sources and detectors, the most critical elements of an optical system are the optical and optomechanical components. Many systems have been designed without taking these aspects of the hardware into account, resulting in large losses of investor and taxpayer money.[11] These aspects include the fabrication of optical components such as lenses, mirrors, beam splitters, windows, prisms, and so forth.[15] Once fabricated, the components must be assembled and aligned with respect to each other in complex mechanical assemblies (Fig. 1.11). In Chap. 7 we will see that the alignments must also be maintained under such environmental stresses as vibration and changes in temperature. Testing and calibration of the assembled optical system are also key parts of optical engineering, although they will not be covered in this book.[16]

The process of optical engineering thus consists of design, fabrication, assembly, and test. Figure 1.12 presents a flowchart that summarizes some of the interactions between optical and mechanical design, fabrication, and optical systems engineering tasks. For illustrative purposes these elements appear in separate

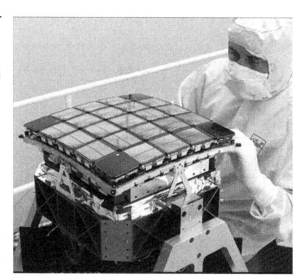

FIGURE 1.10 The Kepler Space Telescope's focal plane array (FPA), which consists of 42 individual focal planes. (*Photo courtesy of Ball Aerospace & Technologies Corp.*)

12 Chapter One

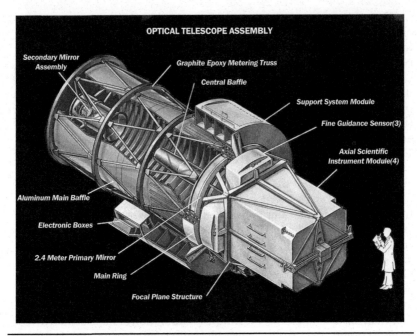

FIGURE 1.11 Mechanical assembly and associated support hardware for the Hubble Space Telescope. (*Credit: NASA, www.nasa.gov.*)

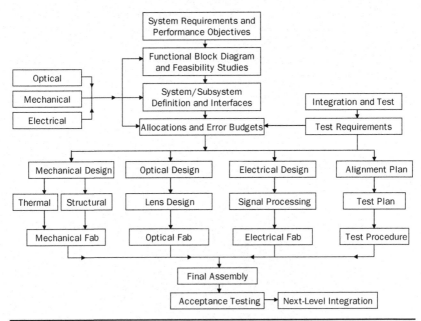

FIGURE 1.12 Summary of the interactions between optical, mechanical, and electrical design as well as fabrication, assembly, and testing. (*Figure developed in collaboration with Paul Forney of Lockheed Martin Corp.*)

"boxes"—an inaccurate simplification given that good design must also include fabrication and assembly considerations and that testing can range from inefficient to impossible if alignment and assembly issues are not taken into account from the beginning of the process.

One can easily be misled into thinking that the ability to manipulate design software and complex formulas is the only skill required for a successful design. A different perspective is given by Einstein and Infeld in *The Evolution of Physics*, where they argue that "Books on physics are full of complicated mathematical formulae. But thought and ideas, not formulae, are the beginning of every physical theory."[17] In an engineering context, Anthony Smart summarizes the issue in a similar manner: "Building systems that actually meet specifications requires more than a knowledge of optics and contemporary design codes. Excellence also comes from a kind of 'street' knowledge, learned not from textbooks but from experience, mostly from failures."[18] Mouroulis gives an excellent summary of the many real-world issues faced by the optical engineer and the optical systems engineer.[19]

For the pessimist, the "experience" mentioned by Smart is synonymous with failure; a more balanced view is that preparation reduces the risk of failure. Such preparation—"Chance favors only the prepared mind," argues Pasteur—can take the form of hands-on experience as well as textbook and classroom learning, none of which is guaranteed to be of the "good" variety. Preparation can also involve looking more closely at the processes that can lead to failure; these may include implicit messages sent by management, the social environment at work, the stress of unrealistic schedules, and even the distraction of background noise in an engineer's work cubicle leading to a million-dollar error in calculation.

These issues may seem far removed from optical engineering, but in fact they are at the core of all engineering—part of the "street knowledge" referred to by Smart. Such knowledge concerns the "soft" issues no less than the "hard" technical content. Combining these two types of knowledge is the task of the optical systems engineer, a key person in the process of hardware development.

1.3 Optical Systems Engineering

Optical systems engineers are really part optical engineer and part project engineer, but *their key role is to take the perspective of the user* to ensure that the hardware meets the customer's requirements. This requires an understanding of optical engineering that's broad enough to ask the right questions and deep enough to estimate the right answers; it also requires an appreciation of the project engineer's schedule, budget, and personnel constraints. The roles of systems engineer and project engineer are often confused with each other. In order to help clarify the distinction, this section focuses on the optical engineer's side of these systems engineering tasks.[20,21]

Taking the perspective of the user requires that the system engineer "flow down" the customer's requirements into subsystem and component performance. These requirements are simply what the customer wants the optical system to do (which may not be the same as what the system *needs* to do). A camera, for example, may only be needed to monitor one's business or property. Yet a camera may also be needed to monitor action on a national level from a low-Earth orbit in space, or it may be needed to detect the presence of planets many light years away. The requirements for these systems are clearly different.

There will be other requirements besides the distance to the object, such as: how good an image is needed? Does it need to be a crisp, high-resolution image that's good enough for publication by the paparazzi? Or is it sufficient to know only that a property line has been crossed, so that a simple motion sensor will do? Will the system be used during the day, when it should be easy to get a bright image, or is it necessary to see through the fog at night? Understanding system requirements is a critical part of the design process because a product that does not meet them will not be useful. Therefore, a system requirements review (SRR) involving the customer and the design team is an important first step in system development.

Once performance requirements are established by the customer and understood by the design team, the next step is to decide what to build—for example, will a camera meet the requirements? If so, will it need a big lens or a small one? Will it need an expensive, high-quality lens, or will an antique monocle found in a secondhand store be sufficient? We think we know what "expensive" means, but what exactly is a "high-quality" lens? To answer such questions we need criteria, which are usually numerical, for evaluating performance and making comparisons with other designs.

Table 1.2 is a list of such criteria for a lightweight, inexpensive digital camera. The table shows the camera specifications—or "specs," a list of properties the system must have to meet the customer's requirements—divided into optical, detector, and mechanical subsystems. Because this particular imager is collecting light from a naturally illuminated (daytime) scene, a source subsystem (e.g., a flash) is not included. Electrical and software subsystems could also be included, but they are not reviewed in this book. The meaning of the listed specs will be developed throughout the course of Chaps. 2 to 7.

Some entries in the spec table come directly from customer requirements; others will be derived (flowed down), with the optical system engineer assigning or allocating the requirements into subsystem and component specs. The process of requirements analysis is often difficult, and performance estimates must often be

Introduction 15

Specification	Value
Optical Subsystem	
Wavelength band, $\Delta\lambda$	0.4–0.7 µm (VIS)
Object distance, s_o	$\infty \rightarrow 1$ m
Effective focal length (EFL)	6.0 mm
Entrance pupil diameter (EPD)	1.7 mm
Field of view (FOV)	43° (H) × 33° (V)
Instantaneous FOV (IFOV)	1.2 mrad
Depth of focus (DOF)	±25 µm
Distortion	< 3%
Throughput, T	> 75%
Relative edge illumination	> 60%
Stray-light suppression	10^{-3}
Out-of-band rejection	N/A
Wavefront error (WFE)	$\lambda/10$ RMS
Modulation transfer function (MTF)	
at 20 lp/mm	> 0.9 ($\lambda = 0.55$ µm)
at 60 lp/mm	> 0.7 ($\lambda = 0.55$ µm)
Ensquared energy (EE)	> 75%
Detector Subsystem	
FPA type	CMOS
Detector material	Silicon
Number of pixels, N_p	640 (H) × 480 (V)
Pixel pitch, x_p	7.4 µm
Pixel size, d	7.4 µm
Responsivity (peak), $\mathcal{R}$	0.5 A/W
Nonuniformity (corrected)	0.1%
Operability	99%
Integration time, t_{int}	30 msec
Frame rate	1 Hz
Dynamic range	12 bits
Mechanical Subsystem	
Operating temperatures, ΔT	−10 to +50°C
Track length, L	10 mm
Volume, V	< 30 mm^3

TABLE 1.2 Typical specifications for an inexpensive digital camera; the units given are based on common engineering usage.

Specification	Value
Weight, W	< 100 g
Fundamental resonance, ω_0	500 Hz
Transmissibility (at ω_0), Q_f	10
Line-of-sight jitter	100 μrad
Line-of-sight drift	500 μrad
Boresight alignment	1 mrad

TABLE 1.2 Typical specifications for an inexpensive digital camera; the units given are based on common engineering usage (*continued*).

developed by the subsystem designers before the actual designs are complete. Determining whether the system can be built as requested requires that the interrelationships between subsystems be known as early as possible in the design process—even if they are based on estimates and not on "hard" design.

The spec table is thus a performance summary which ignores a large amount of design detail. That is, Table 1.2 gives no indication of *how* the specs will be met; it indicates only that these specs must be satisfied to meet customer requirements. The flowdown process is intended to capture the "how" aspect, so that errors can be caught early on in system development. For instance, meeting the specs for distortion and color correction for a professional-quality camera requires a large number of individual lenses within the optical subsystem (see Chap. 3), information that is not obvious from reading a spec table. Flowdown attempts to allocate the acceptable errors for each lens in a way that allows the overall error requirement to be met. As we saw for the HST, optical components can be properly designed yet still have errors in their fabrication; this possibility was properly addressed by the flowdown process but unfortunately was not tested for.

Another property of a spec table is that some specs may be redundant or inconsistent. Redundant (or "overspecified") properties are those that can be derived from other specs. If, for example, the spec for f/# (the ratio of effective focal length to entrance pupil diameter) were also included in Table 1.2, then one of these three specs would be redundant. As long as additional testing is not required to confirm or validate that this spec has been met, there is nothing fundamentally wrong with listing such redundant specs for reference. Problems arise when specs are inconsistent, for in this case the customer requirements may or may not be met depending on which specs were used by the subsystem designers.

Inconsistent requirements can result from subsystem specifications that are interrelated. For example, the optical subsystem must meet its requirements over the temperature range of the mechanical subsystem, and this requires the lens designer to take into account such factors as thermal expansion, change in refractive index with temperature, and depth of focus (see Chap. 7). However, there is no guarantee that all the requirements involving image quality (see Chap. 3) can be met at the same time, given the restrictions entailed by such other requirements as size and cost.

Inconsistencies of this nature must be identified as early as possible in the design process, thereby giving customers the opportunity to reexamine and possibly relax their requirements. Optical, detector, and mechanical engineers will often take the specs given to them and proceed to design their subsystems. But without a bigger picture of the overall system design, they are typically unable to look for inconsistencies between subsystems—a task of the optical systems engineer.

Stepping back a bit from the spec table, it is not always obvious which type of optical system is the best option (or "architecture") to meet the given requirements. Recall that Table 1.2 lists specs for the particular case of an inexpensive camera imaging objects at infinity. By expanding beyond the use of FPAs and including the optical properties of the eye as part of the optical system, hand-held telescopes and binoculars can also be included in the "design space."

Before a spec table can be finalized, then, it is necessary to compare different systems using what is known as a *trade* study. That is, the advantages and disadvantages of different designs are compared ("traded"), and the results are summarized in a trade table. Such tables are much the same as Table 1.2 but include extra columns for each design option. A higher-level summary is often preferred, such as that shown in Table 1.3; it includes key

Spec	Telescope	CMOS Camera	Binoculars
Image quality	High	Moderate	Moderate
Field of view	Small	Large	Large
Instantaneous FOV	Small	Small	Moderate
Size	Large	Small	Large
Weight	High	Moderate	High
Power consumption	Low	Moderate	Low
Cost	Moderate	Low	Moderate
Technology readiness	High	High	High
Risk factors	Low	Low	Low
Reliability	High	High	High
Time-to-market	Short	Short	Short

TABLE 1.3 Trade table summarizing the qualitative strengths and weaknesses of different design options in terms of meeting requirements for a portable instrument to obtain daytime images.

performance metrics such as FOV, IFOV, image quality; size, weight, and power;[22] and project engineering criteria such as cost, risk, reliability, and time-to-market. Qualitative entries such as "large" and "small" are allowable; after all, they are mainly for comparative purposes—in the case of the trade table, to compare the different options listed.

The process of juggling competing effects is not a straightforward task, and settling on a design option is often the most difficult part of the trade study. If the options in Table 1.3 are intended for hikers and backpackers, for example, portability is a key requirement and size and weight are also important specs. If image detail is critical, then a telescope or binoculars with magnification is required; a camera with zoom is also possible, though a 10× zoom will likely exceed the size, weight, and cost requirements whereas a 4× zoom may not. If image storage is important, then a camera is currently the only option available—with the trade table pointing the way to a new product line of binoculars that can store images.

Table 1.3 also shows that assessment of the risk factors and technology readiness level (TRL) for each design option is an important part of the trade space. Risk factors include items such as a vendor's ability to deliver a critical component, the reliability of an unproven part, and the use of systems (e.g., focus mechanisms) that require moving parts. Should the risk be deemed excessive, it may be possible to retain the affected option if it can be shown that the risk can be reduced ("burned down").

Once the number of design options have been reduced to one (or sometimes two), detailed subsystem design is used to develop error budgets. These are required because of nature's insistence that *everything* in an optical system vary with mechanical dimensions, temperature, wavelength, angle of incidence, polarization, stress, and so on. Such dependencies can transform an optical system that works perfectly well on a vibration-isolated table in a climate-controlled building into a system that quickly becomes useless once it is moved out to the field and encounters environmental effects such as temperature changes, shock, vibration, humidity, salt spray, vacuum, radiation, and so forth. In other words, engineering is as much about how things *don't* work as it is about how they do. As we will see in Chap. 7, the question to be answered by the error budgets is: How much change is possible before the system requirements are no longer met?

For detailed subsystem designs, interfaces between them must also be identified. Typical examples include the voltage and current levels fed to the FPA from the electrical subsystem as well as the data format that is sent back from the FPA. An important aspect of this process is understanding the relationships *between* subsystems: has anything been overlooked that could not be seen by designers of the individual subsystems? Once characterized, these interfaces are summarized in a report known as an interface control document (ICD).

In parallel with the subsystem design and interfaces, it is important to develop a test plan with an eye toward how each specification in the spec table will be verified and validated (i.e., demonstrated to the customer that the spec has been met). Three methods are common: test, analysis, and similarity to other hardware. Testing is the most expensive but has the potential to be the most accurate; as we have seen with the Hubble, this potential is not always realized. Analysis and similarity are used when testing is prohibitively expensive and the methods of analysis are mature.

Figure 1.13 summarizes the steps of the optical systems engineering process in a more-or-less chronological fashion. Unfortunately, much of the process has evolved into a formalism with very little content, with systems engineers specializing in "requirements management"—for example, writing ICDs, making verification plans, and establishing procedures for tracking requirements and error budgets—without actually having worked

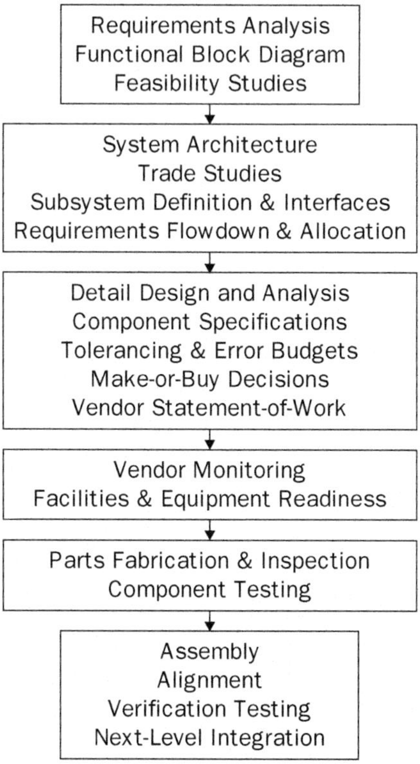

FIGURE 1.13 Summary of the optical systems engineering process. (*Figure developed in collaboration with Paul Forney of Lockheed Martin Corp.*)

```
Requirements Analysis
Functional Block Diagram
Feasibility Studies
        ↓
System Architecture
Trade Studies
Subsystem Definition & Interfaces
Requirements Flowdown & Allocation
        ↓
Detail Design and Analysis
Component Specifications
Tolerancing & Error Budgets
Make-or-Buy Decisions
Vendor Statement-of-Work
        ↓
Vendor Monitoring
Facilities & Equipment Readiness
        ↓
Parts Fabrication & Inspection
Component Testing
        ↓
Assembly
Alignment
Verification Testing
Next-Level Integration
```

- Vague, "blue-sky" concepts
- Qualitative assessments dominate: "small," "lightweight," "low cost,"...
- No quantitative evaluation or comparison of requirements with product specs
- No trade studies comparing design options
- No realistic schematic of hardware implementation or interfaces
- No discussion of manufacturability, test, cost, or schedule
- Promises of success: "In Phase 3, there will be profits!"

Figure 1.14 Summary of the viewgraph engineering "process."

on development of the hardware itself. Such tasks are often delegated to entry-level engineers known as "spec-retaries"; not surprisingly, many soon develop a strong distaste for engineering without ever having an opportunity to practice it.

For those that remain in engineering, the focus on hardware is often lost and replaced by *viewgraph engineering*—vague, "blue sky" concepts sketched out without regard to quantitative requirements, design trade-offs, implementation issues, interfaces, detailed component specs, fabrication, assembly, or testing (Fig. 1.14). This is a difficult situation to turn around; doing so usually requires a failure such as Hubble to force a renewed focus on the most important skill of systems engineering: a solid understanding of optics and of optical system fundamentals.

Problems

1.1 Obtain copies of product catalogs from the following vendors: Newport (www.newport.com), Edmund Optics (www.edmundoptics.com), CVI Melles Griot (www.cvimellesgriot.com), and Thorlabs (www.thorlabs.com). Spend some time browsing through each catalog. Do you see anything interesting?

1.2 Take a look at the following Web sites: SPIE, the International Society for Optical Engineering (www.spie.org); the Optical Society of America (www.osa.org); and the Lasers and Electro-Optics Society of the Institute of Electronic and Electrical Engineers (www.ieee.org/portal/site/leos/). Spend some time browsing through each site, including the short-course descriptions and tutorial texts. Do you see anything interesting?

1.3 Take a look at the Web sites for the following optical software companies: Zemax (www.zemax.com); Code V (www.opticalres.com); ASAP (www.breault.com); and FRED (www.photonengr.com). Try using any of software that is free to download. Are there things you see that you would like to learn more about?

Notes and References

1. Francis A. Jenkins and Harvey E. White, *Fundamentals of Optics*, 4th ed., McGraw-Hill (www.mcgraw-hill.com), 1976.
2. Eugene Hecht, *Optics*, 4th ed., Addison-Wesley (www.aw.com), 2001.
3. Warren J. Smith, *Modern Optical Engineering*, 4th ed., McGraw-Hill (www.mcgraw-hill.com), 2008.
4. Robert E. Fischer, B. Tadic-Galeb, and Paul Yoder, *Optical System Design*, 2nd ed., McGraw-Hill (www.mcgraw-hill.com), 2008.
5. Ed Friedman and J. L. Miller, *Photonics Rules of Thumb*, 2nd ed., McGraw-Hill (www.mcgraw-hill.com), 2004.
6. Philip C. D. Hobbs, *Building Electro-Optical Systems*, Wiley (www.wiley.com), 2000.
7. Frank S. Crawford Jr., *Waves*, McGraw-Hill (www.mcgraw-hill.com), 1968.
8. David DeVorkin and Robert W. Smith, *Hubble: Imaging Space and Time*, National Geographic Society (www.nationalgeographic.com), 2008.
9. L. Allen, *The Hubble Space Telescope Optical System Failure Report*, NASA (www.nasa.gov), 1990.
10. The irony of not having a second set of eyes to review an optical system did not go unnoticed.
11. Philip Taubman, "Death of a Spy Satellite Program," *New York Times*, 11 November 2007.
12. Scientific and technical Academy Awards (Oscars) are given each year for advancements in camera design. The camera designers (to date, almost all men) receive recognition for their technical contributions to the movie industry—which includes having their picture taken with a beautiful actress.
13. Warren J. Smith, *Modern Lens Design*, 2nd ed., McGraw-Hill (www.mcgraw-hill.com), 2005.
14. R. R. Shannon, *The Art and Science of Optical Design*, Cambridge University Press (www.cambridge.org), 1997.
15. Daniel Malacara, *Optical Shop Testing*, 3rd ed., Wiley (www.johnwiley.com), 2007.
16. Joseph M. Geary, *Introduction to Optical Testing*, SPIE Press (www.spie.org), 1993.
17. Albert Einstein and Leopold Infeld, *The Evolution of Physics*, Simon & Schuster (www.simonandschuster.com), 1966.
18. Anthony E. Smart, "Folk Wisdom in Optical Design," *Applied Optics*, 1 December 1994, pp. 8130–8132.
19. Pantazis Mouroulis, "Optical Design and Engineering: Lessons Learned," *Proceedings of the SPIE*, vol. 5865, 2005.

20. Perhaps the most important rule of project engineering is this: *Nine women cannot produce a baby in one month.* Biological systems need time to develop, and optical systems are no different, so there are many situations where "throwing bodies" at a problem does not accelerate a project's completion. The uncertainty comes in when there isn't a common reference (such as nine months) and debates occur regarding how many months it will, in fact, take to design and build a system. Without a historical record, engineers must resort to educated guesses; but even with good estimates, someone will always try to shorten (say) a nine-month schedule to eight or seven—or to whatever is needed to win the contract, although doing so may mean making unrealistic promises to the customer.
21. For an excellent discussion of project management issues, see W. C. Gibson, "A Common Sense Approach to Project Management," NASA Project Management Conference, March 2004.
22. Size, weight, and power (electrical) are often abbreviated as "SWaP." To see how size and weight are affected by simple "design" choices, compare a U.S. dollar bill with 100 pennies. Both options have the same monetary value, but one approach is much larger and heavier than the other. The electrical power consumption required to fabricate these two options is also dramatically different.

CHAPTER 2
Geometrical Optics

With Google Earth providing high-resolution images of large portions of the planet and with NASA and European Space Agency (ESA) space telescopes doing the same over a minute fraction of the universe, the telescope has clearly evolved in importance since its use as a spyglass. The telescope was not Galileo's invention, but pointing one at the heavens for the first time allowed him to see further and with better resolution, initiating a chain of scientific discoveries by himself, Kepler, Newton, and many others that continued through the more recent relativistic and quantum theories.[1] As space telescopes such as NASA's Hubble, Spitzer, James Webb, and Kepler and the ESA's Herschel continue to reveal the size, complexity, and beauty of the universe, remote sensing telescopes pointed back toward the planet also provide a better understanding of Earth's weather patterns (Fig. 2.1), ocean currents, ice flows, and even archeological history.

The basic idea behind any telescope is to collect light from a distant object and create an image on an optical detector such as photographic film, an array of photodetectors, or the retina at the back of the eyeball. In the case of the space telescopes, the objects are often stars many light years away, so an extremely large, high-quality telescope is required to place bright, crisp images of these stars onto a photodetector array. For objects on Earth, the optical instruments of choice are smaller telescopes, cameras, and binoculars. For smaller objects, microscopes are required to reveal yet another world that would otherwise be beyond our reach.

Conceptually, the camera is the simplest of these instruments; it consists of an imaging lens and an optically sensitive detector, with the "lens" often containing many individual lenses (or elements) working together. Knowledge of geometrical optics is fundamental to understanding the lenses in cameras and all optical systems. Snell's law is the foundation on which geometrical optics is built. It is summarized by the equation $n_1 \sin \theta_1 = n_2 \sin \theta_2$, where the angle θ is measured perpendicular to the surface; the equation shows that a larger index ratio (n_2/n_1) bends light more.[2,3] By applying Snell's law to curved surfaces, image formation follows naturally as a change in wavefront curvature (Sec. 2.1). This chapter reviews the basic concepts

FIGURE 2.1 Composite satellite image showing cloud patterns across the planet. This image was collected using the MODIS instrument, an imaging spectroradiometer. (*Credit: NASA, www.nasa.gov.*) (*See also color insert.*)

of image formation as well as important properties of imaging and radiometric systems; these properties include field of view (Sec. 2.2), relative aperture (Sec. 2.3), the concepts of stops and pupils (Sec. 2.8), and afocal systems (Sec. 2.9).

Three techniques are often used to simplify the understanding of optical systems: the use of distant objects at infinity; thin lenses; and the *paraxial* approximation (also known as *first-order* optics), whereby Snell's law is approximated as $n_1 \sin \theta_1 \approx n_1 \theta_1 \approx n_2 \theta_2$ for "small" θ. These tools are limited in their ability to analyze many systems—high-resolution microscopes, for example—so the concept of *finite-conjugate imaging* is introduced in Sec. 2.4 to understand these systems. Combinations of thin lenses that can no longer be considered to be thin, such as the multielement lenses of digital single-lens reflex (DSLR) cameras, are reviewed in Sec. 2.5. In Sec. 2.6 the paraxial approximation is expanded into a form that is useful for tracing rays through an optical system with an arbitrary number of thin lenses or surfaces. This form is then applied in Sec. 2.7 to thick lenses, leading to a reexamination of the concepts of object distance and focal length.

Optical systems known as "afocal" are commonly used in telescopes, binoculars, and microscopes. As the name implies, these systems do not bring the rays from an object directly to a focus; instead, they change the beam diameter to produce an afocal (nonfocal) output with no distinct focal length. A common example is the amateur telescope, where the combination of telescope and eyepiece creates a reduced afocal beam of smaller diameter that the eye then brings into focus. This class of systems, along with the associated relay and field lenses, completes this chapter's overview of geometrical optics.

2.1 Imaging

We know from experience that pointing a white card at a mountain range doesn't put an image of the range on the card. Of course, some light from the mountains will end up on the card, as will light from the sun, sky, moon, nearby buildings, and so forth. With light from these many objects mixed together, the card does not show any specific features such as an epic mountain shot. Something else is needed to obtain an *image* of the mountain, and that something is usually a lens.[4]

Figure 2.2 illustrates how a lens puts an image on the card, where a distant mountain (located at "infinity") is imaged at a certain distance past the lens. This distance is called the *focal length* of the lens; as we shall show, the focal length is determined by the geometry and material of the lens. For a single-element lens, the image is upside down (inverted), a property described in Sec. 2.2.

To understand how the lens creates the image, Fig. 2.2 also shows spherical wavefronts emitted by a point on the mountain that is straight away from the center of the camera. These wavefronts are created when light hits the mountain and is reflected (reemitted by oscillating electrons), resulting in spherical waves emitted from each point in the scene. The distance between wavefronts (d_1) is "the" wavelength of the reflected light in air, which for now we show as only a single wavelength and not as the range of wavelengths actually present in visible light (0.4–0.7 µm; see Table 1.1).

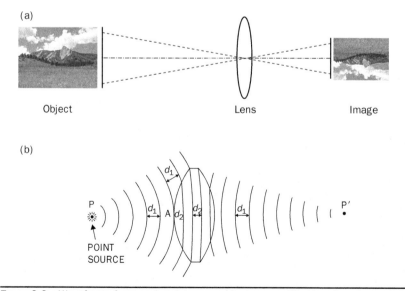

FIGURE 2.2 Wavefronts from each point on an object are modified by a lens to create an image of the scene. (*Credits: (a) photo by Mr. Brian Marotta, Louisville, Colorado; (b) Warren J. Smith*, Modern Optical Engineering, *McGraw-Hill, www.mcgraw-hill.com.*)

By the time the spherical wavefronts from the far-away mountain reach the lens, they are more or less planar. Because of how the lens is shaped, it bends these planar wavefronts into spherical wavefronts on the other side of the lens, where their convergence at the focal length is described by Snell's law. Figure 2.2(b) illustrates this effect for the simple case of an emitting point that is relatively close to the lens. The wavefronts incident on the first surface are bent as they pass into the lens. As indicated by the smaller distance between wavefronts (d_2), they move more slowly in the lens than in the air; this is a result of the lens's refractive index n (which is greater than air's index of approximately 1.0).

Because the lens is thicker at the center than at the edges, the outer part of the wavefront exits the lens first; this allows the outer part to pull ahead of the center parts, which are still slowed down by the lens's refractive index. If the lens surface is spherical, then the wavefronts coming out of the lens are also spherical (with curvature *opposite* that of the surface) on a path that converges at the image distance. If the "lens" were instead planar on both sides, then the wavefronts would continue on through as planar waves and the "lens" would be not a lens but just a flat plate of glass.

An equation developed by lens grinders describes how much curvature is needed for a lens that is thick compared with its focal length (here and throughout the book, the physical units are given in brackets following the equation):

$$\frac{1}{f} = (n-1)\left(\frac{1}{R_1} - \frac{1}{R_2} + \frac{n-1}{n}\frac{CT}{R_1 R_2}\right) \; [1/\text{mm}] \qquad (2.1)$$

where CT is the center thickness of the lens. By setting the center thickness to zero, we see that a larger radius R (closer to infinite or planar) on either surface is less effective at bending planar wavefronts and so gives the lens a longer focal length. At the same time, a larger index difference between the lens and air ($n - 1$) slows down the wavefronts in the center of the lens even more compared with the edges, imparting more curvature to the wavefront and resulting in a higher-power lens with a shorter focal length. Such a lens, whose center thickness is small compared with the focal length, is known as a *thin lens*.

Equation (2.1) assumes an algebraic sign convention for the radii R_1 and R_2. As shown in Fig. 2.3, the convention is that any radius centered to the left of the lens is negative whereas any radius centered to the right is positive. Using different signs in Eq. (2.1) results in different conventions for the radii; Example 2.1 illustrates the convention used in this book.

Example 2.1. To design a lens with a focal length of $f = 100$ mm, there are infinitely many possible combinations of front- and rear-surface radii. For a

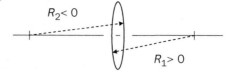

FIGURE 2.3 Sign conventions used in Eq. (2.1) for the front- and rear-surface radii (R_1 and R_2), illustrated for a *positive* lens.

refractive index $n = 1.5$ and a planar front surface ($R_1 = \infty$), Eq. (2.1) for a thin lens (CT = 0) shows that the second surface radius $R_2 = -50$ mm, where the negative radius means, by convention, that the second surface is centered to the left of the lens.

The lens can also be turned around, in which case $R_2 = \infty$ and $R_1 = +50$ mm, with a positive radius for a surface that is centered to the right of the lens. Using Eq. (2.1), the focal length $f = R_1/(n-1) = (50 \text{ mm})/0.5 = 100$ mm; as expected, there is no change in focal length simply by turning the lens around. As shown in Chap. 3, however, the image quality is better when the curved side faces toward the distant mountain.

Given a lens of a certain focal length, the imaging equation can be used to calculate where the image is located for a thin lens in air:

$$\frac{1}{f} = \frac{1}{s_o} + \frac{1}{s_i} \quad [1/\text{mm}] \tag{2.2}$$

This expression states that, for a lens with a positive focal length f (or positive lens power $\phi = 1/f$), a smaller distance from the mountain (or "object") to the lens (= s_o) results in a larger distance from the lens to the image (= s_i). This principle is illustrated graphically in Fig. 2.4.

The figure shows that, as objects are moved closer to the lens, the wavefronts incident on the lens from the left have more curvature ($C = 1/s_o$). Although the lens has enough refractive power to converge the planar wavefronts to an image of distance $s_i = f$ away, it does not have enough power to redirect the diverging spherical wavefronts to the same image distance. As a result, the image is farther away from the lens.

Because the refractive power of the lens is a constant, Eq. (2.2) shows that the curvature of the object and image wavefronts must sum to a constant (equal to $1/f$). As the object is moved closer to the lens, the wavefronts that fall on it have more curvature; the wavefronts that leave the lens must then have less curvature and so converge to an image farther away.

A lens, then, does nothing more than change the curvature of wavefronts. As shown in Chap. 3, this conversion isn't perfect even if the lens surfaces and initial wavefront are perfect. Nevertheless, from the perspective of geometrical optics, Eq. (2.2) is useful for predicting, more or less, where the image will be for a given lens and object distance.

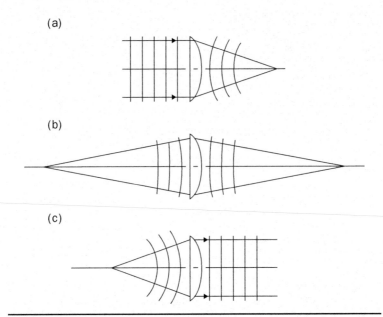

FIGURE 2.4 Effects of a lens on wavefront curvature. As the object is moved closer to the lens, the curvature of the wavefront incident on the lens ($1/s_o$) is bigger and that of the wavefront exiting the lens ($1/s_i$) is smaller.

Example 2.2. In this example, Eq. (2.2) is used to find the image distances for the cases shown in Fig. 2.4. Figure 2.4(a) is already familiar, and Eq. (2.2) confirms that $1/f = 1/s_i$ for a far-away object at infinity ($s_o = \infty$), so the wavefronts converge at an image distance s_i that equals the focal length of the lens. As the object is moved closer to the lens (to an object distance $s_o = 2f$), Eq. (2.2) shows that $1/f - 1/2f = 1/2f$ [Fig. 2.4(b)] and so the image distance $s_i = 2f$. This is a special case where the wavefront radii ($= 2f$) are the same on both sides of the lens.

As the object is moved even closer to the lens, the image continues to move further away. When the object is at the front focal length of the lens we have $s_o = f$, and Eq. (2.2) shows that $1/s_i = 0$, or $s_i = \infty$. The wavefronts on the right side of the lens are now planar [Fig. 2.4(c)] and the wavefront is said to be "collimated"; hence the so-called image is now "at infinity."

Example 2.2 and Eq. (2.2) assume algebraic sign conventions associated with the object and image distances. Specifically, an object distance to the left of the lens is positive and that to the right of the lens is negative. At the same time, image distances to the left of the lens are negative and those to the right are positive. For example, the positive object distance $s_o = +2f$ (to the left of the lens) in Example 2.2 results in a positive image distance of $s_i = +2f$ (to the right of the lens) for a positive focal length f. These conventions are summarized in Table 2.1.

These sign conventions also lead to a distinction between what are known as "positive" and "negative" lenses. A positive lens has a focal length $f > 0$, and it brings an object at infinity to a converging

Location	Object	Image
Left of lens	+ (Real)	− (Virtual)
Right of lens	− (Virtual)	+ (Real)

TABLE 2.1 Object and image sign conventions used in Eq. (2.2) for light propagating from left to right.

focus at a positive location (to the right of the lens); a negative lens has a focal length $f < 0$, and it brings an object at infinity to a virtual focus at a negative location (to the left of the lens). Typical shapes for these two basic types of lenses are illustrated in Fig. 2.5.

2.2 Field of View

An object or scene usually consists of points other than those that are coming straight down the center of the camera lens—either side-to-side, up-and-down, or any combination of these. How much of this off-axis scene can be imaged depends not only on the focal length of the lens but also on the size of our white card (or any other type of detector). Figure 2.6 shows the wavefronts from the bottom of the mountain and how they are imaged at the top of the card; the card size determines, in part, the size of the imaged field, or *field of view* (FOV).

In order to explain this phenomenon, we introduce the concept of a *ray*. Figure 2.7 shows these rays as lines that indicate the direction in which any part of the wavefront is moving. A ray cannot exist without a wavefront, so the properties of rays are determined by those of the wavefront. For example, the three off-axis rays that have been drawn coming from the mountain in Fig. 2.6 are all parallel, indicating that the wavefronts are planar. On the image side of the lens, however, the center ray has gone straight through the lens while the outer two rays are bent by the lens. Thus the planar wavefront has been converted by the lens into spherical waves centered at the image point, which is where the rays converge.

The ray going straight through the center of a thin lens is commonly used to determine the field angle (or FOV) over which a camera can image. This is illustrated in Fig. 2.8(a), which shows the outer rays and the angle they make with the mountain. Straightforward geometry reveals that larger detectors and shorter focal lengths result in wider fields. The angle depends on the size d of the detector and the focal length f of the lens:

$$\text{FOV} = 2\alpha = 2\tan^{-1}\left(\frac{d}{2f}\right) \approx \frac{d}{f} \text{ [rad]} \quad (2.3)$$

Here the FOV is almost the same as d/f for "small" angles, and we must emphasize the concept of a "full FOV" ($= 2\alpha$) because the factor

30 Chapter Two

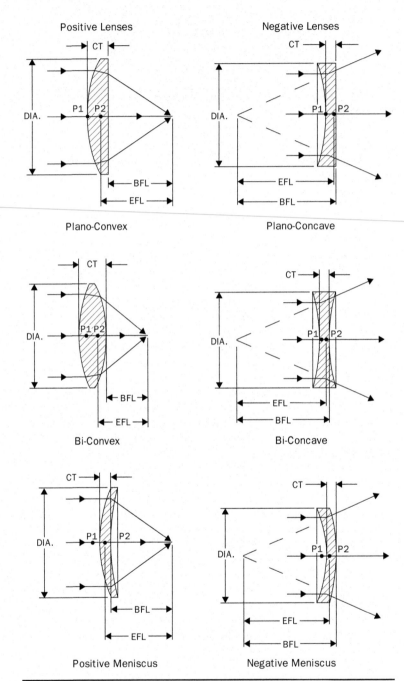

Figure 2.5 Three types of positive and negative lenses. (*Credit: Janos Technology LLC.*)

Geometrical Optics 31

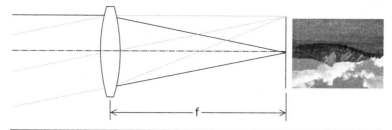

FIGURE 2.6 A single lens recreates a scene as an upside-down (inverted) image, where the field of view is determined by the focal length and detector size. The left side of the image is also switched with the right. (*Photo credit: Mr. Brian Marotta, Louisville, Colorado.*)

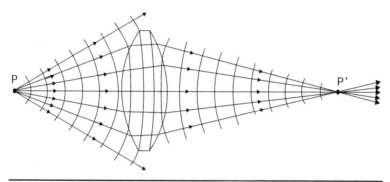

FIGURE 2.7 Rays (the lines with arrows) and their perpendicular relationship to wavefronts. (*Credit: Warren J. Smith*, Modern Optical Engineering, *McGraw-Hill, www.mcgraw-hill.com.*)

of 2 before the arctangent is often not used (in which case we have only a half FOV (or HFOV, also known as semi-FOV). No matter how complex, all lenses can be reduced to the simple case of having a focal length—an "effective focal length"—that can be used in Eq. (2.3) to find the FOV.

Most focal planes are not white cards; instead, as shown in Chap. 6, they are semiconductor devices known as focal plane arrays (FPAs). The individual detectors ("pixels") in these arrays divide the FOV into smaller units known as the instantaneous FOV (IFOV); when the pixels are the same size as their spacing, the IFOV is also called the detector angular subtense (DAS). Figure 2.8(b) illustrates the difference between the FOV and IFOV; the latter can be calculated via Eq. (2.3) if we use the pixel size (rather than the size of the entire FPA) for d.

Snell's law explains why the center ray goes straight through a thin lens (or its equivalent). This is shown in Fig. 2.8(c), where both sides of the center of the lens are approximately parallel over a very small area. If the lens is relatively thin, then the difference in flatness

32 Chapter Two

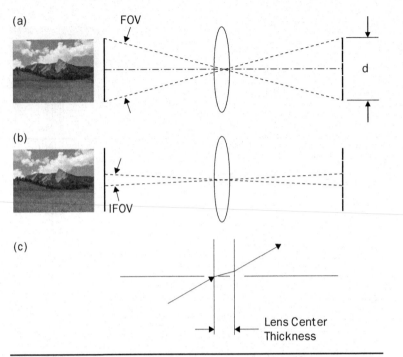

FIGURE 2.8 The field of view (FOV) and instantaneous FOV (IFOV) are determined by a ray that travels through the center of a thin lens without bending. (*Photo credit: Mr. Brian Marotta, Louisville, Colorado.*)

across the center of the lens won't be important; in this case, Snell's law for two flat surfaces then shows that a ray entering the lens from the left will come out at the same angle on the right. If we ignore the small displacement for a thin lens (it is exactly zero for an ideal thin lens of zero thickness), then the ray goes straight through the lens and Eq. (2.3) gives the FOV and IFOV. As will be shown in Sec. 2.7, all lenses can be reduced to an equivalent thin lens.

The use of rays also allows graphical solutions to Eq. (2.2) for basic imaging using the concepts of geometrical optics. Examples are given in Fig. 2.9, where both a positive and negative lens are shown to image an object located at an object distance $s_o = 2f$ from the lens. In both cases, a wavefront is emitted from the top of the object, and the figure shows selected rays that determine the image location and size. The first of these is a ray from the top of the object and drawn through the center of the lens. This is insufficient to locate an image, which requires two intersecting rays. A commonly used second ray is the one parallel to the optical axis; this is equivalent to a collimated wavefront from a point at infinity and thus images at the focal length f of the lens. For a positive lens, this is a real image located to the right

Geometrical Optics

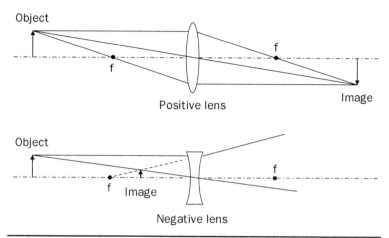

FIGURE 2.9 Graphical method for determining image size and location using rays.

of the lens; for a negative lens, it is a virtual image located to the left of the lens and found by projecting the diverging ray from the lens back to the optical axis.

Example 2.3. For satellite imagers such as Google Earth, the smallest dimension that can be seen on the ground is directly related to the IFOV. This dimension is known as the ground sample distance (GSD), which is equal to the projection of the pixel spacing (sampling) onto a distant surface such as the ground. For FPAs with no dead space between pixels, the GSD equals the projection of the IFOV. For small angles, the GSD is found by multiplying the IFOV by the imager altitude h. In the case of the Google Earth imager, the GSD is approximately 0.5 m; hence Eq. (2.3) gives an IFOV = 2α = GSD/h = (0.5 m)/(500 × 10^3 m) = 1 µrad at an altitude of approximately 500 km. With 35,000 pixels imaging the ground, the FOV = 35,000 × 1 µrad = 35 mrad (~ 2 degrees).

In contrast, commercial cameras often have a much larger FOV (~ 45 degrees), so Eq. (2.3) must be used with caution when relating the IFOV to the FOV. As shown in Table 1.2, such a camera may have a pixel size of d = 7.4 µm and an effective focal length (EFL) of f = 6 mm, from which Eq. (2.3) gives the IFOV as $2\alpha = d/f$ = (7.4 µm)/(0.006 m) = 1.23 mrad. Because of the nonlinearity of the arctangent in Eq. (2.3), the FOV in this case is not found from a simple multiplication of the number of pixels in the FPA (640 in the horizontal direction and 480 in the vertical). Instead, the full size of the FPA must be used in the equation; this yields the FOV in the horizontal direction as follows:

$$\text{FOV}_h = 2\alpha = 2\tan^{-1}\left(\frac{640 \times 7.4 \times 10^{-6} \text{ m}}{2 \times 0.006 \text{ m}}\right) = 0.75 \text{ rad}$$

or 43.1 degrees.

An optical designer is often interested in designing a lens to accommodate the diagonal FOV of the array. Using the specs from Table 1.2 (where we see that the vertical FOV is 33 degrees), we can calculate the diagonal FOV as $(\text{FOV}_h^2 + \text{FOV}_v^2)^{1/2} = (43.1^2 + 33^2)^{1/2} = 54.3$ degrees. Also note that the size of the image is that of the FPA—namely, 640 × 7.4 µm = 4.74 mm in the horizontal direction and 480 × 7.4 µm = 3.55 mm in the vertical direction.

2.3 Relative Aperture

In order to determine the brightness of an image, we must know where it is located and how big it is. For now, we use the term "brightness" in an informal way; intuitively, if an image is too bright (or too dim) then it may not be possible to see detail in the image. Even with a flash on a camera, it is difficult to make some objects (e.g., far-away mountains) brighter; therefore, it's usually best to start out with an image that is too bright and then diminish that brightness via camera settings.

Wavefronts from objects such as mountains are always bigger than the lens, so increasing the lens aperture D collects more light (increases the optical power). A larger lens thus intercepts more of an incident wavefront, with the power collected increasing with the lens area—that is, proportional to D^2. However, Fig. 2.10 shows that a longer focal length yields a bigger image and that the same power distributed over this larger image area looks dimmer. With each side of the image growing with longer f, the area of the image increases with f^2. Thus, the image becomes brighter with a larger *lens* area but dimmer with a larger *image* area.

Our comparison of the aperture area with the image area can be combined into a new concept, known as *relative aperture*. For an object at infinity (and $s_i = f$), the relative aperture is found by taking the ratio of the lens focal length f to its diameter D. This quantity is also known as the *f-number* (written as f/#):

$$f/\# = \frac{f}{D} \tag{2.4}$$

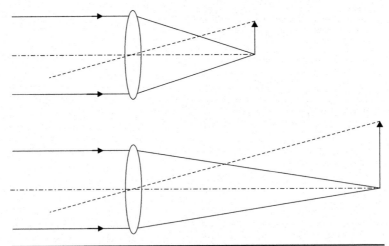

FIGURE 2.10 Longer focal lengths result in dimmer images if the lens diameter is not changed. For a given FOV, the bigger image associated with the longer focal length has the same object energy spread out over a greater area and so is not as bright as the smaller image.

The value obtained is a bit unusual as a performance metric, because a smaller f/# corresponds to a brighter image. Nevertheless, the f/# is commonly used as a measure of the size of the aperture relative to the focal length. Note that, by convention, a positive and negative lens both have a positive f/#.

As the name implies, the f/# can't be used to find the *absolute* brightness of an image, only its *relative* brightness. For example, an image made using an f/2 lens is twice as bright as that made with an f/4; this is because the f/# varies with f/D whereas brightness varies with the square of the f/# (f^2/D^2). The f/2 design is also known as the "faster" lens because the exposure time for FPAs or photographic film is shorter than if an f/4 lens is used.

It is extremely difficult to create a lens whose focal length is short compared with its diameter (see Chap. 7), so a typical number for high-performance cameras is a minimum f/# of about f/1 for the fastest lenses. With the f/D ratio for a lens fixed at the smallest possible f/# of 0.5,[5] Eqs. (2.3) and (2.4) reveal a fundamental relationship between the HFOV and the lens diameter:

$$\alpha \leq \frac{d}{D} \quad [\text{rad}] \tag{2.5}$$

Recall that a larger aperture D must have a longer focal length to keep the f/# the same. This expression, in combination with Eq. (2.3), shows that the longer focal length f (or larger diameter D) will reduce the field angle α if the detector size d is not changed. This law is similar to the second law of thermodynamics in that an equality is the best possible performance. The actual FOV will almost always be smaller, often by a factor of 2 or more, because of f/# limitations and aberrations at the edge of the field (see Chap. 3). Figure 2.11 illustrates the fields and relative apertures (labeled simply as "aperture") available for state-of-the-art designs. Ignoring for now the unfamiliar names, the important point is that—as we work down from the top of the chart through fisheye, retro-focus, double Gauss, Petzval, Schmidt, and so on to bigger, faster lenses with a smaller f/# (larger aperture)—it's much more difficult to design a lens with a large field.

The chart in Fig. 2.11 also shows the use of a related concept: numerical aperture (NA), which is another way of looking at f/#. This value is equal to half the cone angle that the imaged wavefronts make with the optical axis; for paraxial designs with objects at infinity, the NA is given by

$$\text{NA} = \frac{1}{2(f/\#)} \approx \frac{D}{2f} \quad [\text{rad}] \tag{2.6}$$

Here the factor of 2 occurs because the NA is defined using the lens radius whereas the f/# is based on the lens diameter. The inverse expression means that a *larger* NA corresponds to brighter images, which is a more intuitive way to express relative aperture.

36 Chapter Two

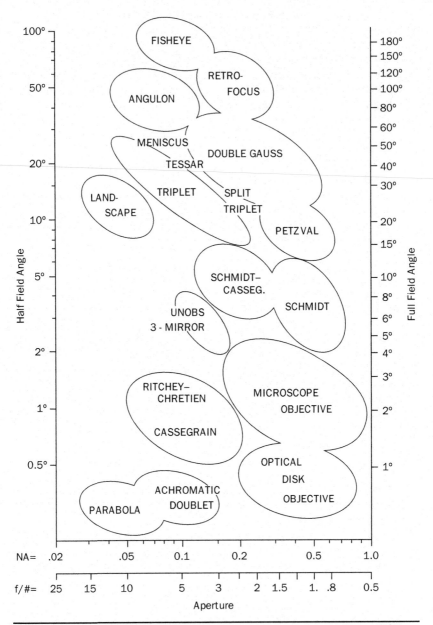

FIGURE 2.11 The combination of aperture (f/# or numerical aperture, NA) and FOV determines the type of lens needed to meet system requirements. (*Credit: Warren J. Smith,* Modern Lens Design, *McGraw-Hill www.mcgraw-hill.com.*)

2.4 Finite Conjugates

Cameras are often used to obtain images of distant objects, but many images are also obtained from nearby objects. A typical example is close-up photography, although microscopes also image objects that are closer than "infinity" (i.e., less than about 10 times the focal length away from the lens and with the wavefront incident on the lens no longer approximately planar). The objects and images in this case are known as *finite conjugates*, and they lead to changes in the calculation of FOV, IFOV, and NA as compared with values for infinite conjugates from distant objects.

As illustrated in Fig. 2.12, a common situation for finite-conjugate imaging is obtained when the object is placed at a distance $s_o = 2f$ from the lens. Equation (2.2) then shows that the image distance $s_i = 2f$, which results in symmetrical wavefronts on both sides of the lens. The FOV in this case is not based on the focal length; instead, the image distance is used with Eq. (2.3) rewritten as $2\alpha \approx d/s_i$ for small angles. Similarly, the calculation for numerical aperture in Eq. (2.6) is rewritten as $NA_i = \sin(D/2s_i) \approx D/2s_i$ on the image side of the lens; numerical aperture can also be defined for the object side as $NA_o \approx D/2s_o$. The use of finite-conjugate image distance can also be generalized to the concept of *working f/#* as given by s_i/D.

More typically, it is the actual size of the image that is used to specify performance with finite conjugates. This leads to the idea of linear magnification, which is the ratio of image size to object size. Similar triangles can be used to show that this ratio is equal to the ratio of image distance to object distance:

$$m = -\frac{s_i}{s_o} \quad (2.7)$$

Here the image may be bigger or smaller than the object, and the negative sign is used because an object is inverted when imaged by a single lens. For the $2f:2f$ object-to-image case shown in Fig. 2.12, the object and image distances are equal and so the linear magnification

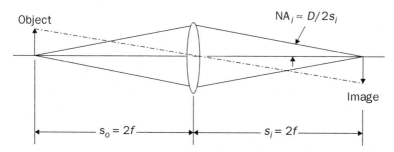

FIGURE 2.12 Finite-conjugate imaging of an object located at an object distance $s_o = 2f$ from the lens. The numerical aperture $NA_i = \sin(D/2s_i) \approx D/2s_i$ for angles less than about 30 degrees.

$m = -1$ (a case known as "unity" conjugates). For infinite conjugates with $s_o = \infty$, Eq. (2.7) shows that the linear magnification $m = 0$; this illustrates the utility of using the angular FOV of Eq. (2.3) for distant objects.

2.5 Combinations of Lenses

It was mentioned at the start of this chapter that, for many cameras, the "lens" actually contains a number of individual lenses. The reason for this is that it's difficult to obtain reasonable image quality over a large aperture and field without a complex combination of lenses. This section reviews how to calculate first-order image locations, sizes, and relative apertures for combinations of thin lenses. Chapter 3 explores how lenses can be combined to obtain good image quality.

The simplest combination is two thin lenses (Fig. 2.13), and we may use the equations from Sec. 2.1 in sequence to establish the resulting image's size and location. This is a powerful graphical technique for laying out a first-order design. However, for quantifying image size and location with a small number of lenses, it is usually easier to use a method based on the *refractive power* of the combination.

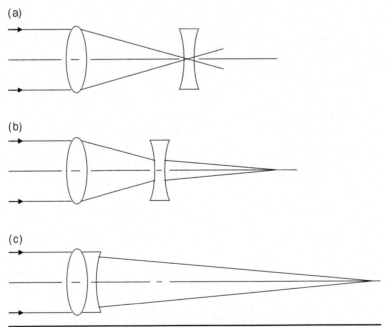

Figure 2.13 Negative lenses can be combined with positive lenses to change the effective focal length. Varying the distance between lenses allows the user to "zoom in" on the object for more detail (smaller IFOV).

We remark that, except for the special case of two thin lenses in contact, the refractive powers of the individual lenses do not simply add when combined. To understand why, Fig. 2.13(a) shows what happens when two thin lenses are separated by the focal length of the first lens. Although it is not obvious what Eq. (2.2) predicts for this case, it is clear that the on-axis rays are not bent as they move through the center of the second lens. As a result, this lens is unable to change the wavefront curvature from an on-axis object at infinity and so the power of the combination is just the power of the first lens.

What happens to off-axis rays is a bit more complex, but the total off-axis power is also the same as that of the first lens. The wavefronts incident on the second lens from points that are either on-axis or off-axis have infinite curvature ($s_o = 0$, giving $1/s_o = \infty$); hence the lens has little effect on this curvature, although off-axis deflection of the entire wavefront still occurs as in a prism (see Sec. 2.9.2).

It is better and thus more common for the first lens not to image directly on the second lens but rather slightly before or after (Chap. 7). Then the rays from the first lens are intercepted off-center by the second lens and are bent as they pass through. The amount of bending depends on the following factors:

- *The refractive power of the first lens*, since it determines the angle at which the incident rays are launched toward the second lens.
- *The spacing between lenses*, which determines how far off-center the intercept occurs at the second lens.
- *The refractive power of the second lens*, which causes the rays to bend when the incident ray is off-center.

When these effects are combined, the total refractive power can be either more or less than the sum of the individual thin-lens powers depending on the distance L between lenses:

$$\phi = \phi_1 + \phi_2 - \phi_1 \phi_2 L = \frac{1}{\text{EFL}} \quad [1/\text{m}] \tag{2.8}$$

Here the individual lens powers ϕ_i are given by Eq. (2.1), and the EFL is the effective focal length of the *combined* lenses.

Example 2.4. Inexpensive digital cameras typically have an optical zoom of about 3–4×, allowing the user to zoom in for a close-up. This is done by lengthening the focal length of a lens combination, yielding a smaller field of view [see Eq. (2.3)]. Changing the focal length of a single lens is not yet a well-developed technology, so inexpensive consumer cameras typically use two or three lenses to change the power (and therefore the effective focal length EFL = $1/\phi$) of the combination. This example illustrates the use of Eq. (2.8) for the design of a two-element zoom lens.

As shown in Fig. 2.14, a typical two-element zoom consists of a positive and negative lens separated by a variable distance.[5,6] With the positive lens (or

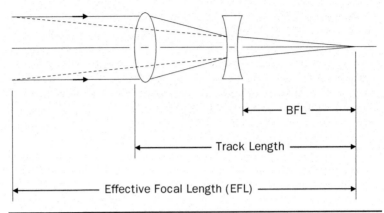

FIGURE 2.14 Typical specs for a telephoto lens: effective focal length (EFL), back focal length (BFL), and track length.

"element") in front, this particular type of zoom is known as a *telephoto* and allows for a longer focal length and narrower FOV than is possible with the front lens by itself. This design form also results in an overall length ("track length") that is less than the effective focal length, enabling compact designs for consumer and space-based applications that require low weight.

The extremes for the spacing are: (1) $L = f_1$, when the lenses are separated by the focal length f_1 of the positive lens on which wavefronts from the object are incident from the left [Fig. 2.13(a)]; and (2) $L = 0$, when the two lenses are in contact [Fig. 2.13(c)]. At this latter extreme, the two lenses in contact have the *least* total power because the rays bent by the first lens don't have room to drop down toward the center of the negative second lens, and the second lens thus subtracts power from the combination. By the same logic, separating the lenses gives the rays some distance to drop, thereby reducing the effects of the second lens to zero when the rays are incident on its center.

Using Eq. (2.8) to quantify the power change, we assume focal lengths of $f_1 = +100$ mm and $f_2 = -150$ mm. The equation shows that the refractive power of the combination for $L = 0$ is $\phi = \phi_1 + \phi_2 = 1/100 - 1/150 = 1/300$, giving EFL = $1/\phi$ = 300 mm for the thin lenses in contact [Fig. 2.13(c)]. When the lenses are separated so that $L = f_1$, Eq. (2.8) gives $\phi = \phi_1 + \phi_2 - \phi_1 \phi_2 f_1 = \phi_1$ and so the EFL = f_1 = 100 mm, as expected from physical reasoning. This particular separation would not produce a practical design—since it's difficult to put a detector right on the back of the second lens—but it does give one a feel for the range of focal lengths available from a zoom lens.

In short, moving a positive and negative lens closer together increases the combination's EFL, enabling a zoom effect. The result is more detail from imaging a smaller FOV onto the same number of pixels, yielding a smaller IFOV.

2.6 Ray Tracing

Equation (2.8) is efficient for two lenses, but it can become tedious for more than that. Fortunately, there's a better method available to find image location, image size, and effective focal length. In its simplest form, the method traces rays through thin lenses and across surfaces using a paraxial form of Snell's law.

This paraxial form, which works for "small" angles when $\sin\theta \approx \theta$ (i.e., for θ less than about 30 degrees), is useful for deriving an algebraic expression that captures how small-angle rays work their way through optical systems. With the paraxial approximation, the resulting expressions are not that useful for exact analysis yet still provide physical insight into first-order optics and optical system design.

The first ray-trace equation describes how rays are bent at a refractive or reflective surface or element such as a thin lens:[7]

$$n'_i u'_i = n_i u_i - y_i \phi_i \quad [\text{rad}] \tag{2.9}$$

Here, as shown in Fig. 2.15, n and u are (respectively) the refractive index and angle of the ray before it hits a surface or thin lens (indexed by the subscript i), and n' and u' are the index and angle after being refracted. This equation is a modification of Snell's law in that u and u' are measured with respect to the horizontal optical axis and not to the surface normal. The modification that results is the $y\phi$ term on

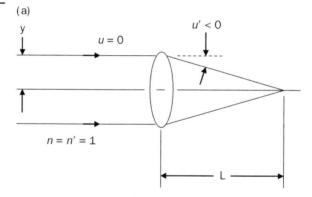

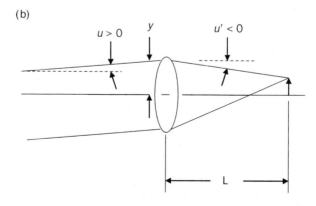

Figure 2.15 Ray tracing using the refraction equation [Eq. (2.9)] for (a) on-axis, and (b) off-axis planar wavefronts incident on a thin lens. Positive angles (u and u') are those with a positive slope; negative angles have a negative slope.

the right-hand side of Eq. (2.9), which is the product of the ray height y on the surface and the surface refractive power ϕ. The derivation of this additional term for the u–u' horizontal axis angle system is given in Chap. 3 of Kingslake.[8]

As shown in Fig. 2.16, the $y\phi$ term has the following physical interpretation: if a ray hits a surface on its optical axis ($y_i = 0$), then that axis is also the surface normal and so the ray bends according to the first-order form of Snell's law ($n'u' = nu$). When rays hit the lens off-axis ($y_i \neq 0$), the positive surface bends the ray more toward the horizontal with a contribution that increases in proportion to the surface power and the height of the ray. This is also shown in Fig. 2.16, where the increase in incident angle of refraction ($u + \Delta u$) depends on the ray height as well as the surface power. The result is a smaller refracted angle ($\theta' = u' + \Delta u'$) than that of the optical axis rays.

After bending at a surface, the ray is then traced to the next surface or thin lens (numbered $i + 1$) using the second ray-trace equation, which is nothing more than an equation for a straight line:

$$y_{i+1} = y_i + u'_i L \quad \text{[mm]} \tag{2.10}$$

Here L is the distance from one surface to the next, and y_{i+1} is the height of the ray where it intercepts the next surface. Observe that the sign convention for a positive angle u' increases the ray height on the next surface.

Thus, we can use Eqs. (2.9) and (2.10) for paraxial ray tracing to follow a first-order ray through an optical system for any number of surfaces and thin lenses. The following example illustrates this procedure for two thin lenses.

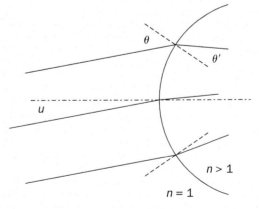

Figure 2.16 Refraction at the surface of a thick lens, which illustrates the dependence of refractive angle on the ray height y_i in Eq. (2.9); here θ and θ' are measured with respect to the surface normal (dashed lines).

Example 2.5. This example illustrates how Eqs. (2.9) and (2.10) can be used to find the effective focal length of the two-element zoom lens analyzed in Example 2.4. As before, the two extremes for the lens spacing are: (1) $L = f_1$, for when the lenses are separated by the focal length f_1 of the positive lens [Fig. 2.13(a)]; and (2) $L = 0$ for when the two lenses are in contact [Fig. 2.13(c)]. Planar wavefronts are incident on the first lens from the left, and the lenses again have focal lengths $f_1 = +100$ mm and $f_2 = -150$ mm. Each lens has the same diameter D.

For case (1), the incident angle $u_1 = 0$ because an on-axis object at infinity is being imaged. The incident height y_1 is also selected as the maximum possible (half the diameter), so $y_1 = D/2$. In addition, the thin lenses are in air, so $n_1 = n_1' = 1$ on both sides of the lens. Equation (2.9) then shows that the refracted angle on the back side of the first lens is $u_1' = 0 - (D/2)\phi_1 = -(D/2f_1)$. As shown in Fig. 2.15, the negative sign indicates that the angle has a negative slope with respect to the horizontal (a positive sign would indicate a positive slope). The intercept with the second lens is given by Eq. (2.10) with $L = f_1$, so that $y_2 = D/2 - f_1(D/2f_1) = 0$. Applying Eq. (2.9) at the second lens gives a refracted angle of $u_2' = u_2$, so the rays go through the center of the lens without bending. Since the second lens has no effective power for these rays, the effective focal length of the combination is that of the first lens—the same result found in Example 2.4.

For case (2), Eq. (2.9) gives the same result for refraction at the first lens, where $u_1' = -D/2f_1$. Because the two lenses are thin and in contact [$L = 0$ in Eq. (2.10)], the intercept with the second lens is the same as that of the first lens; thus we have $y_2 = D/2$. Using Eq. (2.9) for refraction across the second lens—and since $u_2 = u_1'$ (or $u_{i+1} = u_i'$ in general)—it follows that

$$u_2' = u_2 - y_2\phi_2 = -\frac{D}{2f_1} - \left(\frac{D}{2}\right)\phi_2 = -\left[\frac{1}{f_1} + \frac{1}{f_2}\right]\left(\frac{D}{2}\right)$$

Transferring this slope to the third "surface" at focus (where $y_3 = 0$), we obtain $1/\text{EFL} = 1/f_1 + 1/f_2$, which is also identical to what we found in Example 2.4.

Therefore, if we have just two thin lenses then Eq. (2.8) is less work than Eqs. (2.9) and (2.10). However, as shown in what follows, the ray-trace equations are easier to work with for three or more thin lenses or surfaces and also for thick lenses.

Examples 2.4 and 2.5 illustrate how a ray that hits the center of a thin lens goes straight through without bending. Hence the lens has no power for these rays, and it does not influence the effective focal length of a two-lens combination if the second lens is at the focal length of the first.

That being said, lens power doesn't change just because we have selected a certain ray going through it. An unlimited number of additional rays that do *not* go through the center of the lens could also be chosen, and these rays would be bent. Should we therefore conclude that the calculated EFL depends on which rays are chosen? Or that the lens power ϕ is not simply $1/f$ and instead depends on the choice of ray?

Clearly not. Selecting certain rays simply makes it easier to think the problem through. If one ray is chosen that simplifies the problem, the physics doesn't change for a different ray; it just becomes more difficult to understand what's going on. Typically, rays are chosen to help find the image position and FOV; such rays are called the *marginal* ray and the *chief* (or *principal*) ray, respectively. These two rays are used to understand almost every optical system.

Returning to the simple case of the single thin lens shown in Fig. 2.6, the on-axis marginal rays are those parallel to the optical axis and that place the image at the lens's focal length for an object at infinity. The chief ray is the off-axis ray that images at the top of the detector array; it is brought into focus at the maximum image height allowed by the focal length and detector size (and therefore determines the FOV). Note that there are two rays at the outer diameter of the lens that are also brought to focus; these are called *off-axis* marginal rays and the *on-axis* marginal ray (sometimes called the *axial* ray).

For the zoom lens of Example 2.5, the second lens in case (2) increases the focal length of the combination by bending the marginal ray away from the optical axis. However, in case (1) of Example 2.5, the marginal ray is imaged at the center of the negative lens; hence this lens has no effect on the focal length. If this second lens is moved slightly toward the first lens (in order to narrow the FOV) then the marginal ray will be bent away from the optical axis by the negative lens, increasing the effective focal length.

Most practical cases are found between the two extremes of Example 2.5; in other words, the second lens is located between the first lens and its focal length, bending both the chief and marginal rays. It is easy enough to use the ray-trace equations to calculate the EFL for such combinations. What is not clear, though, is just where the EFL is measured from. It may seem that it could be measured from the first lens, but as shown in the next section, this rarely occurs for lens combinations.

2.7 Thick Lenses

The combination of thin lenses in Example 2.5 acts like a thick lens, and that combination forces us to think about imaging a little bit differently. It's also possible to have a curved surface followed by "thick" glass, in which case our assumptions about thin lenses do not apply, even though the same methods used to understand combinations of thin lenses continue to work.

One important difference between thin and thick lenses is locating the EFL. For a single thin lens, it's straightforward to measure (with relatively small error) the focal length from either the front or back of the lens. But for a thick lens or a combination of thin lenses, the EFL is usually not measured from either the front or the back of any surface or lens. In this section, we take a look at how to locate the EFL and how to calculate other distances needed to develop a practical understanding of imagers containing multiple lens elements.

One distance that is a reliable reference is the length from the back of the last lens to the image plane. Known as the back focal length (BFL), it will vary depending on the number of lenses and surfaces, their thickness, their indices of refraction, and so on. The BFL can be

Geometrical Optics   45

calculated using the ray-trace equations from Sec. 2.5. Figure 2.17 shows that the value obtained depends on the angle u'_f of the marginal ray as it exits the back surface of the last lens: $BFL = -y_f / u'_f$, where y_f is the marginal ray height at the back surface.

The EFL also can be found using this same back-surface angle as it intercepts the original height of the marginal ray when it comes into the thick lens. Thus $EFL = -y_i / u'_f$, where y_i is the initial marginal ray height at the front surface of the first lens. This is also seen in Fig. 2.17, which shows the *principal plane*; this plane perpendicular (normal) to the optical axis and is where the parallel marginal rays coming in from infinity appear to start bending on their way toward focus (even though they do not actually bend until they hit the first lens). So to locate the EFL, first the BFL is used to find the image plane with respect to the back surface of the last lens and then the EFL is measured with respect to this image plane. Example 2.6 illustrates the specifics.

Example 2.6. The combination of two thin lenses can be considered a thick lens; in this example, the EFL and BFL are found for the zoom lens in Example 2.4 with the following modification: the negative lens is now located 50 mm to the right of the first lens (or midway between the first lens and its focal length). Calculating the EFL for two lenses is straightforward using Eq. (2.8): $\phi = \phi_1 + \phi_2 - \phi_1\phi_2 L = 1/100 - 1/150 + 50(1/100)(1/150)$, so that the EFL = $1/\phi$ = 150 mm. Yet we need to establish exactly where the EFL is measured from; it is often assumed to be with respect to the first lens, but this is generally not the case.

The BFL is located first; for this we use Eqs. (2.9) and (2.10) to find the angle of the marginal ray as it exits the second lens. Results are shown in spreadsheet form in Table 2.2. Going through the numbers, the angle

$$u'_1 = 0 - \left(\frac{D}{2}\right)\phi_1 = -\frac{D}{2f_1} = \frac{(-20 \text{ mm})}{(2 \times 100 \text{ mm})} = -0.1$$

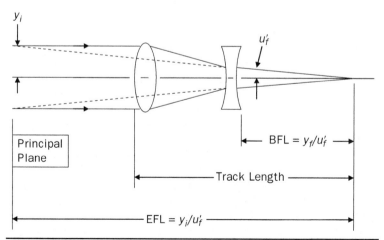

FIGURE 2.17 Determining the BFL and EFL using the ray-trace variables u'_f (the ray angle exiting the final surface), y_i (the marginal ray height on the initial surface), and y_f (the ray height exiting the final surface).

Surface	n	u	n'	y [mm]	f [mm]	φ [1/mm]	u'	L [mm]	BFL [mm]	EFL [mm]
1	1.0	0.00	1.0	10	100	0.01	−0.10	50		
2	1.0	−0.10	1.0	5	−150	−0.006666	−0.066666		75	150

TABLE 2.2 Spreadsheet illustrating the use of the ray-trace equations for on-axis marginal rays ($u = 0$) to determine the image location for the telephoto zoom lens in Example 2.4.

Geometrical Optics

after the first lens (of diameter $D = 20$ mm). The intercept height with the second lens occurs at $y_2 = D/2 - (f_1/2)(D/2f_1) = D/4 = 5$ mm. Using that $u_2 = u'_1$ (or, more generally, $u_{i+1} = u'_i$), the slope of the ray after passing through the second lens is $u'_2 = u_2 - y_2\phi_2 = -D/2f_1 - (D/4)\phi_2 = -2/30$ rad. This makes sense from a physical standpoint given that the negative lens has bent the ray from the first lens away from the optical axis, thereby giving it an angle closer to zero.

Transferring this slope to the third "surface" at the image plane (where $y_3 = 0$), or by using the definition of paraxial angles ($-u'_2 = y_2/$ BFL), the BFL $= (D/4)(30/2) = 75$ mm. This distance is measured from the back surface of the center of the negative lens. These results are also summarized in Table 2.2.

The EFL is thus located 150 mm to the left of the image plane, or 25 mm to the left of the first lens. This is the location of the principal plane, where the two lenses act as a single thin lens and bring collimated light to a focus at a plane 150 mm to the right. This value can be checked by using the same paraxial convergence angle that was used to calculate the BFL and extending it up to the maximum height of the collimated ray; doing so yields EFL $= (D/2)(30/2) = 150$ mm, which is consistent with Eq. (2.8).

The use of the principal plane also illustrates the benefits of using the positive–negative combination of the telephoto lens. Specifically, the track length of 125 mm is *less* than the effective focal length of 150 mm. The principal plane in front of the first lens allows imagers to be designed with a smaller overall length than the EFL, a feature not possible with individual lenses. The reduction in length is quantified by the telephoto ratio, which is equal to the ratio of track length to EFL; in this case, (125 mm)/(150 mm) = 0.83. Depending on the aperture and FOV (and the resulting aberrations), the ratio cannot generally be much smaller than this, yet the telephoto concept remains important in weight-critical applications such as consumer cameras and space-based telescopes.

In addition to combinations of thin lenses, the concepts of BFL and EFL also apply to individual thick lenses. This is shown in Fig. 2.5, which sketches the BFL and EFL locations for six lenses, including a plano-convex lens (a lens of positive focal length that converges an incident wavefront from infinity) and a plano-concave lens (a lens of negative focal length that diverges an incident wavefront from infinity). The principal plane used for imaging objects at infinity is located at the principal point P2.

Figure 2.5 also shows another principal plane (located at point P1), which can be used in conjunction with P2 to trace the chief and marginal rays of finite conjugates. Point P1 is found by turning the lens around and then tracing a new marginal ray; for a symmetric lens, the two planes will be at the same point. Both principal planes are needed for finite-conjugate imaging; this is illustrated in Fig. 2.18, where the object distance S is measured from the object to the first principal plane P_1 and the image distance S' is measured from the second principal plane P_2 to the detector or FPA.

We refer to Table 2.3 to determine whether a lens is "thick" or "thin"; the table shows the reverse engineering of a plano-convex lens manufactured by CVI Melles Griot and illustrates calculations for the BFL and EFL. The lens is made from Schott SF-11 glass, which

Surface	n	u	n'	y [mm]	R [mm]	φ 1/mm	u'	L [mm]	BFL [mm]	EFL [mm]
1	1.0	0.00	1.76312	10	83.98	0.0090869	−0.05	4		
2	1.76312	−0.05	1.0	9.79384	1E+10	−7.63E−11	−0.09087		107.8	110.0

TABLE 2.3 Spreadsheet illustrating the use of the ray-trace equations for a plano-convex lens manufactured by CVI Melles Griot (Product Number 06-LXP-009). The second surface is flat with an infinite radius of curvature, but is entered in the spreadsheet as $R_2 = 10^{10}$ mm.

Geometrical Optics

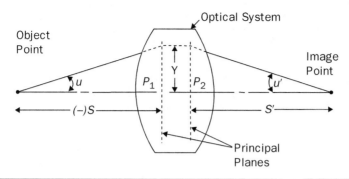

FIGURE 2.18 Finite-conjugate imaging using the principal planes of a thick lens. (*Credit: Warren J. Smith, Modern Optical Engineering, McGraw-Hill, www.mcgraw-hill.com.*)

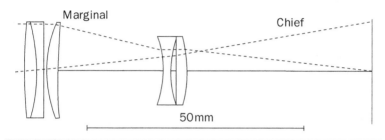

FIGURE 2.19 Chief and marginal rays for a telephoto lens. The positive lens consists of the three-element group on the left; the negative lens is the two-element group in the middle. (*Credit: Warren J. Smith, Modern Optical Engineering, McGraw-Hill, www.mcgraw-hill.com.*)

has refractive index $n = 1.76312$ at a wavelength $\lambda = 830$ nm. Note that—instead of a thin-lens focal length—a radius of curvature R [as in Eq. (2.1)] is now associated with each surface, so that the surface power $\phi = (n' - n)/R$. Also note that the length L is the center thickness of the lens (CT in Fig. 2.5). When calculating EFL and BFL, we change the radius of curvature for the first surface (R_1) until the EFL and BFL match the values given in the catalog.

Note that the BFL and EFL are not significantly different for this lens. Because this difference is a small fraction of the EFL, the thin-lens approximation is valid in this case. In practice, the thin-lens concept is used to establish the first-order design. Such designs are then expanded into thick lenses (to correct aberrations) and propagated through the optical system using a commercial software package such as Zemax or Code V. Figure 2.19 shows the example of a telephoto lens assembly that splits the positive and negative lens elements into multiple elements to correct for aberrations. The transformation from geometrical optics to aberration-corrected designs will be reviewed in Chap. 3.

2.8 Stops, Pupils, and Windows

In addition to focal length, the use of thick lenses changes our perspective on a number of properties of an optical system; these properties include f/# and FOV as well as object and image distance. This section reviews how the f/# and FOV for a thick lens differ from the simple thin lens.

2.8.1 Aperture Stops and Pupils

Section 2.3 showed how the f/# for a thin lens was determined by lens diameter and focal length, and we also saw in Sec. 2.7 that an effective focal length is used for thick lenses or combinations of thin lenses. So it seems that the EFL is necessary to calculate the f/#. Although this is certainly true, there is another complexity to be included—namely, for a system with more than one lens, it's not obvious which lens diameter should be used for D in the equation for calculating f/#.

Figure 2.20 shows what happens when the diameter of the negative lens in Example 2.6 is changed. In Fig. 2.20(a), both lenses are the same size, and the diameter of the first lens controls the amount of light that enters the system. The first lens is thus known as the aperture stop (or often just "the stop"—a physical aperture that controls the amount of light collected by an optical system), and its diameter is what we use to calculate the f/# of the telephoto.

In Fig. 2.20(b), the second lens eventually becomes the aperture stop as its transmitting aperture is reduced, limiting the off-axis rays. However, its diameter cannot be used to calculate f/#. The reason is that wavefronts coming from the object will not "see" the diameter of the second lens; rather, they will be limited by the size of the *image* of the second lens. Just as a lens creates an image of an object at infinity, so it also creates an image of anything else that is sending wavefronts its way—including light scattered by other lenses. In Fig. 2.20(b), the image of the second lens is what controls the amount of light entering the telephoto; this lens is called the entrance pupil (EP). Returning to Fig. 2.20(a), there is no image of the aperture stop, only the stop itself, and in that case the stop is *also* the entrance pupil. In any case, the entrance pupil diameter (EPD) that's seen from the object determines the amount of light entering an optical system, and the f/# can then be calculated from the ratio EFL/EPD.

Similarly, when we stand at the image and look back through the second lens, the diameter that controls the amount of light leaving the zoom is known as the exit pupil (XP). So in Fig. 2.20(a) the exit pupil is the image of the aperture stop (the first lens) created by the second lens, and in Fig. 2.20(b) the exit pupil is the aperture stop itself (which is now the second lens), since there are no lenses after the stop. The following example illustrates these concepts.

(a)

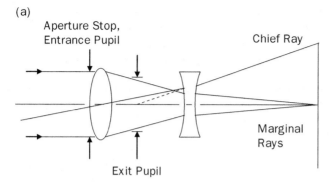

(b)

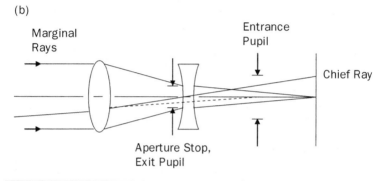

FIGURE 2.20 Change in entrance and exit pupils in response to reduction of the transmitting aperture of the negative lens in a two-element telephoto.

Example 2.7. Figure 2.20(a) shows the first lens as both an aperture stop and entrance pupil, with the f/# for the system given by f/# = EFL/EPD = (150 mm)/(50 mm) = 3 for a 50-mm lens diameter. The exit pupil is the image of the aperture stop by any lenses after the stop (the negative lens between the stop and the detector). If we place the lenses 50 mm apart, then the imaging equation locates the image of the aperture stop for $s_o = 50$ mm and $f = -150$ mm; substituting these values into Eq. (2.2), we find that the exit pupil is located at $s_i = -37.5$ mm (to the left of the second lens).

Figure 2.20(b) shows a different situation: now the second lens is the aperture stop and the exit pupil. This means that the diameter of the clear aperture of the second lens controls the amount of light getting into the system. The on-axis marginal-ray trace in Table 2.2 shows that a 50 mm diameter for the first lens requires a diameter of at least 25 mm for the second lens. By using an aperture with a smaller diameter (e.g., 20 mm) for this lens, it becomes the aperture stop.

In this case, the entrance pupil is the image of the aperture stop by any lenses before the stop (the positive lens between the stop and the object). Because looking into the telephoto from the left requires wavefronts to propagate from right to left, the sign conventions in Table 2.1 must be reversed [or Fig. 2.20(b) can be turned upside down, in which case the conventions are not changed]. Again placing the lenses 50 mm apart, the lens equation is used to find the image location. This gives $s_o = 50$ mm and $f = 100$ mm; when substituted in Eq. (2.2), these values result in an entrance pupil located at $s_i = -100$ mm (to the *right* of the first lens, or 50 mm to the right of the negative lens).

The size of the entrance pupil is given by $m = -s_i/s_o = 2$ times the size of the second lens, so the EPD = 40 mm. The f/# of the telephoto in this case is f/# = EFL/EPD = (150 mm)/(40 mm) = 3.75 at this particular zoom setting. Note that the f/# of the first lens is not the same as that of the telephoto, since the f/# of the first *lens* (or element) is $f_1/D_1 = 100/50 = 2$. We should expect the system to have a larger f/# because the negative element has increased the overall focal length, which makes the system slower in comparison with the positive element by itself.

To verify that the sign conventions for the entrance pupil have been modified properly, left-to-right ray traces for the chief ray are shown in Table 2.4 for Figs. 2.20(a) and 2.20(b). The lens in Fig. 2.20(a) is an aperture stop, so the trace in Table 2.4(a) shows the chief ray going through the center of the first lens ($y_1 = 0$) and the wavefronts from all field angles must go through its center (as explained in what follows). The XP location is then determined by projecting the slope of the chief ray exiting the second lens ($u'_2 = 0.02327$ rad) back to the optical axis, since this is the point from which the center of the wavefronts from all field angles appear to be exiting the telephoto.

The table thus illustrates that chief-ray tracing allows the location of the pupils to be determined; they are found where the chief ray intercepts the optical axis. As we have seen, chief-ray tracing is also used to calculate the size of the image. In sum: the chief ray determines the image size and stop location (as it intersects the optical axis at the center of the stop); and the marginal ray determines the stop size and image location (as it comes to focus at the center of the detector, which is usually also the field stop). There may be various pupils other than the entrance and exit; these are known as intermediate pupils.

The second trace shows the location of the entrance pupil with respect to the first lens, which is found by maintaining the same chief-ray field angle (of 1 degree) into the first lens. This is calculated in Table 2.4(b) by varying the chief-ray y intercept on the positive lens until the intercept on the negative lens is zero (since the negative lens is now the aperture stop). The EP location is then determined by projecting the slope of the chief ray entering the first lens ($u_1 = 0.01745$ rad) up to the optical axis, since this is the point from which the center of the wavefronts appear to be entering the system.

It was mentioned in Example 2.7 that the aperture stop is where the wavefronts from all field angles enter the optical system. Physically, this means that all the light that makes its way to the image must first pass through the aperture stop. Certainly some of that light will be absorbed or reflected and thus never make it to the image; still, to have any chance at all of adding to the image, light must pass through a real aperture stop. Apertures stops, then, are like a Main Street for wavefronts, a bottleneck through which all rays must pass before they are redirected by the optics to their place in the image.

Since the entrance and exit pupils are images of the stop—the word "conjugate" is often used instead of "images"—any ray directed toward the center of the entrance pupil will go through the center of the stop itself and appear to leave the lens from the center of the exit

Surface	n	u	n′	y [mm]	f [mm]	φ [1/mm]	u′	L [mm]	XP [mm]
1	1.0	0.01745	1.0	0.0	100	0.01	0.01745	50	
2	1.0	0.01745	1.0	0.8725	-150	-0.006666	0.02327		-37.5

TABLE 2.4(A) Left-to-right ray trace, for the chief ray in a two-lens zoom, used to determine the location of the exit pupil with respect to the second lens (37.5 mm to its left). As in Fig. 2.20(a), the positive lens is the aperture stop and entrance pupil.

Surface	n	u	n′	y [mm]	f [mm]	φ 1/mm]	u′	L [mm]	EP [mm]
1	1.0	0.01745	1.0	-1.745	100	0.01	0.03490	50	
2	1.0	0.03490	1.0	0.0	-150	-0.006666	0.03490		100

TABLE 2.4(B) Left-to-right ray trace, for the chief ray in a two-lens zoom, used to determine the location of the entrance pupil with respect to the first lens (100 mm to its right). As in Fig. 2.20(b), the negative lens is the aperture stop and exit pupil.

pupil. Since the center of the pupils is the optical axis, Example 2.7 shows that the chief-ray angles can be projected to (or from) the optical axis to find the pupil locations. As a result, whatever happens at the entrance to the system must be "mirrored" (imaged) at the exit.

Hence there are two ways of looking at aperture stops and their pupils: (1) the entrance and exit pupils are images of the aperture stop; or (2) the chief ray appears to enter the system through the center of the entrance pupil, passes through the center of the aperture stop, and appears to leave through the center of the exit pupil. Note that this second perspective is a consequence of the first.

In addition to using the lens diameter as a stop, it is also common to use a separate piece of metal or plastic with a hole in it. In this case, the stop is not restricted to being located at a lens and can be placed somewhere between lenses, or even in front of or behind them. The motivation for using a separate aperture is that it facilitates the control of aberrations that reduce image quality (see Chap. 3). Figure 2.21 illustrates the use of such an aperture stop for a three-element lens.

Finally, objects at infinity were used in Example 2.7 to determine which lens is the stop. When the object is closer (as with close-ups taken with a digital camera), the lens or aperture that acts as the stop can be different. The stop will always be the lens or aperture diameter that has the smallest-size image when seen from the object, just as it is for an object at infinity. For more details, see Sec. 7.7 of Jenkins and White.[2]

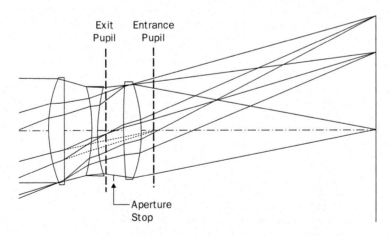

FIGURE 2.21 Graphical method for locating the entrance and exit pupils of a lens. The aperture stop is the physical aperture located between the second and third lens, just to the right of the exit pupil. (*Credit: R. R. Shannon,* The Art and Science of Lens Design, *Cambridge University Press, 1997, reprinted with permission.*)

2.8.2 Field Stops and Windows

The previous sections have shown how the size of an aperture stop is part of what determines the f/# of an optical system. Another, equally important stop is the *field stop*, which is a physical aperture that controls the size of the imaged field. This aperture is almost always placed at an image plane and may simply be the detector itself; for thin lenses, the FOV is then determined by the detector size and the focal length of the lens [Eq. (2.3)]. This result may be compared with the typical field angles shown in Fig. 2.11 in order to confirm that the optical system is not being specified to perform the impossible.

For thick lenses and lens combinations such as the telephoto zoom, the details of how rays work their way through the system can be ignored, and the effective focal length can be used in Eq. (2.3) to find the FOV. For example, given an FPA consisting of 512 × 512 pixels of size 10.2 μm each, the detector size d = 512 × 0.0102 mm; this is equivalent to a square 5.22 mm × 5.22 mm (7.39-mm diagonal). These are also the dimensions of the field stop, for which Eq. (2.3) then gives an FOV ≈ d/EFL = 7.39 mm/150 mm = 0.049 rad (2.8 degrees) if we use the chief-ray angle u_1 = 0.0246 rad for the telephoto in Example 2.4. More generally, the FOV of the object is determined by the ray from the object to the center of the entrance pupil.

Placing a field stop at an intermediate image (Fig. 2.22), as is often done to keep stray light from other parts of the system being reimaged onto the FPA by the relay, may unintentionally restrict the field to a smaller FOV than allowed by the field stop at the FPA. To prevent this, the stop locations *and shapes* must be conjugates, with the relay lens transferring the intermediate stop to the image plane (or vice versa). Intermediate field stops are sometimes made circular even though the FPA is square or rectangular. This clearly restricts the FOV differently than does the FPA, introducing losses or allowing stray light into the system from the edges or corners of the field.

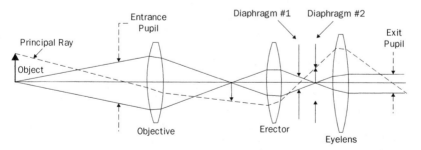

Figure 2.22 The stop that limits the FOV in a multielement system may be either a field stop at an intermediate image (diaphragm #2) or the size of the FPA (not shown for this afocal system). Also shown is the aperture stop at diaphragm #1. (*Credit: Warren J. Smith,* Modern Optical Engineering, *McGraw-Hill, www.mcgraw-hill.com.*)

The field stop, like the aperture stop, has both object and image conjugates created by the elements to the left and right of the stop. Such conjugates are called "windows," as in a window through which the optical system looks out into the world. The image of the field stop by all the lenses before the stop (between it and the object) is known as the entrance window, and the image made by any lenses after the stop is the exit window. When the field stop is located at an image plane, the entrance window is located at the object and the exit window is located at the detector (the image of the object). Because this is usually the case, the term "window" is rarely used and instead reference is usually made to the field stop itself. The entrance window can also be used in conjunction with the entrance pupil to determine the FOV.[5]

2.9 Afocal Telescopes

Some imagers are not as compact as the digital-camera zoom used as a design example in the previous sections because they require a larger aperture to collect more light. These larger imagers have sometimes been called telescopes, and their bigger size entails several important design differences compared with cameras, such as a longer EFL and a smaller FOV. In the past these imagers were restricted by the use of photographic film as an image medium, but the development of FPAs has allowed imaging telescopes to become much more useful.

A second type of telescope is the so-called *afocal* telescope, which has a number of important uses in optical systems. As the name implies, these telescopes do not bring the rays from an object directly to a focus; instead, they change the beam diameter and so produce an afocal (nonfocal, or collimated) output. This is shown in Fig. 2.23, where the incoming wavefronts from a distant object are collected by a relatively large aperture lens or mirror and then reduced to some smaller diameter by a second lens (or mirror). After the diameter is reduced, the rays are brought to a final image by additional lenses or mirrors later on in the optical path.

There are two main reasons for reducing the diameter in this way. First, the lens used for the final imaging may be a human eye, which works well with a collimated beam about 5 to 8 mm in diameter. The afocal telescope is perfect for this application, and it is commonly used in binoculars, riflescopes, and hobby telescopes for backyard astronomy. Second, space-based telescopes often use the smaller beam diameter to scan the heavens with a moving mirror; this technique involves significantly less volume, weight, power, and money than scanning a larger beam or moving the entire telescope.

(a)

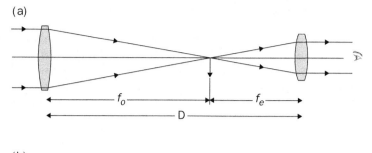

(b)

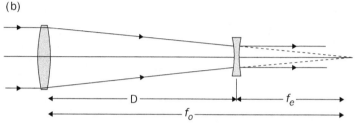

FIGURE 2.23 The two types of afocal telescopes: *(a)* the astronomical (or Keplerian) telescope consists of two positive lenses with an inverted intermediate image located at the focal length of the first lens; *(b)* the Galilean telescope consists of a positive and negative lens and has no intermediate image. (*Credit: Warren J. Smith*, Modern Optical Engineering, *McGraw-Hill, www.mcgraw-hill.com.*)

2.9.1 First-Order Optical Design

In previous sections, the distance between two lenses was changed to zoom in on an object. In this section, the wavefronts after the second lens in an afocal telescope are not converging to an image, but are instead collimated. As shown in Fig. 2.23, this happens when the second lens (known as the "eyepiece" to amateur astronomers) is placed at the focus of the first lens (the "objective"). This arrangement allows the beam, which is converging after the focus of the objective, to become diverging; the second lens takes these diverging wavefronts and converts them to planar wavefronts (this process is the reverse of the convergent focusing of planar wavefronts, reviewed in Sec. 2.1).

There are two different design forms for afocal telescopes. Galileo used a positive lens followed by a negative lens; that is, a telephoto arrangement with an infinite focal length. Unfortunately, this combination has a small field;[5] hence it useful mainly for laser applications and low-performance telescopes such as opera glasses. Kepler invented a more versatile design by using two positive lenses to give a bigger field of view, which can be made even bigger by using a third lens (known as a *field* lens). Although a Galilean telescope is shorter and lighter than its Keplerian counterpart, the latter's larger field and accessible exit pupil have made it more popular for many imaging applications.

Because the afocal telescope does not bring an object at infinity to a focus, it does not have any refractive (or reflective) power. Another way to look at this is that each individual lens has a focal length but their combination does not. Equation (2.8) demonstrates this mathematically: the telescope's power $\phi = \phi_o + \phi_e - \phi_o \phi_e L$ (where the subscripts o and e denote "objective" and "eyepiece"). Since $L = f_o + f_e$, we find that $\phi = 0$ as expected.

Just as with imaging telescopes, the size of the entrance pupil affects image brightness. Suppose we have a telescope with an objective focal length of 200 mm and an eyepiece with a smaller focal length (say, 20 mm); it's straightforward to see that, for an object at infinity, the objective is the entrance pupil. The exit pupil is then the image of the objective formed by the eyepiece. Using Eq. (2.2), we calculate that it is located 22 mm to the right of the eyepiece and although it is smaller than the objective by the transverse magnification factor $m = -s_i/s_o = -22/220 = -1/10$, the brightness is controlled by the size of either the entrance or exit pupil (see Chap. 4).

It's also interesting to see what happens to the chief ray as it works its way through an afocal telescope. Starting with a FOV angle of 1 degree (for example) incident on the objective from an object at infinity, the chief ray goes through the center of the objective before intercepting the eyepiece at a height y_e = 220 mm × 0.01745 rad = 3.84 mm. The eyepiece then bends the chief ray according to Eq. (2.8), which gives the angle after the eyepiece as $u'_e = u_e - y_e \phi_e = 0.01745 - 3.84/20 = -0.1745$ rad, or −10 degrees for an angular magnification (or magnifying power) $M = u'_e / u_o = -10$.

Using the "similar triangles" geometry shown in Fig. 2.23(a) yields the same result, which shows that the angular magnification is also given by the ratio of focal lengths: $M = -f_o/f_e$. It is thus no accident that the angle after the eyepiece is exactly −10× the incident angle for this example. Looking through the exit pupil from the eyepiece side, any object in the field is magnified to appear 10 times bigger (in this case), with the exiting chief ray in general M times larger than what's coming in through the entrance pupil (see Fig. 2.22). If this 10× afocal is turned around, then the reverse is true and the object now appears smaller, as can easily be seen by looking into the objective lens of a binocular. This change in chief ray angle is the reason for calling M an *angular* magnification factor, distinct from the transverse magnification m used for finite-conjugate optics.

Looking into the smaller pupil of an afocal telescope always gives a magnified image, and looking into the larger pupil always gives a "minified" image. Mathematically, this is illustrated by our definitions for M. Physically, it is a result of the different chief ray angle through the exit pupil, though explaining *why* this is the case is difficult using only the concepts discussed so far in this book. Briefly: in the same way that a shorter pulse has more high-frequency components (in time), a smaller pupil has more high-angle components (in space). As shown in Chap. 4, this phenomenon—an optical engineering version of the Heisenberg uncertainty principle—is described by a constant known as the optical invariant, so smaller pupils have larger chief-ray angles (and vice versa). It is a basic physical principle that holds for all optical systems, afocal or imaging, as the chief-ray angles in Table 2.4(b) also illustrate for the imaging (focal) telephoto.

The negative sign in the angular magnifications shows that the angle exiting the eyepiece is in the opposite direction of the angle incident on the objective, so a field of view of +1 degree through the EP becomes –10 degrees through the XP. This results in an upside-down image unless another lens, known as a *relay* lens, is added to correct for the inversion. This is not an issue with telescopes used for astronomy, since the concepts of "up" and "down" are meaningless in outer space, but the correction is necessary when designing binoculars or a riflescope for terrestrial use.

Another engineering issue that must be addressed when designing afocal telescopes for use with the eye is the distance from the eyepiece to the exit pupil. To see the entire field, the eye's (entrance) pupil must be at the telescope's exit pupil. If the XP is too close to the eyepiece then it's impossible to get the eye close enough to the XP to see the entire field, as anyone who wears glasses knows all too well. This distance is known as the eye relief (ER), and it varies with application and user. For example, whereas an ER of about 10 mm is typical for an astronomy telescope used by someone without glasses, 20 mm or so is needed for someone who wears glasses—and 100 mm or more is needed for a riflescope used on a high-power gun that recoils back toward the eye.

2.9.2 Relay and Field Lenses

In addition to the objective and eyepiece, there are two types of lenses often used in afocal telescopes (as well as many other optical systems). These are relay lenses for transferring an image from one place to another and field lenses for increasing the FOV.

A relay lens used in a riflescope to correct an upside-down image is shown in Fig. 2.24, where the intermediate image created by the objective is now transferred (reimaged) by the relay lens. As shown in previous sections, simple imaging always creates an inverted image, so when the object is imaged twice (first by the objective, and then by the relay), a normal view of the world results.[9] The relayed image is then collimated by the eyepiece, as occurs in the usual Keplerian arrangement. If the relayed image has a linear (transverse)

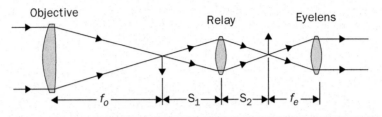

FIGURE 2.24 The relay lens is used to invert the upside-image created by the objective lens into an image suitable for viewing by the eye. (*Credit: Warren J. Smith, Modern Optical Engineering, McGraw-Hill, www.mcgraw-hill.com.*)

magnification of $m = -1$ (unity conjugates with $2f : 2f$ imaging), then the angular magnification is again given by $M = -f_o/f_e$.

Not all afocal telescopes are used with the eye, however, and another common use of a relay lens is in scanning systems. In this case, the relay reimages the entrance pupil onto a scan mirror. Such mirrors are used to either stabilize an image against satellite or platform motion, or to scan across the heavens to collect different images. If the mirror is not placed at a pupil, then any mirror motion would also reduce the FOV as the mirror "view" out through the telescope becomes obstructed by the edges of the objective. With the mirror at an image of the objective, however, the wavefronts that made their way to the exit pupil have all gone through the objective, with the chief ray going through the center of both. As a result, small motions that occur at this image are reproduced at the entrance pupil, and the FOV doesn't change. This is illustrated in Fig. 2.25 and Example 2.8.

Example 2.8. It's common for the so-called "eyepiece" in a space-based telescope to serve also as a relay lens for the objective. That is, the eyepiece simultaneously creates images of both the object (at infinity) and the telescope objective. Since the objective is the aperture stop, its image is the exit pupil. This is shown in Fig. 2.25, where a scan mirror has been placed at an exit pupil and tilted at 45 degrees with respect to the telescope axis.

As the mirror pivots, the (demagnified) angular change at the exit pupil is relayed to the entrance pupil to sweep the telescope's FOV over the sky or ground. With the mirror at a pupil, the chief ray through the center of the objective must reflect off the center of the mirror no matter where the mirror is pointing. If the mirror is not at the exit pupil, then a change in mirror angle will sweep the field across the aperture instead of pivoting through it.

A scan mirror inside the telescope has the benefit of being smaller than a mirror outside, which is especially important when the mirror must be scanned at a high rate. Unfortunately, a 10-degree mirror scan (for example) does not give a 10-degree sweep over the FOV—for the same reason that the angular magnification increases the chief-ray angle inside the telescope. The magnification also makes scanning more difficult when looking out: a ±10-degree sweep inside a 10× afocal telescope (in "image space") is reduced (demagnified) to ±1 degree outside the telescope (in "object space"). This

Geometrical Optics

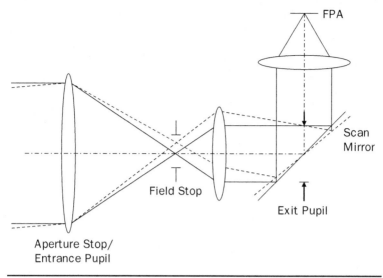

FIGURE 2.25 Placing a scan mirror at a pupil of an afocal system allows the FOV to be scanned without moving the beam across the entrance pupil.

reduction equals the angular magnification M, and the result is that optical system trade-offs must be made among aperture, angular magnification, scan angle, and scan rate for afocal systems.

That the FOV can be swept over a larger field leads to a new concept, the *field of regard* (FOR). This is simply the entire field over which the telescope can collect images. A simple example is the ability of binocular users to be able to look at a 360-degree FOR simply by turning around, even though the FOV of the binoculars may only be 5 degrees. For a space telescope with a 5-degree FOV and an internal scan mirror sweep of ±1 degree in object space (±M degrees in image space), the FOR is only 7 degrees.

In addition to relay lenses for transferring intermediate images, field lenses are commonly used to increase the FOV of optical systems. As shown in Fig. 2.26, rays that would otherwise be lost are bent by a field lens back toward the optical axis, increasing the FOV and allowing a reduction in the size of any optical elements that follow it.

Because it is located at an image plane, the field lens is also a field stop in that its size also affects the FOV. In practice, the field lens is usually moved a small distance away from the intermediate image so that any imperfections in the lens are not reimaged by the following optics onto a retina or FPA. Locating the field lens near an image plane also generates the effect of a prismlike bending of wavefronts without changing their curvature. As with any first-order design, the effects of a field lens are best illustrated using the ray-trace equations.

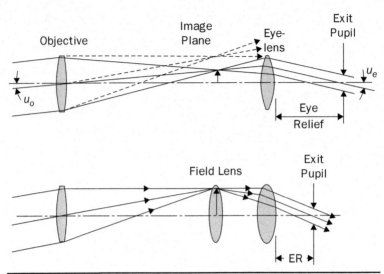

Figure 2.26 A field lens is used at an intermediate image in order to capture rays that would otherwise be lost. In practice, the field lens is placed slightly away from the intermediate image to prevent imperfections in the lens itself from being imaged. (*Credit: Warren J. Smith*, Modern Optical Engineering, McGraw-Hill, www.mcgraw-hill.com.)

Example 2.9. It is instructive to make a critical evaluation of the sales literature for cameras, telescopes, binoculars, and other consumer optics. For example, a reputable recreational equipment store claims that "a bigger exit pupil means more light is reaching your eyes" with its binoculars. This example reviews the inaccuracies of this statement in the context of the trade-offs in binocular design among exit pupil, magnification, and FOV.

Binoculars are typically specified in terms of magnification and entrance pupil; for example, "7 × 50" refers to a 7× afocal magnification with a 50-mm EPD.[10] Based on an afocal Keplerian telescope, the binocular design requirements are as follows:

- Objective-lens diameter $D = 50$ mm
- Angular magnification $M = 7\times$
- Track length $L < 130$ mm
- Eye relief $ER = 15$ mm
- Field of view FOV = 8 degrees (on an object at infinity)

Starting with the objective-lens (entrance pupil) diameter of 50 mm and an angular magnification $M = 7\times$, we calculate the exit pupil diameter as XPD = (50 mm)/7 = 7.1 mm. This EP is close to the size (~ 8 mm) of the eye's pupil in the dark, so aligning to the eye should not be a problem during the day when the eye's pupil is smaller. A larger exit pupil could clearly be obtained—but at the expense of reduced magnification and with little increase in the amount of light entering the eye.

Using the track length and magnification to find the focal lengths of the objective and eye lens yields $L = f_o + f_e = 128$ mm and $f_o = 7f_e$, from which it follows that $f_e = 16$ mm and $f_o = 112$ mm. If the eye lens is the same size as the exit pupil, then the f/# for both lenses is f/# = 112/50 = 16/7.1 = 2.24; this is a little bit on the fast side, a result of the relatively short track length required for a compact, lightweight package.

However, if the eye lens is the same size as the exit pupil (7.1 mm), then the only rays that can get completely through are the on-axis rays. The off-axis marginal rays cannot all get through and so the outer edges of the field are dimmer than the center, an effect known as *vignetting* (see Chap. 4). To avoid this we must use a larger eye lens; increasing the diameter to 16 mm results in an eye-lens f/# = f/D = (16 mm)/(16 mm) = 1.0. This would be a very expensive lens for a commercial product.

Continuing with the analysis to see where this extreme case leads, recall that the FOV depends on the size of the intermediate image. This value is found from the geometry given in Fig. 2.27, which shows the angle of the off-axis marginal ray as it exits the objective. The intercept height on the objective is known ($y_o = -25$ mm), as is the height on the eye lens ($y_e = +8$ mm). The distance from the bottom of the objective to the top of the eye lens is thus 33 mm, which occurs over a track length of 128 mm and gives an exit angle—of the lower marginal ray from the front objective—of $u'_o = 33/128 = 0.2578$ rad (the subscript *o* denotes "objective"; additional subscripts are sometimes added to distinguish between the chief and marginal rays).

Given the focal length of the objective, $f_o = 112$ mm, Eq. (2.9) is next used to find the incident angle of the parallel rays on the left side of the objective. This is the HFOV of the object, and we can use Eq. (2.9) to show that here

$$\text{HFOV} = u_o = u'_o + y_o \phi_o = 0.2578 - \frac{25}{112} = 0.0346 \text{ rad} = 1.98 \text{ degrees}$$

The FOV is then twice the HFOV, giving 3.96 degrees—or only about half of the requirement despite the use of a very fast eye lens.

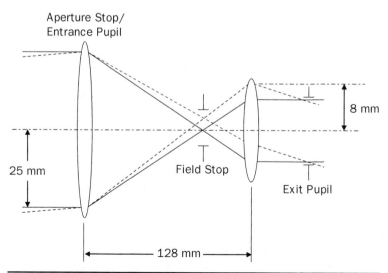

Figure 2.27 Dimensions for ray tracing through the afocal telescope design in Example 2.9.

In principle, a larger eye lens can increase the FOV further. In practice, however, the eye-lens speed of f/1 is a limiting factor in the design and so a new approach must be used. Options include (1) reexamining ("pushing back on") the requirements for the track length and/or the 8-degree FOV or (2) using a field lens.

If we use a field lens with the same focal length as the eye lens, for example, the result is a chief ray exiting from the field lens (or incident on the eye lens) at an angle $u'_f = u_f - y_f \phi_f = 0.0346 - [(3.8752 \text{ mm})/(16 \text{ mm})] = -0.2076$ rad. The chief ray has thus been bent from a positive to negative slope, allowing for an increase in the FOV and/or a smaller (slower, less expensive) eye lens. In practice, both the field and eye lenses have a speed on the order of f/2 or slower, allowing an increase in FOV without an excess increase in cost. Note that adding a field lens also results in a smaller eye relief.[10]

The first-order design of a Keplerian telescope is thus fairly involved and requires trade-offs among various parameters (e.g., f/#, lens size, lens-to-lens distances) in order to meet such requirements as magnification, FOV, and eye relief. Once the first-order design is complete, we must examine what the individual lenses might actually look like. It is not possible to increase the FOV indefinitely without running into some limitations, and these affect the design of the objective, eye lens, and field lens.

One limitation illustrated in Example 2.9 is the f/#. Other limitations, known as *imaging aberrations*, are determined by f/#, focal length, lens diameter, and field angle as well as the shape, surface radii, and refractive index of the lenses. In Figure 2.19 we saw an example of a practical telephoto lens in which the multielement groups for the positive and negative lens clearly differed from a first-order, thin-lens design. These differences are reviewed in detail in the next chapter.

Problems

2.1 Use Eq. (2.3) and f/# = 0.5 to obtain Eq. (2.5). What if an f/1 lens were the fastest possible? What would Eq. (2.5) look like in that case?

2.2 Why does better resolution generally imply a smaller FOV? Referring to Eq. (2.5), why does a bigger aperture D force the FOV to be smaller? **Hint:** What happens to the focal length f as D increases?

2.3 Which thin lens has the biggest FOV: the lens with the most power or that with the least?

2.4 For a Google Earth satellite that can image a GSD of 0.5 m from an altitude of 500 km, what focal length is needed for a pixel size of 10 µm? What if the images are instead collected from an airplane at an altitude of 35,000 feet? What is the IFOV in both cases?

2.5 The paraxial form of Snell's law is valid when the angles are "small." It was stated in the text that this occurs for angles of less than about 30 degrees. Make a table comparing the refraction angle using Snell's law and the paraxial approximation $\sin\theta \approx \tan\theta \approx \theta$ for a range of angles incident (in air) on a flat surface whose refractive index $n = 1.5$. What is the error? Repeat for a surface with $n = 3$. Does the meaning of "small" depend on index?

2.6 For systems with three thin lenses, use Eq. (2.8) in succession to find the EFL. In other words, first reduce two lenses to an equivalent thin lens, and then use this equivalent lens with the third lens to find the EFL of the entire combination. What is the length L when combining the equivalent lens with the third?

2.7 Which has the shortest track length for an imager with a 100-mm focal length: a refractive zoom or a two-mirror reflective such as a Cassegrain?

2.8 The plano-convex lens used in Table 2.3 can be considered a "thin" lens because the thickness (4 mm) is much less than the EFL. When is it necessary to consider the difference between the EFL and BFL? *Hint:* How much accuracy is required in the design?

2.9 In Example 2.7, the two lenses in the zoom were placed 50 mm apart before the entrance and exit pupils were calculated. Where are the pupils located if the lenses are placed 25 mm from each other? What about 75 mm? What does this mean to the designer of the zoom?

2.10 Design an attachment to a Keplerian telescope that makes it easier for people with glasses to use the telescope. The attachment should increase the eye relief from the standard 10 mm to the 20 mm needed for people who wear glasses. It should also fit onto the small-diameter end of the telescope, and it must be as short and as lightweight as possible. Do you think there's a market for such a device?

2.11 In the afocal telescope (binoculars) designed in Example 2.9, the field lens was placed just *after* the intermediate image to avoid relaying any imperfections in the lens to the user's eye. Could the field lens be placed just *before* the intermediate image to avoid the same problem? If so, then how much power would the field lens now need? Before putting the numbers in a spreadsheet, would you expect it to need more or less power?

2.12 In the optical design discussed in Example 2.9, is it possible for the device to have a 10× afocal design with a 20-degree FOV (on the object)? *Hint:* What is the chief ray angle in image space? What would the final imaging lens look like for this angle?

2.13 A typical optical system trade study compares the weights of different telescope designs. Compare the weight of the binoculars in Example 2.9 with a different design that, instead of a field lens, uses a larger eyepiece to increase the field to 7.16 degrees. What is the f/# of the eyepiece?

2.14 Are there any other approaches to increasing the FOV and decreasing the ER besides moving the field lens away from the intermediate image? Would increasing the size of the field lens work—or the size of the eyepiece, or both? Is it possible to increase the size of the lenses and still keep the cost reasonable? Explain your reasoning in words before tracing any rays.

Notes and References

1. Einstein's general theory of relativity was confirmed in the early 20th century by readings from a telescope put together in the African jungle by Sir Arthur Eddington.
2. Francis A. Jenkins and Harvey E. White, *Fundamentals of Optics*, 4th ed., McGraw-Hill (www.mcgraw-hill.com), 1976.
3. Eugene Hecht, *Optics*, 4th ed., Addison-Wesley (www.aw.com), 2001.
4. Pinhole cameras can also produce a basic image, but it will be fuzzy compared with that produced by a decent lens. See Arthur Cox, *Photographic Optics*, Focal Press, 1974.
5. Warren J. Smith, *Modern Optical Engineering*, 4th ed., McGraw-Hill (www.mcgraw-hill.com), 2008.
6. Robert E. Fischer, B. Tadic-Galeb, and Paul Yoder, *Optical System Design*, 2nd ed., McGraw-Hill (www.mcgraw-hill.com), 2008.
7. Warren J. Smith, *Practical Optical System Layout*, McGraw-Hill (www.mcgraw-hill.com), 1997.
8. Rudolf Kingslake, *Lens Design Fundamentals*, Academic Press, 1978.
9. Since the eye is a single-lens imager, it would seem that our "normal" view of the world should be an inverted image. Our brain corrects this inversion for the images we normally receive. Experiments have been performed in which subjects wear glasses that invert the image before it enters the eye; the experimental subjects are able to adapt after a few days of "seeing inverted."
10. William B. Wetherell, "Afocal Systems," in M. Bass, E. W. Van Stryland, D. R. Williams, and W. L. Wolfe (Eds.), *Handbook of Optics*, vol. 2, McGraw-Hill (www.mcgraw-hill.com), 1995, chap. 2.

CHAPTER 3
Aberrations and Image Quality

The problem with the Hubble Space Telescope's original optical system was not one of geometrical image size or location; instead, the issue was that of image quality. The images of stars ("point" objects), for example, were not as small as expected, but blurred into something much larger. It was eventually found that this blurring was due to an incorrectly shaped mirror that resulted in what is known as spherical aberration. The effects of aberrations on image quality are important in many other areas as well—including biomedical optics, where the ability to detect, recognize, and identify tumors can avoid the expense and difficulty of operating on a "false positive" (i.e., an image that looks like a tumor but in fact is not).

The fact that optical systems do not have perfect image quality is not apparent from the material reviewed in Chap. 2. There it was shown that a point images to a point and that the focal length of a lens depends only on its refractive index and surface radii [see Eq. (2.1)]. A point object at infinity thus images to a perfect point at the focal length of the lens.

In practice, the imaged point can actually be a large blur. We saw in Fig. 1.8(b) that an image is reasonably sharp when the convex side of a plano-convex lens points toward an on-axis object; yet when the lens is turned around so that the flat side points toward the object [Fig. 1.8(a)], the image is a fuzzy blur even though the focal length predicted by the geometrical optics of Chap. 2 hasn't changed. This reduction in image quality is a result of spherical aberration induced by the lens; as will be shown in Sec. 3.1, spherical aberration is only one of many types of optical aberration—most all of which reduce the quality of the image.

Fortunately, aberrations can be controlled. An example is shown in Fig. 3.1(a), where a MWIR imager using two thin lenses images an object at infinity onto a cooled focal plane array. The first lens brings the object at infinity to an intermediate image. The second lens relays this image onto the FPA; for reasons described in Chap. 5, it also images the aperture stop (the first lens on the left) to the location of

(a)

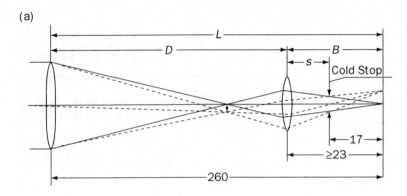

(b)

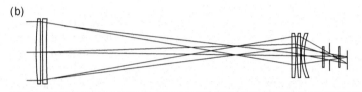

FIGURE 3.1 Comparison of (a) the first-order design of a MWIR imager and (b) a detailed design corrected for aberrations, which also includes the addition of a dewar window and cold filter. (*Credit: Warren J. Smith*, Modern Optical Engineering, *McGraw-Hill, www.mcgraw-hill.com.*)

the cold stop. The calculations of intermediate image location, as well as image and pupil relay, are easily handled using the geometrical optics equations of Chap. 2.

After taking aberrations into account, the design no longer consists of two thin lenses. Figure 3.1(b) shows that the first lens of the MWIR imager has been split into two elements and that the relay lens has been split into three. These splits are necessary to correct and control the aberrations that are unique to this system. As will be shown in Sec. 3.1, lens orientation and splits are just two of the many methods available to improve image quality. Another example is illustrated in Fig. 3.2, where the telephoto lens reviewed in Chap. 2 is shown in both thin lens and aberration-corrected forms. Clearly, aberration control requires an increase in sophistication beyond the simple thin-lens theory of geometrical and first-order optics.

Even with the most complex design, however, aberration control cannot image a point to a perfect point. All images are ultimately limited by a phenomenon known as *diffraction*. This phenomenon is fundamental to all optical systems and is the result of a wavefront spreading after it propagates through an aperture; hence "diffraction-limited" elements minimize the optical blur most of the time. As

Aberrations and Image Quality

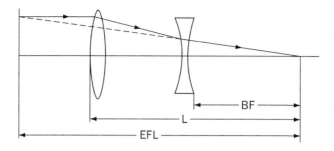

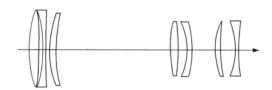

FIGURE 3.2 Comparison of the first-order design of a telephoto imager (upper image) and a detailed design corrected for aberrations (lower image). (*Credit: Warren J. Smith, Modern Optical Engineering, McGraw-Hill, www.mcgraw-hill.com.*)

reviewed in Sec. 3.2, the spreading depends on the wavelength compared with the aperture size (λ/D); this means that there are two system parameters that we may modify to reduce the size of the diffraction-limited blur.

Section 3.3 reviews the effects of aberrations and diffraction on image quality. The most fundamental measure of aberrations is that of wavefront error (WFE), a deviation from the perfect spherical wavefront centered on the ideal image point (Sec. 3.3.1). Additional metrics are also considered, with the modulation transfer function being the most useful to the optical system engineer (Sec. 3.3.3). The FPA also affects image quality; the details are reviewed in Chap. 6.

3.1 Aberrations

The intuitive view of an aberration as "something" that affects image quality is a result of wavefronts that are redirected from their ideal path on the way to the image plane. This redirection is due to the nonlinearity of Snell's law for "large" angles of refraction. That is, the paraxial approximation used for Snell's law in Chap. 2 resulted in a linear relation between incident and refracted angles ($n \sin \theta \approx n\theta \approx n'\theta'$). With larger angles, the paraxial approximation is no longer

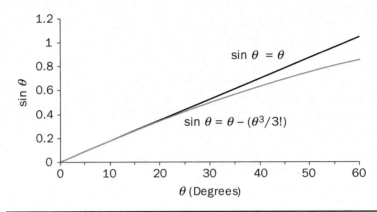

Figure 3.3 Comparison of Eq. (3.1) (lower plot) with the linear approximation (upper plot) used for sin(θ) in geometrical optics.

accurate. Instead, as shown in Fig. 3.3, more terms must be used in the expansion of the sine function:

$$\sin\theta \approx \theta - \frac{\theta^3}{3!} \qquad (3.1)$$

where $3! = 3 \times 2 \times 1$. Here the cubic term leads to the description of aberrations in this chapter as *third-order* optics; the term *first-order* is used to describe the geometrical optics that use only the linear term in Eq. (3.1). There are also higher-order aberrations based on the continued expansion of the sine term ($\theta^5/5!$, $\theta^7/7!$, . . .), but these are not covered in this book.[1,2]

Two types of aberrations result from the third-order expansion in Eq. (3.1). These types result in errors in image location and blur sizes that are larger than the diffraction-limited size; an additional effect related to the index of refraction is reduction of color fidelity. Altogether there are five different third-order aberrations; the one that dominates depends on the aperture size and field of view.

This connection with aperture and field illustrates the key physical dependence of aberrations on local angles of incidence. That is, larger refraction angles have more third-order aberration, as should be expected for rays at steep angles (larger FOV) and far from the optical axis (larger aperture). The first step in reducing aberrations is thus reducing the *first-order* lens powers as much as possible, given the dependence of refractive power on the curvature of the optical surfaces. When this step is insufficient, as it often is, the designer must use the third-order methods reviewed in this chapter—shape factor, orientation factor, achromatic doublets, and so on—in conjunction with aperture and field control.

Depending on the requirements for aperture and field, it can take a fair amount of effort to design a lens that produces a crisp image with little distortion or color blurring. A single lens is almost never good enough, so usually combinations of lenses are used to obtain appropriate image quality. For example, the thin lens used in Chap. 2 as a typical camera lens may actually have as many as six or seven elements so that aberrations are properly corrected.

In addition to camera lenses, other lenses such as telescope objectives, projection lenses, and so on all have specific requirements that lead to unique design forms to minimize aberrations.[3] The compromises and trade-offs in such designs are many, as are the resulting forms. These range from "best form" singlets (optimized for minimum spherical aberration and coma) to achromatic doublets (which remove the chromatic dependence of focal length at specific wavelengths) to something as simple as aperture stop location (to reduce coma and distortion). The full range of design forms is shown in Fig. 2.11, including their dependence on aperture and field.

The purpose of this section is to develop a basic understanding of aberrations, and the methods used to reduce them, without describing in detail the advanced methods used by lens designers.[1-3] "It's just a lens!" is a naive approach that can lead to millions of dollars' worth of trouble. A critical skill of the optical systems engineer is understanding the concepts and subtleties required to develop the first-order designs from Chap. 2 into more realistic optical systems. Toward this end, we first explore the concept of spherical aberration.

3.1.1 Spherical Aberration

This aberration is observed when a lens or mirror with spherical surfaces images an on-axis point; not surprisingly, it is known as spherical aberration (aka SA, SA3, or simply "spherical"). It cannot be avoided except by using surfaces that are not spherical, and it occurs to varying degrees depending on the entrance pupil diameter. The image quality of fast systems with a small relative aperture (f/#) is often limited by the lens designer's ability to minimize spherical.[1]

Figure 3.4 shows an on-axis point at infinity—a star, for example—being focused by a biconvex lens. The figure illustrates that the outer rim ("zone") of the lens brings the wavefronts to a different focus than the central zone near the optical axis. This occurs because the angle between the incident rays and the surface normal of the lens is larger for the outer rays than it is for the central (paraxial) rays near the optical axis. Snell's law shows that the outer rays must bend more; as a result, they are brought to focus closer to the lens while the inner rays are focused farther away. The continuum of zones across the lens thus creates an image blur larger than the crisp point predicted by geometrical optics.

Spherical aberration is not predicted by geometrical optics, where the lenses are treated as a paraxial "black box" with imaging power

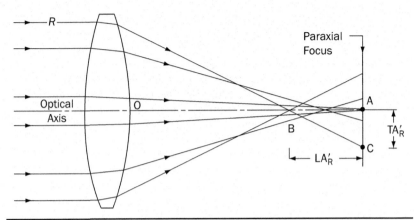

Figure 3.4 Spherical aberration results when the outer rays are refracted more than the central (paraxial) rays. The difference in foci can be measured as either a longitudinal aberration (LA) or a transverse aberration (TA). (*Credit: Warren J. Smith, Modern Optical Engineering, McGraw-Hill, www.mcgraw-hill.com.*)

and the details of how rays are bent by individual surfaces of the lens are ignored. We saw in Fig. 1.8 what happens when a plano-convex lens is turned so that the curved side faces toward the object. The spherical aberration has somehow been reduced, even though the power of the lens hasn't changed (only its orientation has).

By turning the lens around, the spherical is reduced because the incidence angles of the rays have been modified. In Fig. 1.8(a), the surface normals are zero on the planar surface but are steep in the outer zones of the convex surface; this steepness causes severe SA. Figure 1.8(b) shows that facing the spherical surface toward the object reduces the incidence angle on both the convex and planar surfaces; as a result, spherical aberration is much lower. Spherical can thus be reduced by splitting the refraction between the surfaces, rather than relying on one surface to do all the work. This unique aspect of third-order optics stems from the nonlinear dependence of Snell's law on incident angle [Eq. (3.1)].

We have used the *orientation* of a plano-convex lens as an example of how to reduce SA, but spherical also depends on the *shape* of the lens. In addition to plano-convex, other shapes include plano-concave, convex-convex, concave-concave, and so on. With a concave-convex lens whose concave side faces the point at infinity, spherical aberration is greater than with the plano-convex lens in any orientation. In physical terms, the first surface bends the rays in the wrong direction (away from the optical axis) and so the second surface must compensate with even stronger bending in order to obtain a lens with the same focal length as the plano-convex.

The varying degrees of concave and convex surfaces and their effect on spherical aberration can be quantified with the shape factor q:[4]

$$q = \frac{R_2 + R_1}{R_2 - R_1} = \frac{C_1 + C_2}{C_1 - C_2} \tag{3.2}$$

Here $R_1 = 1/C_1$ is the radius of the first surface of the lens (that on the object side) and R_2 is the radius of the second (that on the image side). Keeping in mind the sign convention for surface radii (Fig. 2.3), an equiconvex lens with equal radii on both sides ($R_1 = -R_2$, so $|R_1| = |R_2|$) gives $q = 0$. The shape factor is thus a measure of the deviation from a symmetric lens.

Figure 3.5 graphically illustrates how the shape factor varies with surface radii. Starting with an equiconvex lens, q becomes negative as the radius of the first surface is made larger than the second, with $q = -1$ for a plano-convex lens. Similarly, $q > 0$ when the radius of the second surface is larger than the first. Also observe that reversing the orientation of a lens changes the sign, but not the magnitude, of its shape factor.

The curve plotted in Fig. 3.5 shows how the difference in axial focus varies with surface radii for a point source at infinity. This longitudinal difference is one measure of SA; the size of the blur (lateral spherical aberration) on an image plane is also used. With either metric, the best shape factor for reducing SA for an object at infinity is *not* the equiconvex lens with $q = 0$, given that the object and image distances are not symmetrically placed on either side of the lens. To match the difference between the object and image wavefront curvature, the lens shape with minimum spherical in Fig. 3.5 has a slightly larger radius on the image side of the lens than on the object side; here the shape factor $q \approx +0.714$ for a lens with $n = 1.5$.[4] This is not terribly different from a plano-convex lens with $q = +1$, a useful shape to consider when purchasing off-the-shelf lenses.[5]

The optimum shape for minimum SA thus also depends on the location of the object and image. This is quantified with the position factor $p = (s_i - s_o)/(s_i + s_o)$, where the object and image distances are based on the paraxial focal length. For example, with an object at infinity, $p = (0 - 1)/(0 + 1) = -1$, independent of shape factor. Position factor can be used along with the shape factor in analytical equations to determine the SA;[4] since the advent of lens design software these equations are seldom used, but the concepts are useful for understanding the sources of spherical aberration.

The blur that's created by spherical aberration does not have a clearly defined point of focus. Figure 3.4 shows that it is difficult to define a "best focus" located between the marginal and paraxial foci—that is, an axial location where the blur is "smallest"—leading to the *circle of least confusion* as an acceptable standard for minimum blur.[4] Figure 3.6 is known as a *through-focus spot diagram*, a common

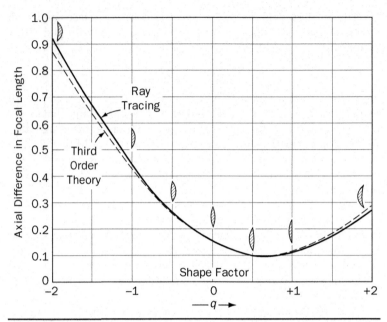

Figure 3.5 Dependence of spherical aberration on the shape factor q. (*Credit: Francis A. Jenkins and Harvey E. White,* Fundamentals of Optics, *McGraw-Hill, www.mcgraw-hill.com.*)

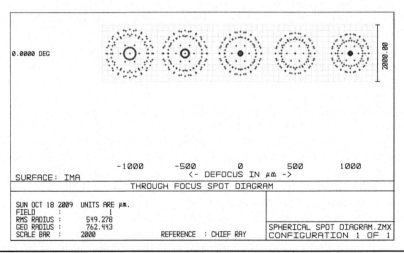

Figure 3.6 Spot diagram for a lens with spherical aberration; it shows the change in spot size and structure as the image plane is moved through focus by ±1000 μm.

plot for illustrating how the blur size (or point spread function, PSF) changes as the image plane moves along the optical axis. This diagram is created by tracing rays through the system and thus does not take into account diffraction, a different contributor to blur size (see Sec. 3.2).

The size of the blur depends on the diameter of the lens because the steepness of the incident angles increases with diameter. The quantitative dependence of blur size on diameter is cubic; for a minimum-spherical singlet, the angular size of the blur at best focus due to transverse (or lateral) SA is[1]

$$\beta_s = \frac{K}{(f/\#)^3} \quad [\text{rad}] \tag{3.3}$$

(where the subscript s denotes "spherical"). Here K is a unitless constant that depends on the lens index, and the physical blur size $B_s = \beta_s f = KD^3/f^2$. For an index $n = 1.5$, Table 3.1 shows that $K = 0.067$; the angular blur for an f/2 lens is thus $\beta_s = K/(f/\#)^3 = 0.067/2^3 = 8.375$ mrad, a relatively large blur. The physical blur for the same f/2 lens with a 100-mm focal length is $B_s = \beta_s f = 8.375$ mrad × 100 mm = 837.5 µm, confirming our guess that the blur due to spherical is large for this lens. Reducing the SA thus requires using a much slower lens (larger f/#), which Eq. (3.3) implies can yield an arbitrarily small SA. This is true up to the point where diffraction begins to dominate (see Sec. 3.2).

Table 3.1 also shows that the lens index affects the spherical. Mathematically, the parameter K in Eq. (3.3) becomes smaller as the index increases. Physically, a larger index allows a smaller surface curvature to obtain the same focal length [see Eq. (2.1)]; a smaller

Index, n	K
1.5	0.0670
2.0	0.0273
2.5	0.0174
3.0	0.0129
3.5	0.0103
4.0	0.0087
Spherical mirror	0.0078
Parabolic mirror	0.0000

TABLE 3.1 The constant K is used in Eq. (3.3) for calculating the minimum spot size due to spherical aberration for an on-axis point at infinity.[1]

curvature in turn implies smaller incident angles and thus a smaller difference between marginal and paraxial rays. Unfortunately, this index cannot be chosen arbitrarily because $n \approx 1.5$ happens to be a common index for materials (e.g., BK7) that transmit well in the visible spectrum. In contrast, infrared materials (e.g., silicon and germanium) typically have $n \approx 3$ to 4, values for which a low-SA design can be readily obtained using a shape that's closer to a meniscus than to a plano-convex lens.[3]

The last entry in Table 3.1 shows that a parabolic surface does not exhibit SA for an on-axis point at infinity. The parabola is a unique aspheric (nonspherical) surface for mirrors in that neither hyperbolic nor elliptical mirrors share this property. As a result, the parabola is a common primary mirror shape for many types of telescopes, including the Newtonian, Cassegrain, and Gregorian (see Sec. 3.1.6). Unfortunately, these instruments have a small field due to off-axis aberrations known as coma and astigmatism (reviewed in Secs. 3.1.2 and 3.1.3, respectively). Lenses with aspheric surfaces are also possible, and although these tend to be more expensive than those fabricated with spherical surfaces, are becoming more and more common.

In summary, Fig. 3.5 shows that it is not possible to completely remove SA using a single lens with two spherical surfaces. This residual SA is often low enough to satisfy system requirements, in which case the use of more expensive approaches is unnecessary. Nevertheless, there are many applications for which the spherical must be brought closer to zero; in these cases, the methods used by the lens designer to reduce or "correct" SA include:

- A lens with a small diameter and long focal length (i.e., large f/#).
- Shape factor q and lens orientation.
- Position factor p determined by the object and image conjugates.
- A large refractive index, if possible (see Table 3.1).
- Splitting one lens into two (or more) lenses in order to slowly "walk" the incident angle from its initial value on the first surface of the lens to the final cone angle set by the f/# or image distance s_i (Fig. 3.7). By using more surfaces, the refraction at each surface can be made smaller, thus reducing the SA in a manner similar to the orientation principle.[3]
- Two lenses can also correct for spherical by balancing the positive spherical of one lens with the negative spherical of the other. This was the method used to correct the SA on the Hubble primary mirror;[6] it is more commonly used in high-

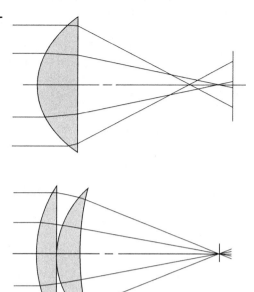

FIGURE 3.7 Splitting a lens into two elements is a common method for reducing spherical aberration; the reduction is obtained by "walking" the rays through smaller refractive angles at each surface. (*Credit: Warren J. Smith*, Modern Optical Engineering, *McGraw-Hill, www.mcgraw-hill.com.*)

index–low-index doublets available as off-the-shelf lenses from a number of manufacturers (Fig. 3.8).

- General aspheric surfaces for lenses; parabolic and general aspheric surfaces for mirrors.

Even when these methods are able to reduce the spherical aberration, the image of a symmetric point source such as a star may appear as an asymmetric blur as the star moves away from the center of the FOV. For small off-axis angles this aberration is known as coma, which we review following Example 3.1.

Example 3.1. As shown in Chap. 2, it is sometimes necessary to relay an image. The relay lens must take into account both the geometric imaging and the aberration corrections. In this example, a 1 : 1 lens combination is used as a relay to reimage an on-axis star with minimum spherical using off-the-shelf lenses.

For a single lens used with 1 : 1 ($2f$: $2f$) image conjugates, the symmetry of the 1 : 1 imaging with a symmetric equiconvex lens (shape factor $q = 0$) is expected to have the smallest SA possible from a "best form" singlet. The resulting size of the imaged point is given by Eq. (3.3), where the image distance is not the focal length f but rather $s_i = 2f$ (so that $B_s = \beta_s s_i = KD^3 s_i / f^3$). However, off-the-shelf doublets are also available for which Eq. (3.3) and Table 3.1 do *not* apply but that are advertised as being "corrected" for spherical, coma, and chromatic aberrations.[7] It seems that the problem is thus easily solved, and the choice of a large-f/# doublet may meet the requirements.

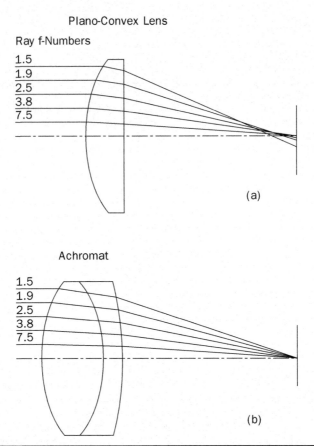

Figure 3.8 *(a)* A plano-convex lens with the curved side pointed toward an object at infinity can still have unacceptably large spherical aberration. *(b)* Adding a second lens with a different index results in a two-element achromat that considerably reduces the SA. *(Credit: CVI Melles Griot.)*

Unfortunately, this conclusion is wrong on two counts: (1) such doublets are usually designed for infinite conjugates (i.e., for an object or image at infinity); and (2) the term "fully corrected" does not mean that rays from *all* annular zones meet at the same focus. Most often it is only the paraxial and marginal rays that have a common focus; the rays from the annular zones between these extremes typically focus closer to the lens, a condition known as *undercorrected zonal* spherical.[1] Less often, the zones between the paraxial and marginal are overcorrected and so image beyond the paraxial focus. In either case, another lens must be added to the system to reduce the zonal SA.

The second lens is added in the configuration shown in Fig. 3.9. Two doublets are used in a back-to-back architecture that enables their use with infinite conjugates: a reimaging lens on the right, and a collimating lens on the left to convert the diverging wavefronts from the point source into plane waves. The lenses are used in their correct orientation (i.e., with the larger radius facing the source or its image); corrected doublets have the smaller zonal spherical.

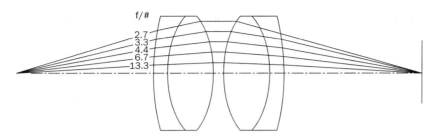

FIGURE 3.9 Infinite-conjugate achromats used in a back-to-back configuration allows SA to be corrected for finite-conjugate images. (*Credit: CVI Melles Griot.*)

3.1.2 Coma

When cataloging stars in the 18th century, Charles Messier was often misled into thinking that bright objects in the heavens were actually comets.[8] This was due to his telescope—a poor-quality instrument by the standards of even the most inexpensive of modern tools—imaging stars into a cometlike shape. It is now known that this was a result of a lens aberration called *coma*, an imaging defect that can be reduced by appropriate lens design techniques.

As illustrated in Fig. 3.10, the images Messier saw had comet-like tails pointing radially. These tails are first seen at small off-axis angles—on the order of 5 degrees for an f/16 lens, for example—with the size of the "comet" being directly proportional to the angle. Geometrical optics predicts the location of the tip of the cone, which is at the intersection of the chief ray with the paraxial image plane; the unusual shape of the blur spreading out from the chief ray has origins similar to those of spherical aberration.

Specifically, the lower marginal rays shown in Fig. 3.11(a) are bent more than the upper rays, a result of the larger angle of incidence on the second lens surface. Compared with the spherical aberration of an on-axis point, the upper marginal ray bends less while the lower marginal ray bends more. The opposite situation, where the upper ray bends more and the lower ray bends less, can also occur. In either case, the focus (intersection AB in Fig. 3.11) of these two rays occurs at a distance that is displaced from the chief ray imaged at point P.

Figure 3.11(a) shows only the two marginal rays (A and B) in the plane of the paper (the tangential or meridional plane) imaging to a point; of course, there are an infinite number of these rays around the circumference of the lens. Each upper–lower pair of rays images in the same way as those in the tangential plane, with the result being that a point object images to a ring [Fig. 3.11(b)]. The marginal focus thus results in the circle at the large end of the cone, and the chief ray determines the point at the tip.

At the same time, there are an infinite number of annular zones besides the marginal; these consist of smaller and smaller concentric

Figure 3.10 Image of a point source, such as a star, as seen through a telescope with a large amount of coma. (*Credit: Warren J. Smith,* Modern Optical Engineering, *McGraw-Hill, www.mcgraw-hill.com.*)

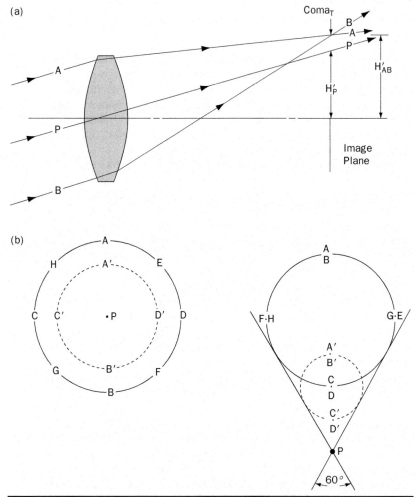

Figure 3.11 Coma results when the off-axis upper marginal ray bends more or less than the lower marginal ray. The paraxial rays intersect the image plane at a height H'_P while the marginal rays intersect at a height H'_{AB}. (*Credit: Warren J. Smith,* Modern Optical Engineering, *McGraw-Hill, www.mcgraw-hill.com.*)

zones in the entrance pupil, each imaging a point object into a smaller and smaller ring in the image plane [circle A'B'C'D' in Fig. 3.11(b), for example]. The upper rays from the smaller zones do not have as large an incident angle on the lens as those in the marginal zone. Therefore, the difference in incident angle between the upper and lower rays becomes smaller as the selected zone moves closer to the chief ray; rays from these zones are thus not bent as much and so image closer to the chief ray. The superposition of an infinite number of these circles from each infinitesimal zone creates the conical shape of the comatic blur seen in Fig. 3.10.

The size of the marginal ring depends on the entrance pupil diameter and field angle. Larger apertures are associated with a larger difference in incident angles between the upper and lower rays, as are larger field angles. Using the lateral size of the marginal ("sagittal") ring as a metric, the angular size of the blur in the image plane for a thin lens *shaped for minimum SA* is given by[1]

$$\beta_c = \frac{\theta}{16(n+2)(f/\#)^2} \quad [\text{rad}] \tag{3.4}$$

(where the subscript c denotes "coma"). Here θ is the half-angle of the field measured in radians, the aperture stop is at the lens, and the size of the blur increases linearly with angle θ up to the half-FOV angle. The physical blur size $B_c = \beta_c f = \theta D^2/16(n+2)f$; for an f/2, minimum-spherical lens with refractive index $n = 1.5$ and a 100-mm EFL, it follows that $\beta_c = (0.0349 \text{ rad})/(16 \times 3.5 \times 2^2) \approx 0.156$ mrad at a field angle of 2 degrees (34.9 mrad). Even at 5 times the field angle, this is significantly smaller than the blur size found for on-axis spherical with the same lens; see Fig. 3.12.

Figure 3.12 also shows that coma can be decreased to zero with the appropriate shape factor, $q = 0.8$ for a lens with $n = 1.5$ and an object at infinity—not much different than the shape factor ($q = 0.714$) for minimum spherical. This shape is also close to plano-convex, which is another reason for the popularity of these lenses as off-the-shelf components. The residual coma for a plano-convex lens (or other shape with $q \neq 0$) can be reduced by using small fields and apertures, if possible. Alternatively, the aperture size can be traded against the field, keeping the coma fixed at a sufficiently small value that meets blur-size or other image quality requirements (Sec. 3.3).

In practice, Eq. (3.4) is only used to approximate the blur size for lateral coma; a more exact analysis relies on traced rays to calculate through-focus spot sizes, as can be done with lens design software. As illustrated in Fig. 3.13, the plots for coma also include across-field results at 0, 3.5, and 5 degrees; this reflects a field dependence that was not present in the through-focus plots for spherical aberration.

In addition to altering the shape factor, position factor, f/#, refractive index, and field angle, another technique for reducing coma is appropriate placement of the aperture stop. As illustrated in

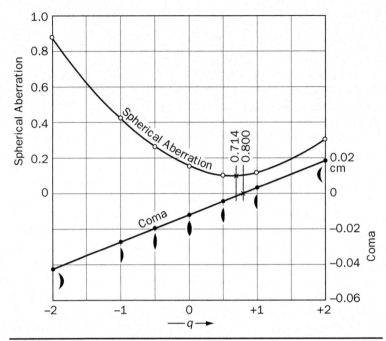

Figure 3.12 Comparison of blurs caused by spherical and coma; note the 10× smaller scale for coma. (*Credit: Francis A. Jenkins and Harvey E. White, Fundamentals of Optics, McGraw-Hill, www.mcgraw-hill.com.*)

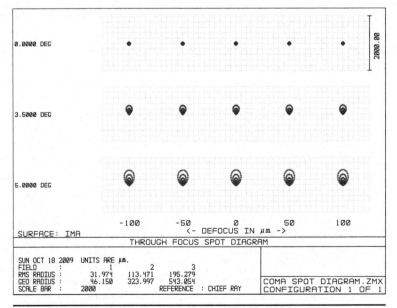

Figure 3.13 Spot diagram for a lens with coma; it shows the change in spot size and structure as the image plane is moved ±100 μm through focus and as the object's off-axis location is changed from 0 to 5 degrees.

Aberrations and Image Quality

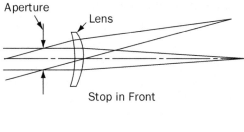

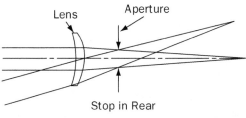

FIGURE 3.14 Stop position can be used to control off-axis aberrations such as coma. (*Credit: Warren J. Smith, Modern Optical Engineering, McGraw-Hill, www.mcgraw-hill.com.*)

Fig. 3.14, stop position controls which part of the lens the wavefronts are incident on and hence the incident angles of refraction. We have seen that the *difference* in refractive bending between upper and lower rays is one approach to understanding coma. Thus, moving the aperture stop can increase or decrease the size of the coma blur depending on the shape factor and how the incident angles are modified. For a thin lens, coma can be reduced to zero by moving the stop in front of the lens (between the lens and the object) to a location—known as the *natural* position—that depends on shape factor, field angle, and so on.[1] Stop shifting is a common technique used by lens designers to reduce certain off-axis aberrations, including coma and astigmatism (reviewed in Sec. 3.1.3).

Example 3.2. The back-to-back arrangement used in Example 3.1 can also be used to reduce coma. This can be done most efficiently with a best-form singlet or corrected doublet in the minimum-aberration orientation. Here we use plano-convex lenses to illustrate the effects of wavefront distortion on blur size.

If a plane wave is incident on a plano-convex reimaging lens, then Eq. (3.4) can be used to approximate the blur size even though the lens does not have exactly the minimum-SA shape. The blur size for finite conjugates starts with the angular blur size, which Eq. (3.4) gives as $\beta_c = \theta D^2/16(n+2)f^2$. For small angles, the physical blur size is then the angular size multiplied by the image distance f; this yields $B_c = \beta_c f = \theta D^2/16(n+2)f$. Given an object size of 2 degrees ($\theta = 1$ degree = 0.01745 rad) when viewed from the entrance pupil (the lens in this case) and an f/2 lens with index $n = 1.5$ and a 100-mm EFL, we obtain B_c = 0.01745 rad × (0.05 m)²/(16 × 3.5 × 0.1 m) ≈ 7.8 μm, a relatively large blur at visible wavelengths when compared with the diffraction limit (see Sec. 3.2).

Unfortunately, a perfectly collimated plane wave is *not* incident on the reimaging lens; instead, the wavefront has been aberrated by the collimating lens in the back-to-back configuration of Fig. 3.9 and therefore deviates from planar. This deviation is known as wavefront error and results in a focused spot larger than that predicted by Eq. (3.4). Figure 3.15, for example, shows that spherical aberration can be viewed as the difference between the actual and paraxial wavefronts, with any *additional* WFE incident on the lens causing the rays to deviate even more than shown in Fig. 3.15 and thus resulting in a SA blur bigger than that given by Eq. (3.3). A comatic blur larger than that given by Eq. (3.4) is also obtained.

The ray and wavefront perspectives are both valid approaches to understanding aberrations. The effects on the blur are difficult to determine using back-of-the-envelope calculations but are straightforward to derive using the ray-trace methods described in Smith.[1] Qualitatively, one can expect a well-corrected doublet to have little WFE and a singlet used at a less-than-optimum orientation to have much more. This effect is explored in more detail in Sec. 3.3.

3.1.3 Astigmatism

It turns out that there is a slight difference in focal length between coma's tangential and sagittal rays. With larger field angles this difference becomes more noticeable, and eventually there is a clear separation between them. Thus, when measured at different points along the chief ray, the image of an off-axis point is seen to have the shape of a blurred line, which is shown in Fig. 3.16 as the tangential and sagittal images. For a finite-size object such as a bicycle wheel, the two ray types are easily seen in the changing orientation of their associated images as the focal plane is moved from the tangential to sagittal (or radial) focus (Fig. 3.17).

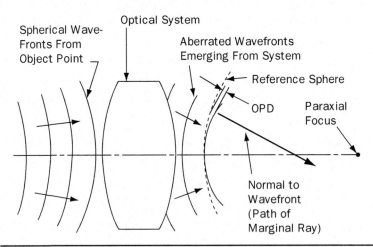

FIGURE 3.15 Aberrations are the result of the difference between the actual and paraxial wavefronts. (*Credit: Warren J. Smith*, Modern Optical Engineering, McGraw-Hill, www.mcgraw-hill.com.)

Aberrations and Image Quality

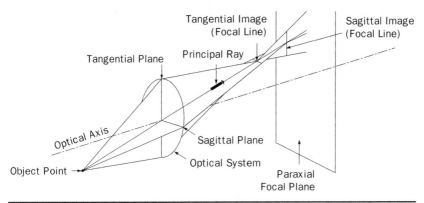

FIGURE 3.16 Large off-axis angles image a point source into two distinct line foci, resulting in a third-order aberration known as astigmatism. (*Credit: CVI Melles Griot.*)

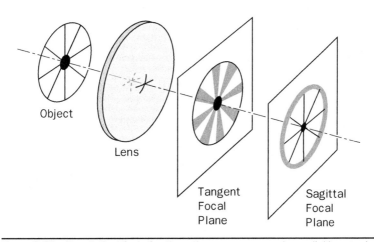

FIGURE 3.17 The distinct foci of astigmatism are seen as crisp radial images in the sagittal plane and perpendicular images in the tangential plane. (*Credit: Francis A. Jenkins and Harvey E. White*, Fundamentals of Optics, *McGraw-Hill, www.mcgraw-hill.com.*)

This aberration is known as *astigmatism*, a common affliction of the human eye and other lenses. With the eye it is a result of either muscular squeezing or an inherent defect in which the eye lens has more power about one axis in the plane of the lens than it does about the other. The extreme case of this situation is a cylindrical lens, which have refractive power about only one axis and thus images a point to a line focus. To correct for astigmatism in nominally spherical lenses, an optometrist will therefore specify the cylindrical power of eyeglasses for both axes; this power is measured in diopters (units of $1/m$ for lens power $\phi = 1/f$ with the focal length f measured in meters).

Spherical lenses do not have a varying curvature across their surface, yet their effect when imaging off-axis points is similar. This is seen in Fig. 3.18, which illustrates cross sections of the center of the lens in the tangential and sagittal planes. The figure shows that the upper marginal rays from an off-axis point in the tangential plane are bent more than those in the sagittal plane; this is a net result of the larger incidence angles (on the left side of the lens for the upper rays and on the right side of the lens for the lower rays). Alternatively, the wavefront perspective shows that the lens surface falls away from the wavefront faster for the tangential rays than for the sagittal—as if the tangential surface had more curvature and refractive power than the sagittal.

As the tangential line image rotates into the sagittal, there is a blur between them known as the *medial focus*; it is circular for a well-fabricated spherical lens but elliptical for a lens with cylinder errors introduced by manufacturing (see Chap. 7). The size of the blur depends on the off-axis field angle and the f/#, with a fast lens with steeper surfaces increasing the (third-order) refractive angle faster for the tangential than for the sagittal rays. These factors result in a larger astigmatic blur, as does a larger field angle. The angular size of the medial blur for a thin lens *shaped for minimum SA* is given by[1]

$$\beta_a = \frac{\theta^2}{2(f/\#)} \quad \text{[rad]} \tag{3.5}$$

(where the subscript a denotes "astigmatic"). Here θ is again the half-angle of the field measured in radians, the aperture stop is at the lens, and the size of the blur increases quadratically with θ up to the half-FOV angle α. The physical blur size $B_a = \beta_a f = \theta^2 D/2$; for an f/2, minimum-spherical lens with index $n = 1.5$ and a 100-mm EFL, we have $\beta_a = (0.0349 \text{ rad})^2/(2 \times 2) \approx 0.305$ mrad at a field angle of 2 degrees

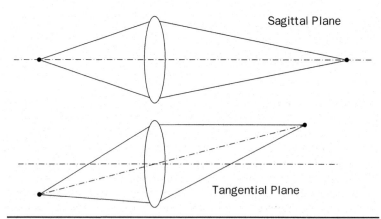

FIGURE 3.18 Marginal rays for a point imaged in the sagittal and tangential planes; the illustration shows how the symmetrical sagittal rays focuses to a different location than do the asymmetric tangential rays.

(34.9 mrad). This is about twice the value found in Sec. 3.1.2 for coma with the same lens, even at the relatively small angle of 2 degrees, which illustrates the trade-off between field angle and f/# in Eq. (3.5). This is the most important trade in an optical system: fast, wide-angle designs require many elements whose number is determined by the allowable aberrations and the lens designer's creativity in correcting for them.

For a singlet with the stop at the lens, the tangential and sagittal lines image onto distinct parabolic surfaces. This is illustrated in Fig. 3.19, where the vertical axis is the image height and the horizontal axis is the distance along the optical axis. As the field angle increases, the focus for both astigmatic lines moves quadratically away from the flat image plane where the paraxial image is located and also farther away from each other.

This curvature results in a large blur for off-axis angles, and the size of the blur increases quadratically with field angle; this is seen in Eq. (3.5) for the medial image, whose dependence on θ also applies to the tangential and sagittal images. Typical through-focus and across-field blur diagrams for a lens with astigmatism are shown in Fig. 3.20.

Even with lenses that are more complex than the singlet, it is difficult to remove astigmatism across the entire field. As with doublets corrected for SA, two zones are typically corrected; these could be the paraxial and marginal rays or the paraxial and $2^{1/2} = 0.707$ times the full field. Such a lens is known as an *anastigmat* ("without astigmatism")—although, as shown in Fig. 3.21, there is

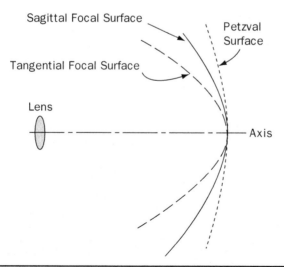

FIGURE 3.19 The image surface for the third-order sagittal and tangential images is not a plane but rather a parabola. (*Credit: Warren J. Smith*, Modern Optical Engineering, *McGraw-Hill, www.mcgraw-hill.com.*)

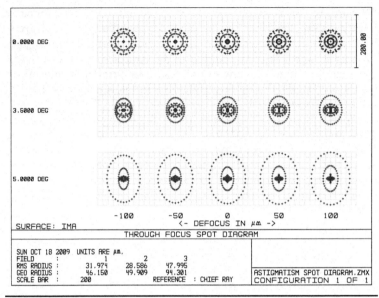

Figure 3.20 Spot diagram for a lens with astigmatism; it shows the change in spot size and structure as the image plane is moved ±100 μm through focus and as the object's location is changed from 0 to 5 degrees.

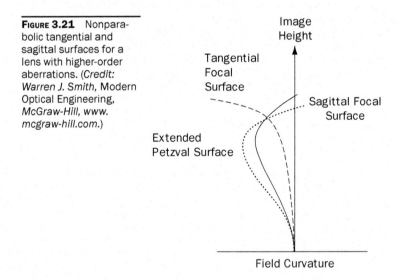

Figure 3.21 Nonparabolic tangential and sagittal surfaces for a lens with higher-order aberrations. (*Credit: Warren J. Smith,* Modern Optical Engineering, *McGraw-Hill, www.mcgraw-hill.com.*)

residual astigmatism at field angles between the two locations where the tangential and sagittal images coincide. A classic design form known as the Cooke-triplet anastigmat is shown in Fig. 3.22; as with all anastigmats, it relies on the principle of separating positive and

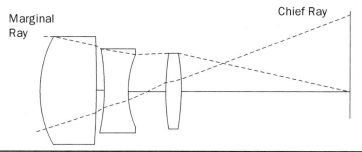

FIGURE 3.22 The Cooke-triplet anastigmat. (*Credit: Warren J. Smith*, Modern Optical Engineering, *McGraw-Hill, www.mcgraw-hill.com.*)

negative lenses so that their inward- and outward-curving image surfaces (pointing toward and away from the lens, respectively) correct astigmatism at a specific field angle.[3]

The residual astigmatism at other field angles can be reduced using many of the same methods that were used for spherical and coma. In addition to the dependence on aperture and field indicated by Eq. (3.5), astigmatism can be reduced using stop shifts. As with coma, the stop forces the rays to be incident on a specific part of the lens, increasing or decreasing the asymmetry of the refracting surfaces with respect to the wavefront and thereby increasing or decreasing the extent of astigmatism.

It is interesting that the residual astigmatism after a shift depends on the amount of SA and coma.[2] For a thin lens, the tangential astigmatism is smallest at the "natural" stop position, which in Fig. 3.23 is seen to be located where coma is zero.[1] Reducing the value of the minimum requires reducing SA using the methods discussed in Sec. 3.1.1—for example, shape factor, f/#, a large refractive index for positive lenses, splitting lenses, and the like. Unfortunately, even if astigmatism can be more or less corrected over the entire field (e.g., over small fields) and the tangential and sagittal lines coincide at the same surface, the image surface can still be parabolic. As explained in Sec. 3.1.4, this phenomenon is due to a fundamental limitation known as Petzval curvature.

Example 3.3. Asymmetry can result in astigmatism not only in lenses but also in flat plates (planar with parallel sides) that do not have optical power. This is commonly seen when beam splitters are used to send a fraction of a beam in one direction, and the remaining fraction in another (Fig. 3.24). If the plate is perpendicular to the beam, then either collimated or converging light will continue through without aberration. Refraction at the flat surfaces also pushes the focus of the converging beam away from the plate by an amount δ that depends on the index and thickness of the plate [$\delta = (n-1)t/n$].

However, if the plate is tilted then collimated light remains free of aberrations while beams converging toward a focus do not. This also is seen in Fig. 3.24; astigmatism is introduced because the asymmetry in refractive angle between

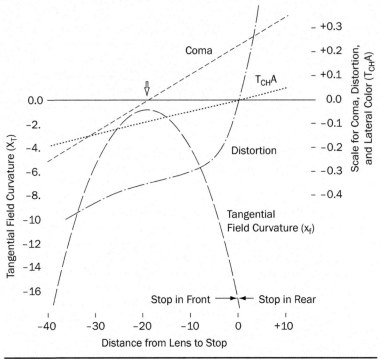

FIGURE 3.23 The effect of a stop shift on off-axis aberrations. (*Credit: Warren J. Smith*, Modern Optical Engineering, *McGraw-Hill, www.mcgraw-hill.com.*)

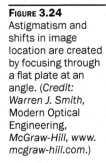
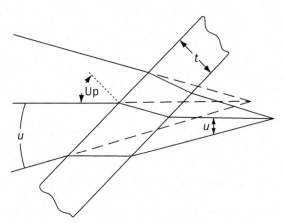

FIGURE 3.24 Astigmatism and shifts in image location are created by focusing through a flat plate at an angle. (*Credit: Warren J. Smith*, Modern Optical Engineering, *McGraw-Hill, www.mcgraw-hill.com.*)

the upper and lower tangential rays converges the beam to a different focus than rays in the sagittal plane. From the wavefront perspective, the situation is that of a planar optic interacting with a curved wavefront; this is the opposite of the conventional case where a curved lens interacts with a planar wavefront. The effect is the same in either case: the difference between the optical surface and the wavefront varies locally with field angle—as if the tangential wavefront had more curvature than the sagittal.

The amount of astigmatism, which is defined formally as the longitudinal difference in focus between the sagittal and tangential image, depends on the tilt angle. Smith gives the approximate (third-order) result as[1]

$$L_a \approx -\left(\frac{(n-1)t}{n}\right)\left(\frac{n+1}{n^2}\right)\theta_p^2 \quad [mm]$$

This expression shows that astigmatism depends on the focus shift of the untilted plate $[(n-1)t/n]$ and the plate's tilt angle θ_p. Observe that there is no dependence on f/#, so astigmatism with an f/10 lens is the same as that with an f/5 (within the limits of the third-order assumptions of this equation).

A typical beam-splitter tilt angle is 45 degrees (0.7854 rad), which sends a beam in two perpendicular directions. The required thickness t of the beam splitter depends on the beam diameter (Sec. 7.1.4); 10 mm is a standard thickness for many off-the-shelf plates. Substituting into Smith's equation for a visible-wavelength beam splitter with $n = 1.5$ gives $|L_a| = [(0.5 \times 10 \text{ mm})/1.5](2.5/1.5^2)$ $(0.7854 \text{ rad})^2 = 2.28$ mm, or approximately one-quarter of the plate thickness. Under the same conditions, for an IR-wavelength beam splitter with $n = 4$ we find that $|L_a| = [(3 \times 10 \text{ mm})/4](5/4^2)(0.7854 \text{ rad})^2 = 1.45$ mm; this value is still large, but it's an improvement over the visible design.

3.1.4 Field Curvature

Even when spherical, coma, and astigmatism have been corrected, still another third-order aberration can give blurred images. Figure 3.19 shows the image of a flat object not as a flat plane, but a curved surface bending inward toward the lens. Although this surface is similar to the tangential and sagittal surfaces of astigmatism for a positive lens, this aberration is known as *field curvature* (or *Petzval curvature*); it occurs, in part, because off-axis object points are slightly farther away from the lens and so their images are a bit closer to it [see Eq. (2.2)]. As a result, the rays on a flat focal plane do not create a crisp image point; instead they create a blur whose size depends on the off-axis angle, the curvature of the image surface, and the convergence angle (or f/#) of the off-axis rays.[9] Reduction of field curvature is thus an essential part of designing lenses for wide-angle photography.

Petzval curvature is typically measured not in terms of blur size but rather as the distance from the image surface to the paraxial image plane. The calculation of the field curvature is not as straightforward as the previously used object- and image-distance heuristic would suggest; physically, the distance Δz_p depends on the index and focal length of the lens[1]

$$\Delta z_p = \frac{y^2}{2nf} \approx \frac{\theta^2 f}{2n} \quad [\text{mm}] \tag{3.6}$$

(where the subscript p denotes "Petzval"). For a given image height, a paraxial approximation yields $y \approx \theta f$; the field curvature is smaller for longer focal lengths and with less departure from a flat image plane. In contrast, shorter focal lengths are preferred for a fixed field angle θ (because the image height is smaller with a shorter focal length at a given field angle). In both cases, a larger refractive index improves the design by increasing the Petzval radius nf, and high-index lenses are favored in the design of flat-field systems. Note that a high-index material has two competing effects that affect field curvature: the first effect is through the presence of n in Eq. (3.6), and the second is through the dependence of focal length f on n [Eq. (2.1)].

Equation (3.6) shows that the field curvature at any given angle depends only on the power and index of the lens, so stop shifts cannot reduce field curvature.[2] However, this curvature can be reduced by combining positive and negative elements. Positive lenses have an inward-curving field (pointing toward the lens) whereas the Petzval radius for a negative lens is outward curving [$f < 0$ in Eq. (3.6)]. As shown in Example 3.4, this balancing of positive with negative lenses employs a term from Eq. (3.6) known as the *Petzval sum*, which is given by $1/nf$ for a single lens and by $\Sigma(1/n_i f_i)$ for separated elements.[2] For example, a positive lens with index $n = 1.5$ and a focal length of 100 mm has a Petzval sum of $1/(1.5 \times 0.1 \text{ m}) = 6.67$ m^{-1} (diopters). Consequently, positive lenses cannot be arbitrarily added to a system; the addition of a field or relay lens, for instance, increases the field curvature.

It is not always possible to correct field curvature to the degree required. One way to *compensate* (but not correct) for field curvature is to use a collection of tilted image surfaces (Fig. 1.10). Unfortunately, tilted imagers are difficult to manufacture and so are used only in extreme cases, such as when astronomers require large-area imagers to cover a wide FOV.

Alternatively, an improvement over placing the image plane at the paraxial focus of the on-axis point is to place it somewhere between the on-axis and off-axis foci. This results in the smallest blur at some intermediate field angle and in the blur size increasing at both smaller and larger field angles. This phenomenon is illustrated in Fig. 3.19, where the image surfaces for tangential and sagittal astigmatism are also shown for a thin lens with third-order aberrations. In practice, the curves of Fig. 3.19 are useful only as third-order approximations; more realistic design curves for an anastigmat are shown in Fig. 3.21, where the Petzval and sagittal image surfaces change from inward curving to outward curving at some field angle—a result of including higher-order (fifth, seventh, etc.) aberrations in the calculations.

Example 3.4. Positive and negative lenses can be combined to reduce the field curvature, but they must be appropriately spaced so that the net power is greater than zero. In this example, we illustrate the use of a positive and negative lens selected so that the Petzval sum $\Sigma(1/n_j f_j) = 0$.

If the lenses have the same index, then the sum becomes $1/f_1 + 1/f_2 = 0$, or $f_1 = -f_2$ (or $\phi_1 = -\phi_2$). If the lenses are also placed in contact then the net power is zero, given that $\phi = \phi_1 + \phi_2 - \phi_1\phi_2 L \approx \phi_1 + \phi_2 = 0$ [see Eq. (2.8)]. However, as the negative lens is moved away from the positive, the power of the system increases until it reaches a maximum when the negative lens is placed near the focus of the first (Fig. 3.25). The negative lens is now known as a *field flattener*, and the net power $\phi = \phi_1 + \phi_2 - \phi_1\phi_2 L \approx \phi_1 + \phi_2 - \phi_2 = \phi_1$ for $L \approx f_1$.

From the ray perspective, this result is due to the small distance above the optical axis at which the marginal rays intercept the field flattener, reducing the incident angle of refraction and local surface bending [the ray height y_i in Eq. (2.9) is small]. The net effect is that the flattener minimizes the power required for the positive lens and thus the effort needed to reduce its spherical, coma, and astigmatism; the lower power also reduces the sensitivities to fabrication errors and misalignments due to assembly and environmental effects (Chap. 7).

Even though the power of the combination is not changed, the field curvature is now smaller. Physically, each section of the negative lens acts as a small prism. Snell's law shows that the prism pushes the focus away from the prism and also causes the rays to change their direction of propagation. As seen in Example 3.3, a thicker plate pushes the image out further depending on the index of the prism. The shift in image location is given by $\delta = (n-1)t/n$, where t is the plate thickness.[1]

Therefore, a negative lens with its thicker section on the outside will image the rays from the larger field angles closer to the image plane, reducing field curvature. In practice, completely removing Petzval curvature is not useful because doing so tends to result in large astigmatism at intermediate field angles.[10,11] In addition, the edge of the lens acts more like a prism than a flat plate and so the flattener changes the image size slightly by bending the chief ray. This distortion of the image is covered in Sec. 3.1.5.

3.1.5 Distortion

Section 3.1.4 showed that objects are not necessarily imaged onto a flat surface. Even when the field curvature is small, however, there's

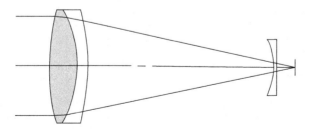

FIGURE 3.25 A field-flattening lens placed near the image plane reduces the Petzval curvature. (*Credit: Warren J. Smith,* Modern Optical Engineering, McGraw-Hill, www.mcgraw-hill.com.)

no guarantee that the image won't also be scrunched or stretched in the flat image plane. This is a problem particularly for aerial surveying or satellite imagers used for Google Earth, given that the eye is extremely sensitive to straight lines that appear curved in the image. This bending is known as image distortion, and it can be found in optical systems even when the image is otherwise perfectly crisp.

Figure 3.26 shows the two different types of distortion. In both cases, a square grid on the object is imaged as a distorted grid. When a stop is placed *in front of* a lens (between the lens and the object), the distortion scrunches the original image into something smaller; this is known as *barrel* distortion. As illustrated in Example 3.5, this is a result of the stop forcing the chief ray to the upper portion of the lens, where the angle of refraction is larger; as a result, the chief ray bends more than it would without the stop and thus creates a slightly smaller image for off-axis points. When a stop is instead placed *after* a lens, the distortion stretches the original image into something bigger; this is known as *pincushion* distortion. Both types of distortion are caused by moving the stop away from the lens (Fig. 3.14).

Placing the stop exactly *at* the lens to remove distortion is not always a convenient design architecture. There may be more than one lens in the system, for example, in which case it is still possible to control distortion via movement of the stop location. Specifically: if the stop is placed fairly symmetrically between two lenses, then the pincushion distortion that is created by the stop located after the first lens can be more or less compensated by the barrel distortion created by the same stop being located also in front of the second lens. This symmetry principle is extremely useful for other reasons as well; it will be explored in more detail in Sec. 3.1.9.

Figure 3.23 shows that, with the stop at the lens, there is no distortion because the chief ray goes through the center of the stop without bending. However, when the stop is shifted away from the

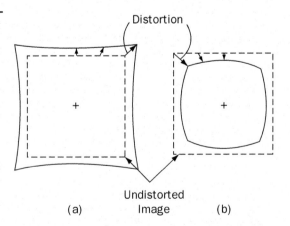

FIGURE 3.26 The image of a rectangular object grid may be imaged as either (a) a pincushion shape or (b) a barrel shape. (*Credit: Warren J. Smith, Modern Optical Engineering, McGraw-Hill, www.mcgraw-hill.com.*)

lens, the incident angle θ is refracted to a different angle θ' on the image side of the entrance pupil, resulting in distortion. As shown in Example 3.5, this distortion has two components: that due to the nonlinearity of the sin θ expansion for Snell's law and that due to nonlinearity of the nonparaxial, $y = f - \tan θ$ image height.

Distortion is thus the difference between the stop-shifted and non–stop-shifted image locations, and it increases with the cube of the field angle ($θ^3$). As a result, the diagonals of the field scrunch or stretch more than the sides, which makes the barrel and pincushion effects more pronounced for wide-angle lenses than for narrow ones. Less than 1 percent distortion at the edges of the FOV is a reasonable design goal for visual applications, although 2 to 3 percent is often acceptable for wide-field imagers. A typical spec for a high-performance camera lens is shown in Fig. 3.27, where it is seen that the type of distortion varies with focal length.

Example 3.5. It was shown in Chap. 2 that the geometrical ray-trace equations bring the chief and marginal rays from an off-axis point to the same focus—at least within the validity of the paraxial approximations for $\sin(u) \approx u$ and $\sin(u') \approx u'$. In this example, the paraxial ray-trace equations are used to show that putting a stop in front of a lens results in image distortion. Although the paraxial equations are not valid for large field angles, they illustrate in a simple way the physical origin of distortion.

There is no distortion for a thin lens with the stop at the lens. The chief ray goes through the center of the lens ($y_1 = 0$), so the image height for a distant object depends on the object's angular size and the image distance. Using a positive lens with a focal length $f = 100$ mm to image a distant object that has an angular size $u = 10$ degrees $= u'$, we obtain an image height of $y = f \tan(u') =$ 100 mm × tan(0.17453) = 17.6327 mm.

Moving the stop out in front of the lens does not change the angular size of the object. However, the stop forces the chief ray to intercept the lens away from

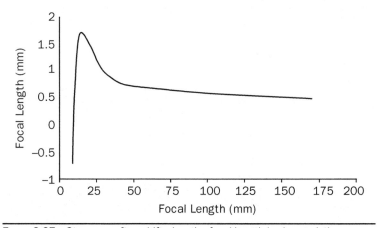

FIGURE 3.27 Stops are often shifted as the focal length is changed, thus changing distortion from barrel to pincushion (distortion > 0). Data is for the Canon YJ20 × 8.5B KRS zoom lens.

its center (Fig. 3.14) at an intercept height $y_1 = \tan(0.17453) \times 10$ mm = 1.7633 mm for a stop that is 10 mm in front of the lens. The curvature of the lens forces the incident angle of the chief ray to be larger at this intercept height; as a result, the chief ray bends more than it did without the stop, creating barrel (negative) distortion. The refracted chief-ray angle after the lens is now smaller ($u' = u - y_1\phi_1 = 0.17453 - 1.7633/100 = 0.1569$ rad) and results in an image height of $h = 1.7633$ mm + 100 mm $\times \tan(0.1569) = 17.5833$ mm. These results are summarized in Table 3.2.

Compared with the stop at the lens, the smaller image height means that the image is now distorted (by 17.5833 mm – 17.6327 mm = –0.0494 mm). Expressed as a percentage of the undistorted height, the distortion is –0.0494 mm/17.6327 mm = –0.0028 (–0.28 percent), a value well within reasonable bounds.

Observe that if we had *not* used $y = f\tan(u')$ for the image height and instead had used the paraxial approximations $\tan(u) \approx u$ and $\tan(u') \approx u'$, then the heights with and without the stop would come out the same (17.453 mm), incorrectly indicating that there is no distortion. The difference in heights for a shifted stop in this example is thus a consequence of using $\tan(u')$ for the calculation of image size; it is not a result of including the component of distortion due to the third-order expansion of Snell's law. The use of this expansion would result in different values for the image height, leading to the general definition of distortion as a percentage variation from the straight-line image shown in Fig. 3.26; it is quantified as $\Delta y_d = (y_s - y)/y$, where y is the image height for the stop at the lens and y_s the image height when the stop is shifted. Both heights use the nonparaxial $\tan(u')$ dependence,. This illustrates that $\tan(u')$ is not a linear function of the refracted angle u' and so, even for a paraxial ray trace, the "lens plus shifted stop" system takes objects that are linear and creates images that are not linear.

3.1.6 Axial Color

We have not yet included the effects of wavelength on image quality. These effects are seen most simply in the rainbow that results when the Sun reflects off raindrops or in the use of a prism to separate visible colors. Of course, there exist other nonvisible "colors" (wavelengths) in the UV and IR portions of the electromagnetic spectrum that can also be separated (or "dispersed").

Color separation is a function of the refractive index. The index of glass in the visible spectrum, for example, is higher for blue wavelengths than it is for red; as a result, a prism designed for color separation bends the blue rays more than the red ("blue bends more"), yielding the characteristic rainbow spectrum.

When glass is used to make a simple lens, different wavelengths for an on-axis point come into focus at different locations along the axis: the higher index for blue wavelengths means that these rays are imaged closer to the lens and that the red wavelengths are imaged farther away [see Eq. (2.1)]. As illustrated in Fig. 3.28, this aberration of axial focus location—known as "longitudinal chromatic," "axial chromatic," or "axial color"—affects image quality by producing a red-plus-green blur surrounding a central blue spot (when the FPA is placed at the blue focus).

Lens	u	Lstop (mm)	y (mm)	f (mm)	φ (mm)	u'	h (mm)
Chief Ray, Stop at Lens							
1	0.17453	0.0	0	100	0.01	0.17453	17.6327
Lens	u		y (mm)	f (mm)	φ (mm)	u'	h (mm)
Chief Ray, Stop Before Lens							
1	0.17453	−10.0	1.7633	10.0	0.01	0.1569	17.5833

TABLE 3.2 Paraxial ray trace for the chief ray through a thin lens with the stop at the lens and with the stop located 10 mm to the left of the lens; reported values reflect use of the exact expression for image height, $y = f \tan(u')$.

98 Chapter Three

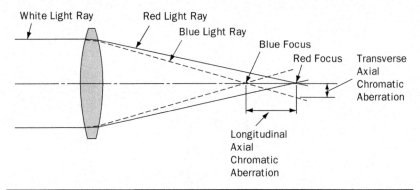

FIGURE 3.28 On-axis points image to different locations along the optical axis as a function of dispersion of the refractive index. (*Credit: Warren J. Smith*, Modern Optical Engineering, *McGraw-Hill, www.mcgraw-hill.com.*)

Many lens designs—for example, those for laser-based systems operating at a specific wavelength—do not require correction of chromatic aberration. However, visual systems can be quite sensitive to lack of color correction; in particular, high-definition television (HDTV) lenses are among the most difficult to design. The refractive index in Eq. (2.1) implies that correcting for chromatic aberration is simply a task of geometrical optics; however, given the dependence of spherical aberration, coma, and so forth on focal length and f/#, wavelength-dependent indices clearly affect these aberrations as well. In order to understand the extent of these effects, we must first quantify the concept of dispersion.

Dispersion and V-Parameter

The term *dispersion* refers to the dependence of a material's refractive index on wavelength. This parameter is plotted in Fig. 3.29 for the common optical glasses N-BK7 and N-SF6. The dominant feature of such curves is that the shorter ("blue") wavelengths have a higher index while the longer ("red") wavelengths have a lower index.[12] The increase in index at shorter wavelengths occurs as the material absorption increases; we shall explore the consequences of absorption in Chap. 4.

Quantitatively, the dispersion curve can be approximated by a power-series expansion that is named after Sellmeier and now used as a standard by the glass manufacturer Schott Glass:[13]

$$n^2(\lambda) - 1 = \frac{B_1 \lambda^2}{\lambda^2 - C_1} + \frac{B_2 \lambda^2}{\lambda^2 - C_2} + \frac{B_3 \lambda^2}{\lambda^2 - C_3} \qquad (3.7)$$

Here the constants (B_1, C_1, . . .) are experimentally measured values, available in handbooks and manufacturer data sheets, and are based on the wavelength λ being measured in units of microns. The curves

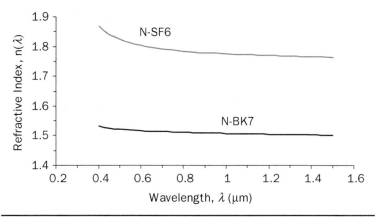

FIGURE 3.29 Refractive index versus wavelength for two common optical materials.

shown in Fig. 3.29 were created using Eq. (3.7) and data from Schott Glass.[13] The six-constant power-series expansion in λ^2 is accurate to 0.000005 over the spectral range 0.36–1.5 μm.[2]

The simplest measure of dispersion is the index difference between the wavelengths at the extremes of the curve. Using the Sellmeier coefficients given in Fig. 3.30, for example, the refractive index for N-BK7 at the so-called F wavelength (λ_F = 486.1 nm, a blue spectral line of hydrogen) is n_F = 1.52238 and while the index at the other end of the visible spectrum (the C wavelength, λ_C = 656.3 nm, a red hydrogen line) is n_C = 1.51432.[13] One measure of dispersion for N-BK7 in the visible band is thus estimated as $n_F - n_C$ = 1.52238 − 1.51432 = 0.00806, very close to the exact value given in the upper right-hand corner of Fig. 3.31.

In designing lenses for color correction, it is useful to include a measure of refractive power in the figure of merit for chromatic aberration. Note that a glass could have a large dispersion as given by the index difference $n_F - n_C$ yet relatively small absolute values of n_F and n_C individually, indicating little refractive power. As a result, a new parameter was introduced by Abbe; it is known as the Abbe number, the V-number (V/#), or the V-parameter:[1,2]

$$V = \frac{n_d - 1}{n_F - n_C} \qquad (3.8)$$

The refractive power of the material [$\phi \approx (n - 1)$] is included in the numerator of this figure of merit comparing refraction with dispersion. The index n_d is measured at a wavelength between the extremes; for visible light, the wavelength used is the yellow emission (d-line) of helium at λ_d = 587.6 nm, which is why the symbol V_d is

	Glass Type				
	BK7 (N-BK7)	SF11	LaSFN9	BaK1 (N-BaK1)	F2
Melt-to-Melt Mean Index Tolerance	±0.001	±0.001	±0.002	±0.001	±0.001
Homogeneity within Melt	±1 × 10^{-4}	±1 × 10^{-4}	±1 × 10^{-4}	±1 × 10^{-4}	±1 × 10^{-4}
Striae Grade (ISO 10110)	A	A	A	A	A
Stress Birefringence, nm/cm, Yellow Light	10	10	10	10	10
Abbé Factor (v_d)	64.17	25.76	32.17	57.55	36.37
Constants of Dispersion Formula:					
B_1	1.03961212	1.73848403	1.97888194	1.12365662	1.34533359
B_2	2.31792344 × 10^{-1}	3.11168974 × 10^{-1}	3.20435298 × 10^{-1}	3.09276848 × 10^{-1}	2.09073176 × 10^{-1}
B_3	1.01046945	1.17490871	1.92900751	8.81511957 × 10^{-1}	9.37357162 × 10^{-1}
C_1	6.00069867 × 10^{-3}	1.36068604 × 10^{-2}	1.18537266 × 10^{-2}	6.44742752 × 10^{-3}	9.97743871 × 10^{-3}
C_2	2.00179144 × 10^{-2}	6.15960463 × 10^{-2}	5.27381770 × 10^{-2}	2.22284402 × 10^{-2}	4.70450767 × 10^{-2}
C_3	1.03560653 × 10^2	1.21922711 × 10^2	1.66256540 × 10^2	1.07297751 × 10^2	1.11836764 × 10^2
Density (g/cm^3)	2.51	4.74	4.44	3.19	3.61
Coefficient of Linear Thermal Expansion (cm/cm/°C):					
−30° to +70°C	7.1 × 10^{-6}	6.1 × 10^{-6}	7.4 × 10^{-6}	7.6 × 10^{-6}	8.2 × 10^{-6}
+20° to +300°C	8.3 × 10^{-6}	6.8 × 10^{-6}	8.4 × 10^{-6}	8.6 × 10^{-6}	9.2 × 10^{-6}
Transition Temperature	557°C	505°C	703°C	592°C	438°C
Young's Modulus (dynes/mm^2)	8.20 × 10^9	6.60 × 10^9	1.09 × 10^{10}	7.30 × 10^9	5.70 × 10^9
Climate Resistance	2	1	2	2	1
Stain Resistance	0	0	0	1	0
Acid Resistance	1.0	1.0	2.0	3.3	1.0
Alkali Resistance	2.0	1.2	1.0	1.2	2.3
Phosphate Resistance	2.3	1.0	1.0	2.0	1.3
Knoop Hardness	610	450	630	530	420
Poisson's Ratio	0.206	0.235	0.286	0.252	0.220

*PAUV coating: 10-nsec pulse at 355 nm; all others: 20-nsec pulse at 1064 nm typical

Figure 3.30 Optical and mechanical properties of common lens materials, including constants for the dispersion equation [Eq. (3.7)]. (*Credit: CVI Melles-Griot.*)

Aberrations and Image Quality

N-BK7
517642.251

n_d = 1.51680		v_d = 64.17		$n_F - n_C$ = 0.008054		
n_e = 1.51872		v_e = 63.96		$n_{F'} - n_{C'}$ = 0.008110		

Refractive Indices		
λ [nm]		
$n_{2325.4}$	2325.4	1.489210
$n_{1970.1}$	1970.1	1.494950
$n_{1529.6}$	1529.6	1.500910
$n_{1060.0}$	1060.0	1.506690
n_t	1014.0	1.507310
n_s	852.1	1.509800
n_r	706.5	1.512890
n_C	656.3	1.514320
$n_{C'}$	643.8	1.514720
$n_{632.8}$	632.8	1.515090
n_D	589.3	1.516730
n_d	587.6	1.516800
n_e	546.1	1.518720
n_F	486.1	1.522380
$n_{F'}$	480.0	1.522830
n_g	435.8	1.526680
n_h	404.7	1.530240
n_i	365.0	1.536270
$n_{334.1}$	334.1	1.542720
$n_{312.6}$	312.6	1.548620
$n_{296.7}$	296.7	
$n_{280.4}$	280.4	
$n_{248.3}$	248.3	

Constants of Dispersion Formula	
B_1	1.03961212
B_2	0.231792344
B_3	1.01046945
C_1	0.00600069867
C_2	0.0200179144
C_3	103.560653

Constants of Dispersion dn/dT	
D_0	$1.86 \cdot 10^{-6}$
D_1	$1.31 \cdot 10^{-8}$
D_2	$-1.37 \cdot 10^{-11}$
E_0	$4.34 \cdot 10^{-7}$
E_1	$6.27 \cdot 10^{-10}$
λ_{TK}[μm]	0.170

Temperature Coefficients of Refractive Index						
	$\Delta n_{rel}/\Delta T[10^{-6}/K]$			$\Delta n_{abs}/\Delta T[10^{-6}/K]$		
[°C]	1060.0	e	g	1060.0	e	g
−40/−20	2.4	2.9	3.3	0.3	0.8	1.2
+20/+40	2.4	3.0	3.5	1.1	1.6	2.1
+60/+80	2.5	3.1	3.7	1.5	2.1	2.7

Internal Transmittance τ_i		
λ [nm]	τ_i (10 mm)	τ_i (25 mm)
2500	0.67	0.36
2325	0.79	0.56
1970	0.933	0.840
1530	0.992	0.980
1060	0.999	0.997
700	0.998	0.996
660	0.998	0.994
620	0.998	0.994
580	0.998	0.995
546	0.998	0.996
500	0.998	0.994
460	0.997	0.993
436	0.997	0.992
420	0.997	0.993
405	0.997	0.993
400	0.997	0.992
390	0.996	0.989
380	0.993	0.983
370	0.991	0.977
365	0.988	0.971
350	0.967	0.920
334	0.905	0.780
320	0.770	0.520
310	0.574	0.250
300	0.290	0.050
290	0.060	
280		
270		
260		
250		

Color Code	
λ_{80}/λ_5	33/29
($*=\lambda_{70}/\lambda_5$)	

Remarks

Relative Partial Dispersion	
$P_{s,t}$	0.3098
$P_{C,s}$	0.5612
$P_{d,C}$	0.3076
$P_{e,d}$	0.2386
$P_{g,F}$	0.5349
$P_{i,h}$	0.7483
$P'_{s,t}$	0.3076
$P'_{C',s}$	0.6062
$P'_{d,c'}$	0.2566
$P'_{e,d}$	0.237
$P'_{g,F'}$	0.4754
$P'_{i,h}$	0.7432

Deviation of Relative Partial Dispersions ΔP from the "Normal Line"	
$\Delta P_{C,t}$	0.0216
$\Delta P_{C,s}$	0.0087
$\Delta P_{F,e}$	−0.0009
$\Delta P_{g,F}$	−0.0009
$\Delta P_{i,g}$	0.0035

Other Properties	
$\alpha_{-30/+70°C}[10^{-6}/K]$	7.1
$\alpha_{+20/+300°C}[10^{-6}/K]$	8.3
T_g[°C]	557
$T_{10}^{13.0}$[°C]	557
$T_{10}^{7.6}$[°C]	719
c_p[J/(g-K)]	0.858
λ [W/(m-K)]	1.114
ρ [g/cm^3]	2.51
E[10^3N/mm^2]	82
μ	0.206
K[10^{-6}mm^2/N]	2.77
$HK_{0.1/20}$	610
HG	3
B	0.00
CR	1
FR	0
SR	1
AR	2.3
PR	2.3

FIGURE 3.31 Typical data sheet of optical and mechanical properties supplied by Schott for a wide variety of optical materials. (*Credit: SCHOTT North America, Inc.*)

sometimes used. Using different wavelengths, the V/# can also be defined for other spectral ranges—for example, the NIR and SWIR bands over which N-BK7 also transmits.

Typical values for common glasses that transmit in the visible range from 20 to about 90; small V-numbers imply a large dispersion, with different wavelengths strongly separated by a lens or prism. A distinction is made between materials with $V = 20$ to 50 and $n > 1.60$ (commonly known as flints) and less dispersive materials with $V = 55$ to 90 and $n < 1.60$ (known as crowns). Depending on the refractive index, the intermediate V-numbers between 50 and 55 can be either crowns or flints: with an index > 1.6 they are crowns and with an index < 1.6 flints.[1] Traditionally, flints were made by replacing the alkaline (fluorine) in crowns with lead oxides;[14] more recently, lead-free flints have been made with approximately the same optical properties but a less hazardous chemical composition.

The distinction between crowns and flints is commonly represented by Schott Glass with the letter "K" for crown (from the German *kron*) and "F" for flint. A summary using this distinction for the materials available from Schott Glass is shown in Table 3.3. Barium (BaSF) and lanthanum (LaSF) flints are also available in dense (S) form, as are phosphate (PSK), lanthanum (LaSK), and titanium (TiSK) dense crowns. A material that does not fit the nomenclature is the crown/flint (KF), presumably an intermediate material between crowns and flints. In the spectral bands where the glasses are not used, the previous quantitative distinction is not made between flints and crowns; hence there are V-numbers as high as 250 for silicon in

Glass Type	Crown	Flint
Crown	K	—
Flint	—	F
Light	—	LF
Very light	—	LLF
Dense	SK	SF
Very dense	SSK	—
Barium	BaK	BaF
Boron	BK	—
Fluorine	FK	—
Lanthanum	LaK	LaF
Phosphate	PK	—
Titanium	TiK	TiF

TABLE 3.3 Selected glass types manufactured by Schott Glass. A recent addition to the nomenclature is the prefix N, as in N-BK7, for glass types that contain no lead, arsenic, or cadmium.

Aberrations and Image Quality 103

the MWIR band and of approximately 1000 for germanium in the LWIR band.[15] There are also materials (e.g., germanium) that are a flint in the MWIR band but a crown in the LWIR band.[15]

In addition to their higher dispersion, flints also tend to have larger refractive indices. This is seen in Fig. 3.29, where the index for N-SF6 is higher than that of N-BK7 across the visible spectrum. It is also illustrated in the glass map shown in Fig. 3.32, where the index is plotted against the V-number (symbol v_d on the horizontal axis of this plot). This is an extremely useful tool for the lens designer because it displays visually the available glasses for reducing chromatic aberration. The dominant trend is that shown along the bottom of the plot, where more dispersion (i.e., smaller v_d toward the right) is associated with a higher refractive index; for example, a barium flint (BaF) has a higher index and dispersion than a barium crown (BaK). With different materials, however, it is possible to find crowns with a higher index than a flint—for instance, the lanthanum crown (LaK), which has a higher index than light flints (LF).

The reason for creating flints with *more* dispersion is that a single crown cannot, by itself, correct for chromatic aberrations. If the dispersion could be made low enough (V/# large enough), then it

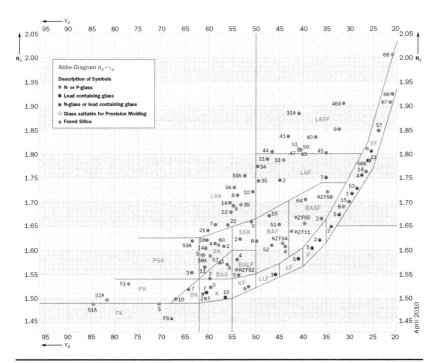

Figure 3.32 Abbe diagram showing refractive index n_d as a function of dispersion v_d for a variety of optical materials. (*Credit: SCHOTT North America, Inc.*) (*See also color insert.*)

would be possible to use a single crown to reduce axial color to acceptable limits. If "acceptable" is based on a strict depth-of-focus criterion (see Sec. 7.2.3), then the maximum V-number of 90 is insufficient for visible designs. Although engineered materials with high V/# are no doubt being developed, for now there are two options for reducing chromatic aberrations: refractive achromats and reflective mirrors.

Achromatic Doublets

A common design for reducing axial color is an achromatic doublet that corrects for axial color at two wavelengths, such as the F- and C-wavelengths of hydrogen. By "correct" we mean that these two colors have the same effective index (and therefore focal length), which Eq. (3.8) shows is equivalent to an infinite V-number *for these two wavelengths*.

Figure 3.33 shows a typical doublet consisting of a positive and negative lens in contact. To obtain a doublet with the net positive power required for imaging, the power of the positive lens must be greater than that of the negative lens [see Eq. (2.8)]. To reduce chromatic aberration, the dispersion of the negative lens must be larger than that of the positive lens. For this reason, the invention of high-dispersion flints was critical for development of the achromat.

The axial chromatic aberration, which is measured as the change in focal length between the two wavelengths that are used to measure the V-number, depends on both the focal length and the V/#:[3]

$$\Delta f \approx -\frac{f}{V} \quad [\text{mm}] \tag{3.9}$$

Here the benefits of a large V/# are quantified, and for a given requirement on Δf, larger values are required for longer focal lengths. Physically, the longer focal length has a greater separation between the "red" and "blue" wavelengths; a larger V/# is thus needed to equalize the difference in focus.

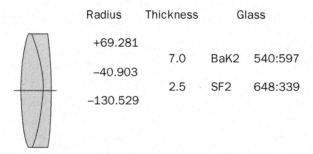

FIGURE 3.33 Optical prescription for a typical achromatic doublet consisting of BaK2 and SF2 glasses. (*Credit: Warren J. Smith,* Modern Optical Engineering, *McGraw-Hill, www.mcgraw-hill.com.*)

Example 3.6. To illustrate the process of designing an achromatic doublet, Eq. (2.8) is used in conjunction with Eq. (3.9) to determine how much dispersion is required for each element. To obtain zero chromatic aberration at the two wavelengths that define the V-number, Eq. (3.9) shows that $\Delta f = f_1/V_1 + f_2/V_2 = 0$. By rearranging we can see that the ratio of V-numbers depends on the ratio of focal lengths: $f_1/f_2 = -V_1/V_2$.

To determine the ratio of focal lengths, Eq. (2.8) shows that the net power $\phi = \phi_1 + \phi_2 - \phi_1\phi_2 L \approx \phi_1 + \phi_2$ when the two lenses are in contact. To obtain a lens with a focal length of 100 mm, any combination of positive and negative elements is possible—for example, $f_1 = 50$ mm and $f_2 = -100$ mm. The focal ratio in this case is thus $f_1/f_2 = -0.5$, so the dispersion ratio $V_1/V_2 = 0.5$. Physically, the dispersion of the negative flint must be higher than that of the positive crown and so the V/# of the flint must be lower by a factor of 2.

The glass map in Fig. 3.32 shows that BaK2 and SF2 have approximately this ratio: BaK2 has a V-number of 59.7, and SF2 disperses to V/33.9. There is no guarantee that lens materials with the correct dispersion ratio are available that will exactly balance the ratio of focal lengths selected; using these glasses will require reiterating the design so that $f_1 = 43.2$ mm and $f_2 = -76.1$ mm, thus matching the ratio of focal lengths to V-numbers, and giving an EFL = 100 mm.

In addition to axial color and the glass map, there are other issues that must be considered when designing an achromatic doublet. These issues include control of spherical aberration by optimizing the index difference (and V-number) between the crown and the flint;[1] shape factor for additional control of spherical and coma; stop location; and environmental stability, availability, and cost of the materials selected.

In Example 3.6, the combination of a low-index crown and a high-index flint resulted in a relatively strong field curvature. Recall from Example 3.4 that the Petzval sum equals $\Sigma(1/n_i f_i) = 1/n_1 f_1 + 1/n_2 f_2$ for two lenses that are not in contact. Assuming a slight separation between the crown and flint to illustrate the concept, the combination of a low index and short focal length in the crown ensures that $1/n_1 f_1$ is large. If instead we use a higher-index material for the positive element, the Petzval is reduced considerably. Such doublets are known as "new" achromats and may incorporate very dense (SSK) or lanthanum (LaK) crowns in combination with a light flint (LF) to simultaneously correct for axial color and Petzval.[3]

Even new achromats cannot remove axial color over a wide range of wavelengths. Because an achromat is designed to correct for only two wavelengths, there is residual chromatic aberration at the wavelengths between for which the achromat is designed. This residual is known as *secondary color* and is measured as the axial distance between the corrected and uncorrected wavelengths. Depending on the magnitude of this residual, it may be necessary to use lenses designed to correct for three or four wavelengths (known as "apochromats" and "super-achromats," respectively). These complex and expensive lenses are unnecessary where instead reflective optics can be used to remove axial color over entire wavebands.

Reflective Optics

Even a quick look at an object "in" a flat mirror reveals that different wavelengths are reflected without dispersion; there is no chromatic aberration because there is no index of *refraction* for the reflection angle. As a result, all wavelengths reflect off the mirror with the same angle (with respect to the local perpendicular) at which they are incident and without chromatic spread.

Curved mirrors share some properties with lenses; these properties include the redirection of collimated wavefronts into converging or diverging beams. A spherical mirror, for example, brings an on-axis point at infinity to a focus at a distance equal to half its radius of curvature ($f = 0.5R$). Like flat mirrors, curved reflectors exhibit no dispersion. Hence they are used for imaging, radiometric, and spectroscopic systems that require color correction across a wide range of wavelengths.

Examples of reflective telescopes using different conic sections are shown in Fig. 3.34.[16] The most common configuration is the Cassegrain telescope, which consists of a paraboloid primary mirror and a hyperboloid secondary mirror that are curved to act as a positive and negative lens, respectively; the Cassegrain is thus the reflective equivalent of the telephoto explored in Chap. 2. Some properties of this and other types of reflective 'scopes' are summarized in Table 3.4. Note that afocal telescopes are also possible.

In addition to being inherently achromatic, reflective optics have the additional advantage of relatively small size and low weight. Because reflective surfaces allow rays to bounce back toward the source, the overall track length can be shorter than in a refractive system; for example, with mirrors an EFL of 1000 mm can be folded into a track length of 500 mm or less. Also, the weight of a multielement lens is almost always greater than that of an equivalent reflective system, especially for entrance pupils on the order of 150 mm or larger. This is why apertures of this size are designed almost exclusively as reflective optics. Exceptions, such as the 1-meter refractor at the Yerkes Observatory, have larger FOVs than are available with reflective optics.

One of the disadvantages of not having a refractive index is that reflective systems are thus more difficult to correct for nonchromatic aberrations. It was shown in Sec. 3.1.1, for example, that the refractive index can be used to reduce spherical aberration; moreover, lenses have two surfaces with which to control aberrations whereas mirrors have only one surface. As a result, reflective systems are limited off-axis by coma and astigmatism, and they generally have much smaller field coverage for a given f/# (or a larger f/# for a given FOV). The reflective systems shown in Fig. 2.11, for example—the parabola, Cassegrain, Ritchey–Chretien, unobscured three-mirror, Schmidt, and Schmidt–Cassegrain (described below)—have a maximum FOV of only 10 degrees or so.

FIGURE 3.34 Three types of reflective telescopes used to remove chromatic aberration. (*Credit: Warren J. Smith*, Modern Optical Engineering, *McGraw-Hill, www.mcgraw-hill.com.*)

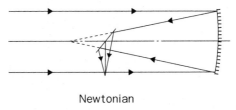
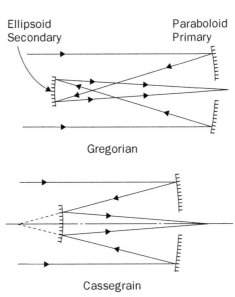

Telescope Type	Primary	Secondary
Cassegrain	Parabolic	Hyperbolic
Gregorian	Parabolic	Elliptical
Newtonian	Parabolic	Flat
Ritchey–Chretien	Hyperbolic	Hyperbolic
Dall–Kirkham	Elliptical	Spherical

TABLE 3.4 Shapes used for the primary and secondary mirrors of two-mirror reflective imagers. (The Newtonian imager is *not* the same as the Newtonian afocal configuration in Fig. 2.23.)

Reflective optics have other disadvantages as well:

- Most reflective designs require aspheric surfaces (parabolas, hyperbolas, etc.), which are relatively difficult to fabricate, align, and test.

- Central obscurations, such as those shown in Fig. 3.34, block a fraction of the incident photons and reduce contrast between bright and dark parts of the image (see Sec. 3.3.1).
- Although unobscured three- and four-mirror systems (e.g., the three-mirror anastigmat, TMA) can be used to increase the number of surfaces available to correct for such aberrations as astigmatism, these systems tend to be expensive and difficult to align. Hence they are used only when their other advantages—for example, suppression of stray light—cannot be met with any other architecture (see Chap. 4).

Hybrid refractive/reflective forms known as *catadioptrics* (or "cats") add some type of refractive element for aberration correction. The most common is the Schmidt–Cassegrain, which uses spherical mirrors and an aspherically shaped front window placed at the aperture stop. The placement of the stop at the center of curvature of the spherical primary mirror removes coma and astigmatism, and the aspheric window placed at the stop controls spherical aberration.[3] This reflective configuration is used not for color correction but rather because such telescopes are compact and lightweight. The addition of the refractive window means that the eyepiece design must control chromatic aberrations, at the expense of an increase in complexity and cost.[3] The price of such telescopes is nonetheless relatively low, providing a good entry point for moderate-aperture amateur astronomy (Fig. 3.35).

3.1.7 Lateral Color

The effects of wavelength on image quality can also be seen in a second chromatic aberration known as *lateral color*. This is shown in

FIGURE 3.35 Reflective Schmidt–Cassegrain telescopes for amateur astronomy are commonly available from manufacturers such as Celestron and Mead. (*Photo credit: Celestron, LLC.*)

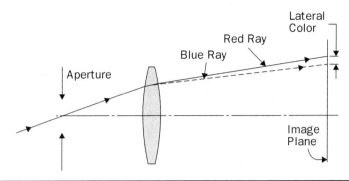

FIGURE 3.36 Off-axis points image to different heights on the focal plane depending on the index dispersion. (*Credit: Warren J. Smith*, Modern Optical Engineering, *McGraw-Hill, www.mcgraw-hill.com.*)

Fig. 3.36, where the larger index of refraction for the shorter wavelengths brings the blue focus for off-axis points closer to the optical axis than the red focus, which creates a color blur in the image. The resulting difference in image heights across the design spectrum is known as lateral (or transverse) color.

Lateral color often controls the design of small-angle systems with a long off-axis image distance over which the colors can disperse. Given that the FOV is often dictated by customer requirements, the position of the aperture stop is commonly used to control this aberration. As shown in Fig. 3.23, placing the stop at the lens removes the lateral color, whereas moving the stop either before the lens (closer to the object) or after the lens (closer to the image) induces significant transverse chromatic. For the stop before the lens, "blue bends more" and places shorter wavelengths closer to the axis. Blue also bends more when the stop is after the lens but, in starting out below the optical axis, the shorter wavelengths end up farther away from the axis than the longer (red) wavelengths.

For multielement lenses, there is no single element at which the stop can be placed to remove lateral color. For these lenses, the symmetry principle is often employed by placing a stop between two elements. This allows the lateral color from the stop after the first element to cancel the lateral color from the same stop, which is located before the second element. This principle is illustrated by the plot in Fig. 3.37 of lateral color as a function of the focal length of a multielement lens. Using symmetry is a common lens design technique, one that is discussed in more detail in Sec. 3.1.8.

3.1.8 Summary

An initial optical design will always create an image of a point that is blurred into something bigger; the imaged points may also be improperly located and may not correctly reproduce the object's

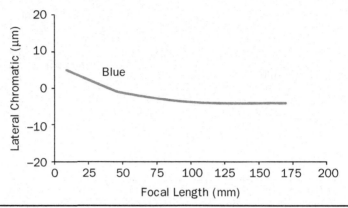

FIGURE 3.37 A lens that is corrected for one wavelength at all foci will likely have lateral chromatic aberrations that vary with focal length at other wavelengths. Data is for the Canon YJ20 × 8.5B KRS zoom lens.

colors. Part of the reason is that the spherical lenses and mirrors used to create images do not do a perfect job, and these imperfections are known as aberrations.

Aberrations can be reduced by decreasing the size of the entrance pupil, the field of view, or the wavelength range. For example, it is relatively easy to design a single-wavelength telescope with a narrow field and small aperture; on the other hand, designing a fast, wide-band, wide-field design presents a real challenge. Table 3.5 summarizes these effects; note the trade-off between aperture and field for the off-axis aberrations, which is consistent with the trend revealed in Fig. 2.11.

Unfortunately, it's not always possible to make arbitrary changes to the entrance pupil, field, or design wavelengths. The entrance aperture, for example, may be based on how much power the system needs to collect (see Chap. 4). The FOV and wavelengths may be requirements that have been decided by the customer, in which case they cannot be modified.

In these cases, the only option is to look for other ways to control the aberrations: shape factor to reduce spherical and coma, use of a doublet to control chromatics, and so on. One of the most challenging careers in optics is that of the lens designer (also known as a "ray bender"), who must know how to modify the lens power, surface radii, spacings, indices, stop locations, et cetera to reduce or balance the aberrations. The designer's results are summarized in a table known as an *optical prescription*, an example of which is given in Table 3.6 for the double-Gauss form shown in Fig. 3.38. The table details the material, thickness, radii, and clear aperture of the elements required to produce an "acceptable" design. The Radius column lists the radius of curvature of each surface, and the Thickness

Aberrations and Image Quality

Aberration	Aperture	Field Angle
Spherical	D^3	—
Coma	D^2	θ
Astigmatism	D	θ^2
Field curvature	D	θ^2
Distortion (linear)	—	θ^3
Axial color	D	—
Lateral color	—	θ

TABLE 3.5 The effects of aperture D and field angle θ on third-order aberrations. All aberrations are measured in the plane of the image (transverse or lateral aberrations), not along the optical axis.[1]

Surface	Radius	Thickness	Glass	Semidiameter
OBJ	Infinity	Infinity		
1	81.40	9.30	LaF15	40.0
2	265.12	0.19		40.0
3	56.2	12.02	LaF20	32.9
4	271.32	6.59	SF15	31.5
5	33.41	16.33		24.6
STO	Infinity	17.00		23.9
7	−32.27	4.65	SF1	23.3
8	−857.27	16.47	LaK12	28.5
9	−49.53	0.58		30.8
10	−232.56	9.69	LaK8	32.0
11	−66.1	0.19		33.5
12	158.91	6.20	LaK8	36.0
13	−890.28	74.557		36.0
IMA	Infinity			

TABLE 3.6 Optical prescription for a double-Gauss lens consisting of seven elements.[1] All dimensions are measured in units of millimeters. The object (OBJ) is an infinite distance away from surface 1, and the image (IMA) is located 74.557 mm past surface 13. The semidiameter is the clear-aperture radius required for the chief and marginal rays to propagate through the lens; blank cells in the Glass column represent air; and STO denotes the aperture stop.

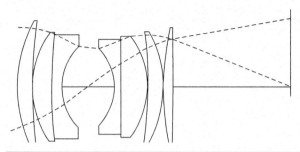

FIGURE 3.38 Double-Gauss lens corresponding to the prescription given in Table 3.6. (*Credit: Warren J. Smith, Modern Optical Engineering, McGraw-Hill, www.mcgraw-hill.com.*)

column lists the propagation distance—center thickness (CT) for a lens—*along the optical centerline*. Aspheric elements would have additional columns listing the conic constant or the coefficients required to define the aspheric surface.[1,2]

In one variation or another, the double-Gauss form is widely used for 35-mm camera lenses. It relies on many of the concepts reviewed in the previous sections to reduce aberrations and is a six- or seven-element version of the geometrical thin lens. Particularly important is the symmetry principle introduced in Sec. 3.1.7, where the lateral color introduced by the lenses on one side of the stop is more or less canceled by the lenses on the other side. Note that the stop (STO) in Table 3.6 is located 16.33 mm to the right of surface 5 and 17 mm to the left of surface 6—that is, almost exactly between them.

This placement of the stop also removes coma (at one field angle) and distortion because these aberrations also have positive and negative values that can be balanced by stop location (see Fig. 3.15). Exact symmetry requires that the lens be used in a 1 : 1 ($2f : 2f$) configuration, in which case the coma, distortion, and lateral color can be completely removed by the stop location. More typically the object is at infinity, in which case these aberrations are reduced considerably but are not exactly zero. Because spherical aberration, Petzval curvature, and axial color are not affected by the stop shift, they must be corrected by modifying the individual elements using the methods reviewed previously.

The information in the prescription is used by the optical and optomechanical engineer to create the individual element drawings used by the optical shop for fabrication (see Chap. 7). The first element in Fig. 3.38, for example, consists of a lanthanum flint, LaF15 [with a radius of its first surface (surface 1 in Table 3.6) of 81.4 mm and a radius of its second surface (surface 2 in Table 3.6) of 265.12 mm], that is 9.3 mm thick when measured along the optical axis and requires a clear aperture of 80 mm. One of the tasks of the optical systems engineer is to work with the lens designer and the optomechanical engineer to determine the criteria for acceptable performance from a given design. Spot size and distortion are reasonable criteria to start with, but how small a blur is required, and what is the corresponding amount of residual aberration? We take up these topics in the next two sections.

3.2 Diffraction

As the lens designer juggles the surface radii, refractive indices, and other elements to reduce aberration, the spot size reaches a point where it is limited not by aberrations but rather by a wave property known as *diffraction*. Ray propagation through an optical system implies that light travels straight through an aperture. Although most of the light does go straight through the aperture, there is some spreading for the part of the wavefront that reradiates from the aperture's edge as a spherical wave (Fig. 3.39). This spreading is called diffraction.

If a lens is used to image the collimated and diffracted wavefronts from a point at infinity, the blur in the image plane looks like that shown in Fig. 3.40. For a circular aperture, the blur angle β (measured in radians) can be obtained from the first zero of the Bessel function J_1 that describes the blur. The first dark ring of this function defines a portion of the diffraction-limited blur known as the *Airy disk*; its angular size is given by

$$\beta = \frac{2.44\lambda}{D} \tag{3.10}$$

Here D is the entrance pupil diameter of the system, and the wavelength λ illustrates why optical systems have so much better resolution than radio-frequency (RF) systems. A useful rule of thumb that follows from Eq. (3.10) for visible wavelengths is that $2.44\lambda \approx 1$ micron, so the angular blur (in microradians) is approximately $1/D$ when the aperture D is measured in meters. For example, a telescope with a 100-mm aperture has a diffraction-limited blur angle of 10 μrad at a wavelength of 0.5 μm, or 10,000 times smaller than the same-size radio-frequency imager operating at a wavelength of 5 mm.

The blur diameter for an object at infinity is the blur angle projected over the effective focal length:

$$B = \beta f = \frac{2.44\lambda f}{D} = 2.44\lambda \times (f/\#) \tag{3.11}$$

The rule of thumb that results from this equation for visible wavelengths is that the spot size (in microns) is more or less equal to the f/#. For example, an f/5 system has a diffraction-limited spot size of approximately 5 microns in the visible.

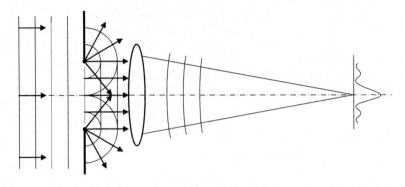

FIGURE 3.39 Diffraction at an aperture consists of both straight-through and reradiated components from the edges of the aperture; interference between these components produces the blur shape shown.

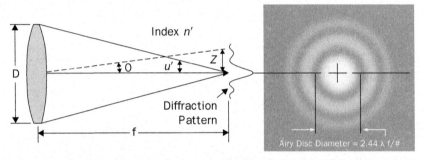

FIGURE 3.40 Diffraction pattern at the focus of a circular lens produces a blur spot in the shape of the Bessel function J_1. The blur size decreases as the interference angle u' (or numerical aperture) increases. (Credit: Warren J. Smith, Modern Optical Engineering, McGraw-Hill, www.mcgraw-hill.com.)

Example 3.7. Equation (3.11) implies that the spot size can be made arbitrarily small simply by reducing the f/# for a diffraction-limited lens. There are two problems with this argument: (1) the smallest possible f/# for a lens is 0.5;[11] and (2) aberration-limited blurs are usually larger than the theoretical diffraction-limited spot for small f/#. Recall from Sec. 3.1.1 that the smallest possible spot size due to spherical aberration is bigger for a faster lens because the angular blur β_s varies inversely with $(f/\#)^3$. For diffraction, we see from Eq. (3.11) that the diffraction-limited spot size varies directly with f/#.

Figure 3.41 plots these two competing trends, showing how the diffraction-limited [Eq. (3.11)] and spherical-limited [Eq. (3.3)] spot sizes vary with f/#. The figure incorporates four assumptions: (1) a thin lens is used in the orientation and bending for minimum spherical spot size; (2) the index $n = 1.5$ ($K = 0.067$) for a lens that transmits in the visible spectrum; (3) a wavelength $\lambda = 0.5$ µm; and (4) a focal length $f = 100$ mm. The plot shows that the smallest spot size occurs at an optimum f/# ≈ 8.5. Spherical aberration dominates with faster lenses, and it is not possible to design a single-element thin lens with these parameters and a diffraction-limited spot size at these speeds.

Aberrations and Image Quality

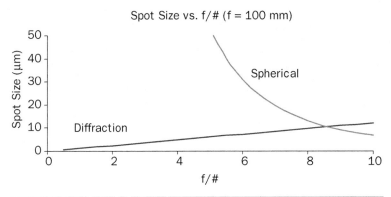

FIGURE 3.41 Plot of Eq. (3.3) and Eq. (3.11), showing how the spherical-limited and diffraction-limited spot sizes vary with f/# for a thin lens with index $n = 1.5$; the spot size obtained in practice is the larger of the two.

The optimum f/# can be calculated by equating the diffraction-limited and spherical-limited blurs. Thus we have $(f/\#)^4 = Kf/2.44\lambda$, and plugging in the preceding values for a visible lens yields the optimum f/# = 8.61. Designing a fast (f/1 to f/2) lens for the visible is thus a difficult task—limited in part by the large spherical aberration. However, for a LWIR lens with index $n \approx 4$, the K-parameter from Table 3.1 is 0.0087 (approximately 8 times smaller than for the visible lens) and the wavelength $\lambda = 10$ μm (20 times larger than for the visible lens), giving an optimum f/# = 2.44. Longer wavelengths are thus inherently easier to correct for aberrations, making f/1.5 lenses for the LWIR readily available.

The optimum f/# thus depends on the assumptions used for wavelength, field of view (zero in this example), lens index, focal length, and selection of spherical as the dominant aberration. The existence of an optimum f/# implies that bigger, faster lenses are not always better—unless the increase in size is used to reduce aberrations. By using multiple, thick, and/or aspheric lenses, some improvement can be expected over the single-element lens. Even so, a back-of-the-envelope "sanity check" such as that shown in Fig. 3.41 is useful for indicating where such design methods are not likely to be effective. In practice, coma and astigmatism limit the f/# to approximately f/2 to f/3 for off-the-shelf diffraction-limited optics in the visible-to-NIR, with faster lenses typically requiring specialized designs with more elements.

In addition to spot size, another useful metric for evaluating telescope performance is this: How close can two point sources (e.g., two different stars) be to each other and still be imaged as distinct objects? With each star's angular size smaller than that of the Airy disk,[17] their images will each be a diffraction pattern similar to that shown in Fig. 3.40. Even when far apart, some of the energy from each star starts to overlap the image from its neighbor.

As the stars get closer together, they reach a point where the peaks of their Airy disks sit on the first dark ring of their neighbor (Fig. 3.42). At this point, there is still a sufficient drop in peak power from star to star for them to still be distinguishable (or "resolved"). Although there are variations on the precise meaning of "sufficient,"

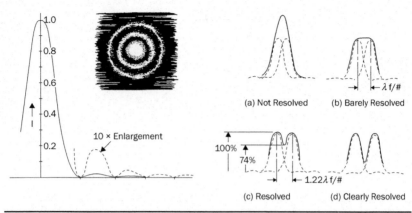

Figure 3.42 The overlap of Airy disks from two adjacent point sources determines the best attainable optical resolution for diffraction-limited systems. The commonly used Rayleigh criterion is illustrated in (c) and the Sparrow criterion in (b). (*Credit: Warren J. Smith, Modern Optical Engineering, McGraw-Hill, www.mcgraw-hill.com.*)

the main idea in this case is that two stars (of equal power) can be resolved if they are separated by at least the angular radius of the Airy disk; if they are any closer, then resolving the two stars becomes difficult. As a result, the resolving power of any diffraction-limited imager is half the angular diameter, or an angular radius of $1.22\lambda/D$. This criterion was first proposed by Lord Rayleigh and has proven to be extremely useful over the years.

As an example of the use of Rayleigh's criterion, two planets reflecting visible light from their star are 10 light-years away from Earth, and the system requirement is to image them with a space-based telescope that has a diffraction-limited, 1-meter aperture. Since 10 light-years $\approx 9.5 \times 10^{16}$ m and since the angular resolution for a 1-meter aperture is $\beta = (1.22 \times 0.5 \times 10^{-6}$ m$)/(1$ m$) = 0.61$ µrad, the planets must be separated by at least 9.5×10^{16} m $\times 0.61$ microradians $= 5.78 \times 10^{10}$ m in order to be seen as separate planets from Earth. This is equivalent to 36 million miles, or less than half the mean distance from the Earth to the Sun. So with a 1-meter telescope, it's difficult to know whether there are one or two (or more) planets across a 36-million-mile swath of universe located 10 light-years away. Answering this question requires a larger telescope with better resolving power or the use of some other technique that does not rely on imaging;[18] Chap. 4 describes one of these methods.

Most imaging is not of two point sources but rather of extended sources such as mountain ranges. An extended source consists of an infinite number of points, and its image is the overlap of all the blurred point images. Diffraction-limited lenses are still important because large blurs have large overlap, which leads to poor image quality. The diffraction-limited spot size and resolving power are still useful for evaluating quality, but there are also other figures of merit—wavefront error, Strehl ratio, modulation transfer function, and encircled energy—to help sort out good images from bad images and everything in between. These metrics are discussed in the next section.

3.3 Image Quality

The concept of image quality is an extensive topic that takes into account diffraction- and aberration-limited optical performance, the focal plane array used to measure this performance, and subjective interpretations of resolution and contrast. As will be seen in Chap. 6, such evaluations ultimately relate to the pixel IFOV and its comparison with the optical blur. Independent of this systems-level comparison, lens designers have developed a number of metrics over the years for the optical performance itself, criteria for a successful optical design that are reviewed in this section.

As an optical design evolves from a rough concept with highly aberrated blurs down to a refined design with diffraction-limited performance, the tools used by lens designers to monitor progress change. For the initial aberrated design, the spot size is determined using third-order ray tracing: it is calculated from the diameter of a blur circle that contains 68 percent of the energy of an imaged point.[1] A full specification of an imager includes both through-focus and full-field spot sizes, as described in Sec. 3.1.

However, as diffraction-limited imaging is approached, the geometrical ray trace predicts much smaller spots than are possible. It was shown in Example 3.7, for example, that at larger f/# the aberrated spot size due to spherical aberration is smaller than that due to diffraction. Diffraction cannot be removed from an optical system, so the larger, diffraction-limited spot is the limiting one in this case. As a result, a second method for monitoring design progress has been developed that takes diffraction into account and relies on a quantity known as wavefront error.

3.3.1 Wavefront Error

Section 3.1 showed that wavefront error (WFE) is the cause of imaging aberrations such as spherical, astigmatism, and coma. In this section WFE is quantified, an important step in using it as a design metric.

Figure 3.15 shows a thin lens imaging an on-axis point at infinity with both an unaberrated and aberrated wavefront on the image side of the lens. The unaberrated wavefront is spherical and centered on the geometrical image point—that is, at the focal length of the lens. This is where an ideal lens with no aberrations brings the wavefront to an image.

The actual wavefront is aberrated and deviates from the paraxial sphere. The specific deviation in Fig. 3.15 is larger at the edge of the lens, bringing the outer parts of the wavefront to a focus closer than the paraxial focal length and so creating spherical aberration. The distance between the aberrated and unaberrated wavefronts is the *optical path difference* (OPD). The wavefront error is the OPD measured in units of wavelength, so the peak OPD (e.g., 0.5 µm) is specified as a WFE of one wavelength at 0.5 µm (or "one wave peak-to-valley at 0.5 µm"), or one-half wave ($\lambda/2$) peak-to-valley at 1 µm, depending on the wavelength used to measure the WFE. A standard wavelength is 632.8 nm, which is transmitted by many materials and is emitted by the common helium-neon (HeNe) laser.

Using the peak WFE at the edge of the lens gives the worst-case deviation, but it does not indicate how the lens performs on average across the aperture. To cover the case where the wavefront deviations across the lens are independent of each other, the average used is a root-mean-square value equal to the standard deviation of the average WFE at 100 or so points across the lens aperture, depending on aperture size. As a statistical measure across the entire aperture, the root-mean-square WFE is usually the better metric of lens quality.

The number of points used also depends on how quickly the wavefront varies across the aperture. For the smoothly varying aberrations reviewed in Sec. 3.1, Table 3.7 summarizes the relation between the peak-to-valley (PV) and root-mean-square (RMS) measures of wavefront error.[2] A common reference is that of defocus, for which $W_{PV} = 3.5 W_{RMS}$; this table also reports another commonly used rule of thumb—namely, $W_{PV} = 5 W_{RMS}$ for random fabrication errors.[19]

Returning to the use of WFE as a design metric, spot sizes for PV wavefront errors on the order of one wave or greater can be found using geometrical ray tracing; anything smaller requires a diffraction

Wavefront aberration	W_{PV}/W_{RMS}
Defocus	3.5
Spherical	13.4
Coma	8.6
Astigmatism	5.0
Random fabrication errors	5.0

TABLE 3.7 Relation between peak-to-valley and root-mean-square wavefront errors for various wavefront aberrations, including fabrication errors for polished optical surfaces.[2-3] A defocus value of 3.5 is often used as a common reference for comparison, so a diffraction-limited lens may be specified as having either 0.071 waves RMS WFE or 0.25 waves PV.

analysis. One finding from diffraction analysis is that a peak-to-valley WFE of one-quarter wave ($\lambda/4$) produces blurs that are only slightly larger than diffraction limited and with a resolution equivalent to the Rayleigh criterion. This is significant because it tells us that *obtaining a diffraction-limited lens does not require a zero WFE*. For this reason, so-called quarter-wave optics are considered the minimum standard for high-quality imagery. Vendors also specify such optics using RMS wavefront error, which Table 3.7 shows us corresponds to $\lambda/(4 \times 3.5) = \lambda/14$ or approximately 0.07 waves RMS.

However, it is necessary for individual elements within a complex lens to have WFE $< \lambda/4$ peak-to-valley for the optical system to maintain diffraction-limited performance, with each element typically on the order of $\lambda/10$. One reason is that the WFE for multiple elements must add to a system WFE $\approx \lambda/4$. Another reason is that additional factors contribute to the overall WFE, including lens-to-lens alignments and fabrication errors such as imperfections in each lens's radius of curvature. Alignment and fabrication issues are examined in Chap. 7. Example 3.8 shows how to use the results from that chapter to create a WFE budget, a commonly used optical engineering tool for assessing the relative contributions to diffraction-limited performance.

Example 3.8. The total WFE for a lens assembly is required to be diffraction limited, with the WFE = 0.071λ ($\lambda/14$) RMS at λ = 633 nm. The assembly incorporates two lenses, both similar in design and fabrication; we therefore initially assign ("allocate") a WFE = 0.05λ for each lens so that the root sum of squares (RSS) of the lenses adds to a total WFE of 0.071 waves for the assembly.

Figure 3.43 shows how the WFE budget for each lens flows down to the components that contribute to the WFE for each lens. Included are WFE entries for design residual, fabrication, initial alignment, and temperature- and vibration-induced misalignments from the operating environment. The *design residual* is the WFE due to remaining aberrations; this is not necessarily the smallest WFE the lens designers can deliver but only the smallest necessary to meet the $\lambda/14$ spec for the assembly. Note that the design WFE for each lens must be significantly smaller than diffraction limited to meet this spec.

The other four contributors to WFE will be covered in Chap. 7. The five components shown in Fig. 3.43 are independent of each other, so they RSS to yield the total WFE for lens 1: $(0.02^2 + 0.04^2 + 0.01^2 + 0.01^2 + 0.01^2)^{1/2} = 0.048$ waves RMS; a similar sum is computed for lens 2, giving 0.052 waves RMS and a total WFE = $(0.048^2 + 0.052^2)^{1/2} = 0.071$ waves for the assembly. The WFE budget is useful because it shows which term dominates the total error; in this example, it's the lens fabrication with WFE = 0.04λ. Therefore, designer time would best be spent on reducing this item if the lens is intended to be used with additional optics. Observe that the small values for alignment, thermal, and vibration terms may require an extremely difficult alignment procedure or optomechanical design to obtain and may be traded off (increased) against possible reductions in the fabrication term.

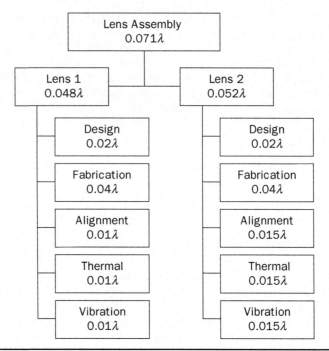

FIGURE 3.43 Allocation of wavefront error budget for a two-lens optical assembly (e.g., an objective and a field flattener) that includes the effects of design, fabrication, alignment, thermal changes, and vibration.

3.3.2 Strehl Ratio

A related metric for evaluating image quality is the Strehl ratio. As shown in Fig. 3.44, an object point imaged with WFE has a lower peak power than one that has no aberrations (and hence zero WFE). For values greater than about 0.6, the Strehl ratio is related to the RMS WFE as follows:[2]

$$S = \exp\{-4\pi^2 \text{WFE}_{\text{RMS}}^2\} \approx 1 - 4\pi^2 \text{WFE}_{\text{RMS}}^2 \qquad (3.12)$$

Here the approximation of the exponential is valid for "small" WFE ($\ll 2\pi$ waves RMS), about 0.1 waves RMS.

The Strehl ratio is typically used to describe overall system performance, not individual lens quality. For example, the system WFE for the lens assembly in Example 3.8 is 0.071 waves ($\sim \lambda/14$) RMS (or 0.25 waves PV), giving $S = 0.80$. The most important aspect of this metric is not that the peak is lower but that the peak is lower because the energy is redistributed by WFE to the outer rings of the optical blur. This type of redistribution is the basis of the modulation transfer function described next.

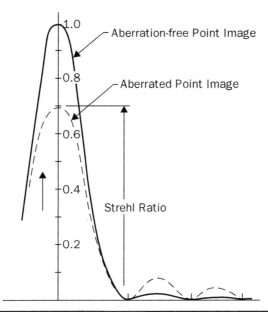

Figure 3.44 The Strehl ratio is a metric for image quality that is given by the ratio of the peak of the aberrated image to that of an image with no WFE. A Strehl ratio of 0.8 is still considered to be diffraction limited. (*Credit: Warren J. Smith*, Modern Optical Engineering, *McGraw-Hill, www.mcgraw-hill.com.*)

3.3.3 Modulation Transfer Function

Spot size, resolving power, WFE, and Strehl are useful to the lens designer and optical system engineer as monitoring metrics, yet these methods are best used for evaluating the image quality of points, not of extended objects such as mountain scenes. The evaluation of extended objects requires new tools: the contrast transfer function (CTF) and modulation transfer function (MTF). These functions describe how well the brightness variations in the scene are transferred to the image for both diffraction-limited and highly aberrated optical systems with asymmetric spot shapes.

The CTF represents extended scenes with an idealized object known as a *bar chart*. Shown as the object in Fig. 3.45, it consists of groups of white lines and dark spaces—a white picket fence or a semiconductor chip with circuit lines are typical examples—where each group has a different spacing between the lines. When the chart is imaged by a lens, the lines become blurred. Diffraction and lens aberrations create this by causing the sharp edge of each line to be imaged into a blurred line called a *line-spread function* (LSF). Like the Strehl ratio, the LSF quantifies the redistribution of image energy from the white lines to the adjacent dark spaces.

To put some numbers on how much energy is redistributed by the lens, the MTF replaces the lines and spaces with sinusoids of the same height and fundamental spatial frequency. This process enables

(a)

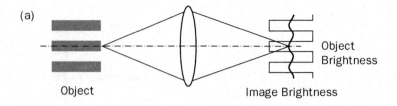

Object   Image Brightness   Object Brightness

(b)

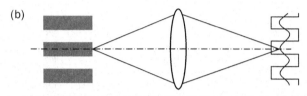

Figure 3.45 The modulation of an object depends on the difference in brightness between light and dark and might not be transferred efficiently by a lens. In panel (b) the lines and spaces are farther apart and the variation between black and white is easier to resolve, giving a greater transfer of the modulation. The object brightness is shown on the image side of the lens for reference only.

definition of a quantity known as *intensity modulation*, which is found by measuring the brightness of the object's sinusoidal peaks (I_{max}) and valleys (I_{min}) and then calculating the ratio of the difference to twice the average:[20]

$$m_o = \frac{(I_{max} - I_{min})_o}{(I_{max} + I_{min})_o} \qquad (3.13)$$

Using a lens to image the sine-wave pattern, it is seen that diffraction and aberrations spread energy from the peaks into the valleys and thereby decrease I_{max} while increasing I_{min}. As a result, the modulation of the image (denoted m_i) is smaller than that of the object, and the ratio of the image modulation to that of the object (m_i/m_o) describes how well the lens transfers the modulation at one specific spatial frequency (line pairs per millimeter of image).

If the spacing between the peak and valleys of the object is changed [Fig. 3.45(b)], thus changing the object's spatial frequency, then a different number is obtained for the modulation transfer. For example, if the spatial frequency is higher then the distance between the peaks and valleys is smaller. The closer overlap of the imaged sinusoids puts more energy from the peaks into the adjacent valleys, reducing the image modulation. A plot of the modulation transfer for a range of spatial frequencies is known as the modulation transfer function curve.

An example MTF curve is shown in Fig. 3.46 for a lens with zero to one-half wave of spherical aberration. The vertical axis is the

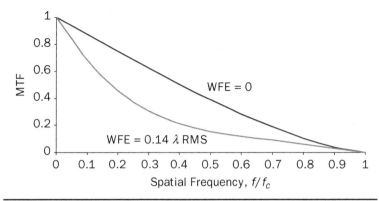

Figure 3.46 Example MTF curves for a lens with no aberrations (WFE = 0) and WFE = 0.14 waves RMS. The curves were obtained by plotting Eq. (3.17).

modulation transfer, and the horizontal axis is the range of spatial frequencies used to calculate or measure that transfer. The units for spatial frequency f_s are line pairs (or cycles) per millimeter in the image plane, where two peaks (two bright lines) constitute a line *pair* [see Fig. 3.42(c)]. This is analogous to time-based frequencies measured in cycles per second (Hz), where each cycle can be selected to start and stop at a peak.

For zero WFE, Fig. 3.46 shows a nearly linear roll-off in modulation from $f_s = 0$ (an image with no variation in intensity across the FOV) to a cutoff frequency that depends on the f/# and the wavelength. An important aspect of the figure is that even a perfect lens (WFE = 0) does not perfectly transfer the modulation for *any* frequency except $f_s = 0$; again, the reason is that diffracted light in the image redistributes the energy from the peaks into the valleys, reducing the modulation for higher frequencies with more closely spaced lines.

The usefulness of the MTF as a measure of image quality goes beyond the artificial scenes represented by lines, spaces, and sine waves. Fourier analysis, for example, shows that any scene can be analyzed in terms of sinusoidal spatial frequencies.[21] The limiting spatial frequency that determines how much detail is visible is called the *cutoff frequency*. The term "cutoff" refers to the case where the MTF = 0; this occurs when the line widths and spacings become comparable to the diffraction-limited blur radius determined by the wavelength and f/#. The Rayleigh criterion in Sec. 3.2 allowed two closely spaced points to be resolved when the distance between them is 1.22λ multiplied by the f/#. The Sparrow criterion in Fig. 3.42(b) also showed that, when the spacing between a pair of point images is reduced to λ × f/# then they are no longer resolvable, so that $I_{max} = I_{min}$ and the intensity *modulation* is zero (even though the *absolute* intensity is, of course, not zero). The corresponding cutoff frequency f_c (line pairs per mm, denoted by lp/mm) that can no longer be transmitted by a lens is then the reciprocal of this distance:[22,23]

$$f_c = \frac{1}{\lambda \times (f/\#)} \quad (3.14)$$

where the f/# is, as usual, the working f/#. Note that this frequency can also be expressed in units of line pairs per milliradian (lp/mrad) by using the angular version of the Sparrow criterion (for which $1/\beta_c = f \times f_c = D/\lambda$ is the angular cutoff frequency). With either set of units, understanding the connection between Sparrow resolution and MTF cutoff frequency allows deeper insight into the limitations of image quality.

Modulation transfer function curves for complex geometries are easily obtained using lens design software. For spreadsheet calculations, a useful expression of the optical MTF for circular, diffraction-limited lenses with zero WFE at a specific wavelength is[2]

$$\text{MTF}_d = \frac{2}{\pi}\left[\cos^{-1}\left(\frac{f_s}{f_c}\right) - \frac{f_s}{f_c}\left(1 - \frac{f_s^2}{f_c^2}\right)^{1/2}\right] \quad (3.15)$$

This expression can be approximated as a straight line for f_s/f_c up to about 0.5:

$$\text{MTF}_d \approx 1 - \frac{4}{\pi}\frac{f_s}{f_c} \quad (3.16)$$

Both equations use the normalized spatial frequency f_s/f_c, where the normalization is with respect to the cutoff frequency and gives an upper range of 1.0 on the MTF horizontal axis.

For aberrations or fabrication errors with small wavefront error (WFE ≤ λ/2 PV), a useful approximation for the optical MTF_{opt} is given by the product of MTFs due to diffraction and aberrations:[2]

$$\text{MTF}_{opt} = \text{MTF}_d \times \text{MTF}_a \quad (3.17)$$

Here the reduction in MTF due to aberrations is given by[2]

$$\text{MTF}_a = 1 - \left(\frac{\text{WFE}_{RMS}}{0.18}\right)^2\left[1 - 4\left(\frac{f_s}{f_c} - \frac{1}{2}\right)^2\right] \quad (3.18)$$

for an RMS WFE measured in units of waves. The WFE in this equation is independent of aberration type. As illustrated in Fig. 3.46, this is a valid approximation for small WFE (up to λ/2 PV or 0.14 waves RMS).

Asymmetric aberrations such as astigmatism have different MTF curves for the tangential and sagittal images. Recall from Sec. 3.1.3 that the asymmetric blur sizes for astigmatism are different in the x

Aberrations and Image Quality

and y directions; as a result, we expect the MTF to be different for the bigger (tangential) blurs than for the smaller (sagittal) blurs in low-spatial frequency regions where they are reasonably distinct. This is seen in the MTF curves of Fig. 3.47 for a lens with astigmatism.

Other factors that can reduce the optical MTF include being out of focus (resulting in bigger blur sizes), lens scattering [which puts stray light into the peaks and valleys, increasing the average in the denominator of Eq. (3.13)], and detector noise (increasing the minimum detectable value of I_{min}). The last two items are covered in more detail in Chaps. 4 and 6, respectively. In any case, the official story is that "the cutoff frequency cannot change." In practice, MTF values of less than about 0.1 are essentially cutoff values, because it is difficult (though not impossible) to extract a useful image with such a low degree of contrast.[24] So-called obscuration MTFs are particularly interesting in this regard: the reduction in effective aperture size reduces the MTF at intermediate spatial frequencies but actually *enhances* it at higher frequencies (Fig. 3.48).

Other factors besides optical MTF contribute to the overall system MTF. Detector MTF is the most critical, a story that will be continued in Chap. 6. Specifications for MTF are often summarized in a table that takes both the optical and detector MTFs into account. For

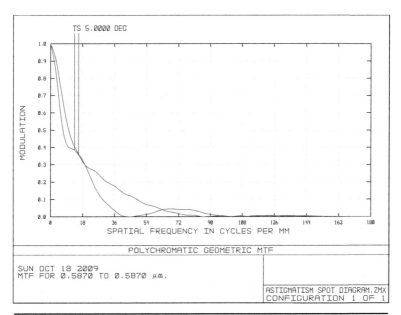

FIGURE 3.47 MTF depends on the presence of aberrations; this plot shows the difference in tangential (T) and sagittal (S) MTF at a field angle of 5 degrees for a lens with astigmatism.

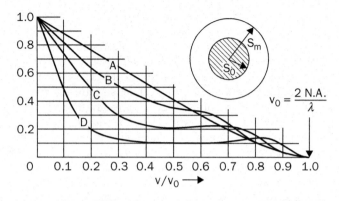

FIGURE 3.48 MTF curves for an obscured aperture such as that found in a Cassegrain telescope. The A, B, C, and D curves represent an obscuration ratio $\varepsilon = s_o/s_m$ of 0, 0.25, 0.50, and 0.75, respectively. (*Credit: Warren J. Smith, Modern Optical Engineering, McGraw-Hill, www.mcgraw-hill.com.*)

example, typical MTF specifications for an inexpensive digital-camera are shown in Table 3.8. The 0° point is the center of the field, and the 15° column refers to the corner of the FOV; the differing values illustrate the effect of aberrations in reducing MTF over the entire field.

3.3.4 Encircled Energy

So far we have ignored the size of the detector used to collect photons. A detector that is smaller than the image blur—whether limited by diffraction or aberrations—cannot measure the energy that is in the outer rings of the blur. Using an array of detectors to capture that energy will spread the blur energy over multiple pixels; using larger detectors captures more energy, but it also increases the IFOV and thus may reduce the perceived image quality.

Table 3.9 shows the fraction of energy contained in various image radii for a perfect lens with no aberrations. The table shows that a circular detector of the same size as the central Airy ring (= 2.44λ × f/#) contains (or "encircles") 83.9 percent of the total energy available; the rest (16.1 percent) is diffracted to the outer rings. The other radii listed correspond to the dark-ring radii in the Bessel function J_1 that describes the Airy pattern. The table also indicates that increasing the detector diameter is progressively less effective at capturing more energy for a perfect lens, and this is true also for one with a quarter-wave PV wavefront error (data not shown).

We have seen throughout this chapter that all designs have some residual aberration. In Sec. 3.3.2, for example, we saw that the peak energy of the Airy disk is reduced to 80 percent of the perfect-lens value with only a quarter-wave of peak-to-valley WFE. The fraction

Spatial frequency, f_s	MTF @ 0°	MTF @ 15°
20 lp/mm	> 0.90	> 0.85
50 lp/mm	> 0.30	> 0.25

TABLE 3.8 Summary of MTF specifications for an inexpensive digital-camera.

Ring diameter	Encircled energy
2.44λ × f/#	83.9%
4.46λ × f/#	91.0%
6.48λ × f/#	93.8%
8.48λ × f/#	95.3%
10.48λ × f/#	96.3%

TABLE 3.9 Comparison of fractional encircled energy (EE) for a zero-WFE lenses of various diameters.[1]

of energy in the central Airy disk in this case is 68 percent; the remaining energy (32 percent) is diffracted and aberrated into the outer rings. As a result, even a diffraction-limited lens with small residual aberrations has twice as much energy in the outer rings as does a perfect lens, and this may push the system design toward a bigger detector.

A common image quality metric, then, is a plot of how the fraction of encircled energy varies with detector radius. Figure 3.49 shows such a plot for a lens with astigmatism, where blur radius is measured with respect to the blur's centroid. Also plotted in the figure is the encircled energy for an unaberrated lens; as expected, the aberrated image requires a bigger detector to collect the same fraction of energy as the perfect lens.

The curves plotted in Fig. 3.49 take both diffraction and aberrations into account, and they can be generated as a standard option in any lens design software such as Zemax or Code V. As with the other measures of image quality, encircled energy varies with field angle and focus position; for asymmetric blurs (coma, astigmatism), the radius is best measured with respect to the blur centroid, not the chief ray. Plots can also made for square detectors, in which case the energy captured by the detector is said to be "ensquared" rather than encircled.

A common design goal is to achieve greater than 80 percent encircled (or ensquared) energy, although more is not always better. With certain types of imagers, for example, it is sometimes useful to

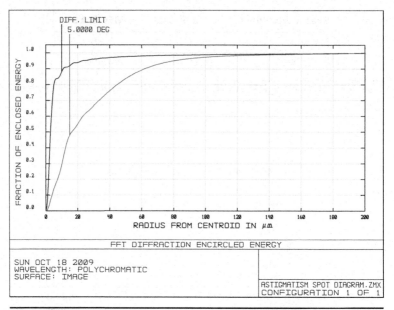

Figure 3.49 Encircled energy as measured from the distance from the centroid of the blur. The astigmatism in this example requires a significantly larger detector size to collect the same energy as a diffraction-limited lens with no WFE, which may increase the IFOV and reduce image quality.

image an Airy disk onto four or more pixels, in which case the ensquared energy on each pixel will be small. This topic is examined in more detail in Chap. 6, where the use of FPAs in system design is reviewed.

3.3.5 Atmospheric Distortion

The motivation for putting telescopes into space is not to establish space-agency empires but rather to avoid imaging through the atmosphere. Because of temperature differences and air motion, the atmosphere's refractive index is not uniform; instead, it varies slightly with temperature, time of day, proximity to the ground, the type of ground (asphalt, sand, etc.), and so on. This latter effect can be seen when looking across hot asphalt, which causes images of objects to shimmer and move; the resulting increase in WFE across the lens aperture can also increase the blur size. This means that the atmosphere is often the limiting factor in image quality.

Wavefronts are free of such atmospheric distortion over a diameter known as the *coherence length*. This phenomenon is captured by an image quality metric called the *Fried parameter* (denoted r_o), which is the aperture size over which the wavefront is not distorted appreciably. The significance of this parameter is that the angular blur can be no better than that determined by λ/r_o.[25] Depending on atmospheric conditions, typical values of r_o are on the order of 100 to 150 mm, for which a VIS-wavelength telescope (λ = 0.5 μm) has an atmospheric-limited resolution $\lambda/r_o \approx$ 3 to 5 μrad.

Comparing this result with the Rayleigh criterion ($\beta = 1.22\lambda/D$), we see that the coherence length more or less determines the largest useful aperture for the resolution of an imaging telescope. Because atmospheric wavefront error occurs over apertures larger than r_o, it is not possible to obtain better image quality with a larger telescope unless specialized techniques—known as *adaptive optics*—to sense and control the wavefront are used.[25] Without these techniques, larger apertures collect more light (see Chap. 4) but the resolution cannot improve. Since the atmosphere determines the limiting resolution that can be seen, this phenomenon is known as atmospheric "seeing."

The value of the coherence length depends on altitude, with large values corresponding to higher altitudes (where there is less air). Large-aperture telescopes are thus often located ("sited") on mountains. The 8.2-meter Subaru Telescope on Mauna Kea in Hawaii features a seeing resolution on the order of 0.4 arcseconds ($\approx$ 2 μrad, given the conversion of 4.85 μrad/arcsec) at a wavelength of 2 μm. Despite the high altitude, the seeing resolution is still approximately 5 times worse than the diffraction-limited resolution, which illustrates the continuing need for space-based imagers or alternatives such as wavefront sensing and control systems.

Problems

3.1 What is the shape factor q for a symmetric negative lens?

3.2 In terms of ray angles and wavefronts, why is the equiconvex lens with $q = 0$ not the best shape for an object at infinity?

3.3 For an object at infinity, is the refraction angle the same at both surfaces of a minimum-spherical lens?

3.4 If an object is placed two focal lengths in front of a positive lens, where is the image located? What lens shape minimizes the spherical aberration for this arrangement? *Hint:* Shape-factor calculations are not necessary; use the symmetry of the object and image distances to determine the shape.

3.5 A 1 : 1 ($2f : 2f$) conjugate system is required, but the project cannot afford to design and fabricate a lens that delivers diffraction-limited performance. Looking through manufacturers catalogs, we find an aplanat lens that is well corrected for spherical, coma, and astigmatism *for infinite conjugates*. Is there a way to use these aplanats for the finite-conjugate design? *Hint:* Two aplanats would be required.

3.6 If you wear eyeglasses, take them off and look through a small pinhole in a piece of aluminum foil. How well can you see looking through the pinhole? Does it depend on how close the pinhole can be placed to your eye? Why? What about the field of view? Does it also depend on how close the pinhole can be placed to your eye?

3.7 An object distance of $s_o = 1$ km was used to calculate the field curvature in Example 3.4. What is the field curvature for an object at 10 m? Does the closer object have more or less distortion? Why is it different?

3.8 A focal length of $f = 100$ mm was used to calculate the field curvature in Example 3.4. What is the field curvature for $f = 50$ mm? Does the shorter focal length have more or less distortion? Why is it different?

3.9 Use the paraxial ray-trace equations to show that a field flattener reduces field curvature. For this demonstration you should employ an object at 1 km, a positive imaging lens with $f = 100$ mm, and a plano-concave field flattener with the planar side facing the detector, a center thickness of 3 mm, an index $n = 1.5$, and a radius of curvature $R = -10$ mm on the concave side.

3.10 Using the same field flattener, use Snell's law to calculate the deviation through the field flattener at a 5-degree field angle. Check your result using the equation for deviation θ through a thin prism: $\theta = (n - 1)\alpha$, where α is the prism angle. What accounts for any differences in the calculations?

3.11 Two thin lenses ($f = 100$ mm) are placed 50 mm apart. Place a stop midway between the two lenses, and use the paraxial ray-trace equations to calculate the distortion. Now move the stop 10 mm toward the first lens and repeat the calculation. Do the results show pincushion or barrel distortion? Why?

3.12 Use the ray-trace equations to calculate the distortion for a distant object if the stop is placed 10 mm after a positive lens. Use the same parameters that were used in Example 3.5.

3.13 Select a reasonable set of wavelengths for the V-number of N-BK7 in the NIR band. How does it compare with the V-number in the visible?

3.14 Create a "glass" map for the common materials used in the MWIR (3–5 μm) and LWIR (8–12 μm) regions. Why is "glass" is quotes?

3.15 On what does the blur size associated with axial color depend?

3.16 What does the diffraction pattern look like for a square? What is the resolution criterion for a square lens?

3.17 Compare the optimal f/# for the lens in Example 3.7, whose focal length was 100 mm, with the optimal f/# for a lens whose focal length is 1000 mm. Assume the same lens material, K-factor, and wavelength.

3.18 Compare the diffraction-limited with the coma-limited spot size for the lens used in Example 3.7. What is the optimal f/# where the spot sizes are equal for field angles of 10 degrees and 30 degrees?

3.19 Compare the diffraction-limited with the astigmatism-limited spot size for the lens used in Example 3.7. What is the optimal f/# where the spot sizes are equal for field angles of 10 degrees and 30 degrees?

3.20 How close must a car be before its headlights can be resolved as two distinct lights? *Hint:* What is the resolving power of the human eye? Is it based on the eye lens, on the array of detectors at the back of the eye, or both?

3.21 A visible imager is designed to image an Airy disc on a detector whose IFOV is 100 μrad. What aperture D is required?

3.22 Why is a helium-neon (HeNe) laser often used to measure WFE? (No calculations are required to answer this question.)

3.23 Why is the blur size determined by the f/# and not by the aperture size D? Equation (3.10) shows that larger apertures have a smaller angular blur, so doesn't that imply that the f/# is not relevant?

3.24 A lens is diffraction limited at a wavelength of 1 μm with a PV WFE of $\lambda/4$. Is this lens also diffraction limited when used at a wavelength of 2 μm? What about 0.5 μm?

3.25 When two point sources are placed a distance $1.22\lambda \times$ f/# apart, the energy pattern in the image peaks at the location of the point sources but drops to 74 percent of the peak halfway between them. What is the modulation of the image in this case?

3.26 What size detector is needed to encircle 95 percent of the energy in the Airy pattern for a perfect lens? What about 100 percent of the energy? Express your answer in units of the Airy disk size.

Notes and References

1. Warren J. Smith, *Modern Optical Engineering*, 4th ed., McGraw-Hill (www.mcgraw-hill.com), 2008.
2. R. R. Shannon, *The Art and Science of Optical Design*, Cambridge University Press (www.cambridge.org), 1997.
3. Warren J. Smith, *Modern Lens Design*, 2nd ed., McGraw-Hill (www.mcgraw-hill.com), 2005.
4. Francis A. Jenkins and Harvey E. White, *Fundamentals of Optics*, 4th ed., McGraw-Hill (www.mcgraw-hill.com), 1976.
5. Warren J. Smith, *Practical Optical System Layout*, McGraw-Hill (www.mcgraw-hill.com), 1997.
6. L. Allen, *The Hubble Space Telescope Optical System Failure Report*, NASA (www.nasa.gov), 1990.
7. See, for example, CVI Melles-Griot data sheets at www.mellesgriot.com.
8. See www.seds.org/Messier for more information on Mr. Messier and his discoveries.
9. Implied by the phrase "reducing Petzval" is "reducing Petzval sum" or "reducing field curvature" but *not* reducing Petzval radius. A longer radius deviates less from a flat image plane and so is the preferred design.

132 Chapter Three

10. Arthur Cox, *Photographic Optics*, 15th ed., Focal Press, 1974.
11. Rudolf Kingslake, *Lens Design Fundamentals*, Academic Press, 1978.
12. The terms "blue" and "red" are also used for UV and IR materials outside the visible spectrum to indicate relative wavelengths (shorter and longer, respectively), not visual response.
13. Data for a wide variety of glasses and material properties are available from the Schott Glass Web site, www.schott.com.
14. Solomon Musikant, *Optical Materials*, Marcel Dekker, 1985.
15. R. E. Fischer, B. Tadic-Galeb, and P. R. Yoder, *Optical System Design*, 2nd ed., McGraw-Hill (www.mcgraw-hill.com), 2008.
16. Daniel J. Schroeder, *Astronomical Optics*, 2nd ed., Academic Press, 2000.
17. This is the definition of a point source. In other words, a point source is any object—stars, extra-solar planets, pinholes, et cetera—whose angular size is smaller than the limiting blur size of the optics. For diffraction-limited optics this limit is taken as the size of the Airy disk, and a bigger star does not generate a bigger image with a diffraction-limited lens unless its angular size exceeds the Airy blur size. Whether or not an object is a point source thus depends on its wavelengths and the entrance pupil diameter of the imaging optics.
18. Barrie W. Jones, *Life in the Solar System and Beyond*, Springer-Verlag (www.springeronline.com), 2004.
19. Ed Friedman and J. L. Miller, *Photonics Rules of Thumb*, 2nd ed., SPIE Press (www.spie.org), 2004.
20. Contrast is defined as $C = I_{max} - I_{min}$, which is the same as the modulation except that this difference is not divided by twice the average. Contrast is related to modulation m by $C = (1 + m)/(1 - m)$.
21. J. F. James, *A Student's Guide to Fourier Transforms*, Cambridge University Press (www.cambridge.org), 1995.
22. Glenn Boreman, *Modulation Transfer Function in Optical and Electro-Optical Systems*, SPIE Press (www.spie.org), 2001.
23. Joseph M. Geary, *Introduction to Optical Testing*, SPIE Press (www.spie.org), 1993.
24. Gregory H. Smith, *Practical Computer-Aided Lens Design*, Willmann-Bell, Inc. (www.willbell.com), 1998.
25. Robert K. Tyson and Benjamin W. Frazier, *Field Guide to Adaptive Optics*, SPIE Press (www.spie.org), 2004.

CHAPTER 4
Radiometry

The Kepler Space Telescope, launched in March 2009, is a NASA instrument designed to look for planets outside our solar system that might support life. The approach used to look for these planets is based on radiometry: when the Moon goes between the Sun and Earth, the skies darken as sunlight is blocked. Similarly, when a faraway planet moves between *its* star and the Earth, our own skies may not darken but there is a small change in the amount of light that reaches us (Fig. 4.1). The Kepler telescope was designed to measure these brightness changes to 1 part in 100,000, with light from other sources—the Sun, reflected sunlight from the Moon, nearby stars, etc.—excluded to one part in 10^6 or so.[1] This is extremely precise radiometry, and even small amounts of "stray" light from other sources can lead to poor results.[2]

Chapters 2 and 3 reviewed how to find where an image is located, its size, and its quality; but yet to be established is the brightness of the image and how it is affected by stray light. A simple example of brightness is the use of a "magnifying" lens to image the Sun onto a piece of paper.[3] We know from experience that the Sun's image is bright enough to ignite the paper, but why can't the paper also be torched with an image of the Moon? Is this deficiency inherent in the Moon's level of brightness, or is a bigger and better lens all that's needed to light up a campfire after the Sun has gone down for the day?

Digital SLR cameras have better lenses than a magnifying glass, yet it's possible to take photos of the Sun without "torching" the camera's pixels. One way to do this is to put an optical filter in front of the lens, thereby eliminating most of the solar power. Stopping down the lens to a slow setting is also useful, but it's necessary to consider the "speed" of the detectors. A camera is "fast" for small f/#, but what does it mean for detectors to be fast? Can these effects be combined to produce a camera that is ultrafast? It is shown in this chapter that the combination of filtering, lens speed, and detector size determines whether there is enough light to obtain a good image or so much light that the FPA will be damaged.

Summarizing these examples, optical systems engineering has been described as understanding how to collect the light that's

134 Chapter Four

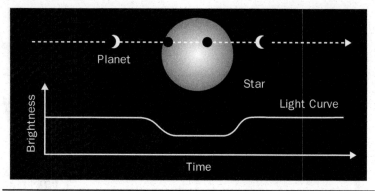

FIGURE 4.1 The Kepler Space Telescope uses changes in a star's apparent brightness to detect planetary transits across the star. (Credit: NASA, www.kepler.nasa.gov.)

needed for an optical system and reject the light that's not (Fig. 1.6). This chapter reviews the tools and trade-offs needed to design a system that meets these general radiometric requirements. The first step is to determine how much power is collected by the system and how much ends up on a single-element detector, pixel, or entire FPA (Sec. 4.1). Transmission losses make these two numbers different, and knowing how to minimize such losses is an important part of the chapter.

The idea of optical power is extended in Secs. 4.2 and 4.3 to include its radiometric variations: power density (irradiance), power density per solid angle (brightness or radiance), and power per solid angle (intensity). All of these concepts are useful in one way or another, depending on the type of source. The emphasis is on the reasons for employing different concepts and on how radiometric performance varies with aperture and field.

Much as with the laws of thermodynamics and quantum mechanics, radiometry has certain fundamental limitations. For example, despite the best efforts of many inventors over the centuries, an image can never be brighter than its source. A simple calculation using conservation of radiance shows why; this conservation law and others are reviewed in Sec. 4.4.

The chapter concludes by discussing the rejection of stray light that's not wanted on the image. Section 4.5 describes a number of design options for controlling stray light, including optical coatings, sunshades, baffles, stops, cleanliness levels, and component surface roughness.

Despite this wide range of topics, the field of radiometry tends to be fairly specialized: the folks who work in radiometry are oftentimes not lens designers, even though there are strong connections between these fields. It was shown in Chap. 3, for example, that a smaller f/#

gives brighter images but that correcting for aberrations becomes more and more difficult for faster optics; as a result, radiometry systems trade-offs can be determined by aberration requirements and geometrical optics. This underscores the need for optical systems engineers to have a strong understanding of the fundamentals and an ability to make order-of-magnitude system trades and "sanity" checks. The goal of this and the following chapters is to provide some practical guidance on these topics.

4.1 Optical Transmission

All of the power collected by a telescope does not make it onto the detector. The power that does make it through is known as the *optical transmission* or *throughput*. This value is usually given as a fraction of power that falls on the entrance pupil, although it's also important to specify the total number of watts that makes its way through the optical instrument.

To determine the total number of watts, start with the simplest case of a point source emitting power in the form of spherical wavefronts. Because the wavefronts are diverging, the power is spread out over a larger and larger area as it propagates. This spherical divergence results in an inverse square $(1/R^2)$ law to describe the power density incident on a surface, which is measured in watts per square meter (W/m^2). For example, the optical power density from the Sun (~ 1000 W/m^2 at noon on a clear day at sea level) sets the upper limit on how much electrical power a solar cell can generate. Moving closer to the source also collects more power in instruments such as video projectors, though this is clearly not an option for distant sources like the Sun.

The total power collected is given by the source's power density [W/m^2] at the telescope multiplied by the area [m^2] of the optical aperture. The optical aperture can be either a nonimaging surface (e.g., for a solar cell) or an entrance pupil diameter (for an imaging instrument). A bigger aperture is required to collect more power, which means that an f/2 imager collects just as much power as an f/1 imager if the entrance pupils are the same size.

Unfortunately, there are optical losses as the collected power works its way through the instrument. The fraction of incident power that eventually hits the detector depends on the transmission of the optics, reflections off the lens surfaces, and the optical design. This section looks at how material properties, optical coatings, and instrument design affect this optical throughput.

4.1.1 Material Properties

When light hits a lens, some of the energy reflects off the lens surfaces, some is absorbed by the lens material, and the rest is transmitted.

Ultimately, the transmitted throughput is the important requirement, since transmission loss (on the way to a solar cell, for example) can't do anything useful (such as generate electricity). There are ways to compensate for the losses—such as using a larger telescope or a more powerful source or a more sensitive detector—all of which add size, weight, and cost to the system. Lost light may also present problems in terms of back-reflections, stray light, and an increase in lens temperature; addressing these problems complicates the system design.

As summarized in Fig. 4.2, conservation of energy shows that there is a connection between the throughput and the reflected and absorbed light:

$$T + R + A = 1 \qquad (4.1)$$

Here T is the fraction of incident power transmitted (often called the external transmission), R is the fraction reflected at a surface, and A is the fraction absorbed in a volume. If two of these quantities are known, then the third may be calculated from Eq. (4.1). The properties

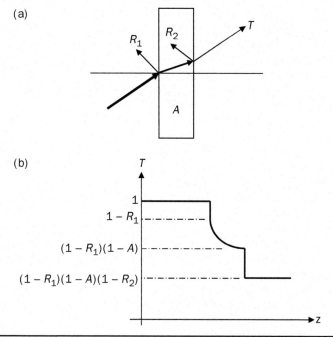

Figure 4.2 The ray diagram in (a) shows reflected, absorbed, and transmitted light propagating through a flat plate; in (b), the remaining fraction of the initial energy is plotted as the rays propagate through the plate (ignoring multiple reflections).

usually measured are the transmission T and reflection R, from which accurate predictions of the absorption A can be obtained.

Reflection from a surface depends on the refractive index of both the surface and the material that is in contact with it. Monsieur Fresnel developed an equation to calculate the reflectivity, which at normal incidence to the surface of a low-absorption material is written as[4]

$$R(\lambda) = \left[\frac{n(\lambda)-1}{n(\lambda)+1}\right]^2 \qquad (4.2)$$

Here the number "1" is used when the material in contact is air (whose index $n(\lambda) \approx 1$); n_m is used when the lens is sitting in something other than air (e.g., $n_m = 1.33$ when the lens is in water). In its more general form, Eq. (4.2) also includes the angle of incidence (AOI) effects, and how the plane of the incident lightwave is oriented (polarized) with respect to the lens surface.[4]

The index $n(\lambda)$ varies with wavelength, but many optical glasses have a refractive index $n \approx 1.5$ in the visible; Eq. (4.2) shows this implies reflection $R = 4$ percent per surface. This can be a serious problem in an instrument with many surfaces, since the throughput will be reduced dramatically. With only five lenses (ten surfaces) and no absorption ($A = 0$), for example, Eq. (4.1) shows that the throughput $T = T_1 T_2 \cdots T_{10} = (1-R)^{10} = (1-0.04)^{10} = 66.5$ percent. Aside from AOI and polarization effects, it seems that this loss is fundamental to the design. As shown in the next section, this is fortunately not the case, and it is possible to reduce the reflection loss to something much less than 4 percent per surface.

However, absorption by the lens material is more fundamental and generally cannot be changed except by using a different material or thickness. The amount of absorption is usually found by measuring the transmission and then using Eq. (4.1) and Eq. (4.2) to remove the reflection losses; what's left is the absorption. A typical curve of how the internal transmission T_{int} for a common lens material (Schott Glass borosilicate #7, or BK7) varies with wavelength is shown in Fig. 4.3.[5] Equation (4.2) is also used to plot the index-limited, external transmission T for two surfaces based on the wavelength-dependent approximation that $T \approx T_{int}[1 - R(\lambda)]^2$. Because there are no Fresnel reflection losses for an internal property, the absorption $A = 1 - T_{int}$. The absorption is extremely small over the wavelength range of 350–1500 nm for BK7, which indicates that this is a high-purity material for visible-to-SWIR optics.

The transmission curves in Fig. 4.3 depend on the thickness of the sample, since a thicker lens absorbs more. Because optical components may be a different thickness than the one used for the measured data, some method is needed to specify material absorption that is independent of the thickness. Such a method exists, and it

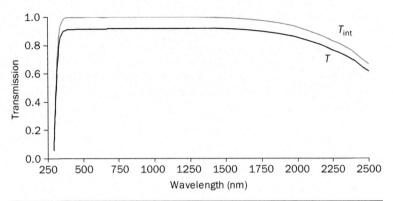

FIGURE 4.3 Internal and external transmission of N-BK7 for a 10-mm sample; the graph illustrates why this material is commonly used for visible, NIR, and SWIR applications.

relies on the concept of an *attenuation coefficient*. Beer's law shows that the material transmission decreases exponentially with thickness:

$$T_{int} = \exp\{-\alpha_m L\} \qquad (4.3)$$

In this expression, a macroscopic property of the optical system (absorption $A = 1 - T_{int}$) is connected with the inherent material loss. This loss is represented by the attenuation coefficient α_m in Eq. (4.3), which takes into account both absorption and scattering. The attenuation coefficient is typically measured in units of 1/cm (or cm^{-1}, "per centimeter"), so that multiplying α_m by the propagation thickness L (in units of cm) gives a unitless number for the exponential and the material transmission. Because the transmission takes only material attenuation into account and not the Fresnel reflections off each surface, it is known as the *internal* transmission. Example 4.1 illustrates the use of these concepts for lenses with different thicknesses than the one at which the transmission is measured.

Example 4.1. The transmission of optical materials is often measured for uncoated substrates of a known thickness (typically 10 mm or 25 mm). However, an optical element, such as a lens or beam splitter, will often have anti-reflection coatings as well as a different thickness than the typical. This example illustrates how to convert measured transmission data into useful design numbers for an optic with a different thickness.
The internal transmission for a 10-mm thickness of N-BK7 is $T_{int} = 96.7$ percent at a wavelength of 350 nm. Because this transmission is an internal material property, it does not account for the Fresnel reflection at the two surfaces of a lens. To include these surfaces, note that the refractive index determines the reflected energy at normal incidence, with Eq. (4.2) giving $R = (0.54/2.54)^2 = 0.045$ per surface for an index $n \approx 1.54$ (from Fig. 3.31, for a wavelength of 350 nm). The first surface (S1) thus transmits $1 - R = 95.5$

percent of the incident light, and if there were no absorption then the second surface (S2) would transmit 95.5 percent of this. Including absorption losses, the external (net) transmission for a lens with an *average* thickness of 10 mm is $T = T_{S1}T_{int}T_{S2} = (1 - R)^2 T_{int} = (1 - 0.045)^2(0.967) = 0.882$, a significant loss of power for a single lens made from relatively low-absorption material.

The lens used, however, has an average thickness of 20 mm, not 10 mm. The exponential form of Beer's law shows that doubling the thickness decreases the internal transmission by the square of the attenuation coefficient—in this case, reducing T_{int} from 0.967 to $0.967^2 = 0.935$ for the 20-mm thickness. [In general, Eq. (4.3) shows that $T_2 = T_1^{(d_2/d_1)}$.] The internal absorption is then $A = 1 - T_{int} = 1 - 0.935 = 0.065$, or approximately twice the internal absorption of the thinner lens. Since the Fresnel reflections do not change, the external transmission is $T = (1 - R)^2 T_{int} = (1 - 0.045)^2(0.935) = 0.853$ for a lens with an average thickness of 20 mm.

In addition to an internal *attenuation* coefficient α_m, it is also possible to define an internal *transmission* coefficient τ. In this case, the internal transmission is now given by

$$T_{int} = \tau^L \qquad (4.4)$$

where L must be measured in the inverse units of the transmission coefficient (such as units of centimeters for L and 1/cm for τ). Physically, the transmission $T = \tau$ for the first centimeter, $T = \tau^2$ for the second centimeter, and so on.

The use of an attenuation coefficient allows us to compare the transmission of different materials independently of their thickness and surface reflectivity. The properties of many of these materials are summarized in Fig. 4.4. Table 4.1 also lists the most common refractive materials with a relatively small attenuation coefficient in each of the bands. The table shows that optical glasses such as BK7 and SF11 are preferred for visible and NIR applications, whereas crystalline materials such as germanium (Ge), silicon (Si), zinc selenide (ZnSe), and zinc sulfide (ZnS) are best for the MWIR and LWIR. Short-wavelength systems requiring high purity for low scatter use a UV-grade form of fused silica, and a water-free grade is available for the SWIR.

For flat optics such as windows and beam splitters, the additional complication of multiple reflections changes the external transmission from that found in Example 4.1. As illustrated in Fig. 4.5, the back-reflections from the second surface are reflected yet again by the first surface, *increasing* the throughput beyond that expected from a two-surface loss. In this case, Eq. (4.2) does not correctly predict the reflection R; instead, the transmission $T = 2n/(n^2 + 1)$ and $R = 1 - T$ for a low-absorption material.

This effect is not significant for low-reflectivity coatings with low back-reflection, but it becomes important for high-reflectivity surfaces such as the high-index materials found in the infrared. For an

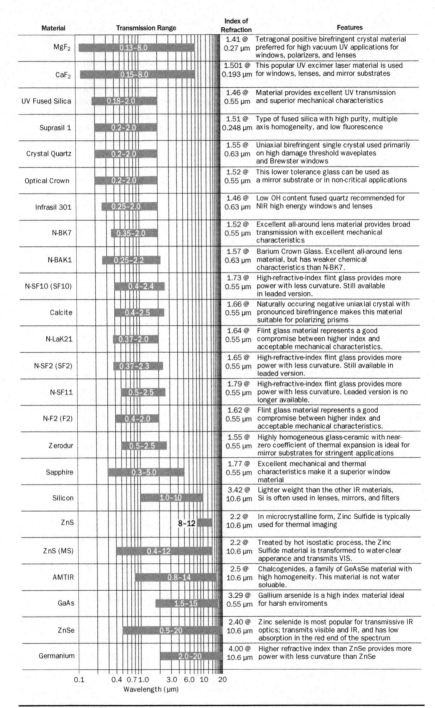

FIGURE 4.4 Properties of common optical materials, including the wavelengths over which they transmit well.[6] (*Credit: CVI Melles Griot.*)

Radiometry

Wavelength Band	Refractive Materials	Wavelengths
Ultraviolet	Fused silica (UV grade), CaF_2, MgF_2	0.2–0.4 µm
Visible	Fused silica, N-BK7, N-SF11, Sapphire, Cleartran™	0.4–0.7 µm
Near infrared	Fused silica, N-BK7, N-SF11, Sapphire, Cleartran™	0.7–1 µm
Shortwave infrared	Fused silica (IR grade), N-BK7, N-SF11, Sapphire, Cleartran™	1–3 µm
Midwave infrared	AMTIR, Ge, Sapphire, Si, ZnS, ZnSe, Cleartran™	3–5 µm
Longwave infrared	AMTIR, Ge, ZnS, ZnSe	8–12 µm

TABLE 4.1 Commonly used materials with high transmission in a given wavelength band.

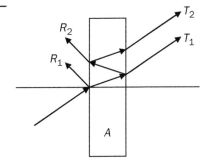

FIGURE 4.5 Multiple reflections from a surface—of which only two are shown in the figure—increase the throughput T over that found from Eq. (4.1) and Eq. (4.2).

uncoated germanium window with $n = 3$, for example, Eq. (4.2) incorrectly predicts a Fresnel reflection $R = [(3 − 1)/(3 + 1)]^2 = 0.25$ per surface [or $T = (1 − R)^2 = 56.3$ percent after including reflections from the front and back surfaces of a low-absorption material]; multiple reflections actually increase the throughput to $T = 2(3)/(3^2 + 1) = 60$ percent, giving a reflectivity $R = 1 − T = 40$ percent for a low-absorption material.

4.1.2 Optical Coatings

The previous section showed that Fresnel reflections reduce the throughput of an optical system. These reflection losses can be large: a 10-surface system will have a throughput of only $(1 − 0.04)^{10} = 66.5$ percent for typical lenses in the visible ($n ≈ 1.5$). In the infrared, the refractive index is often higher ($n ≈ 3$), and the throughput for 10 surfaces is only 5.6 percent in this case! Clearly, then, it would be useful if we could find some way to reduce the Fresnel losses and thereby increase the external transmission T in Eq. (4.1).

It is possible to engineer the inherent reflection of surfaces with thin layers of additional materials. Known as *optical coatings*, these layers usually consist of nonconducting materials (dielectrics) and are deposited on the surface of a lens. If the layers are of a certain index and thickness then they are known as *antireflection* (AR) coatings because they reduce the surface reflection from its natural, index-limited value to something smaller, often on the order of 1 percent or less. The development of such coatings was a major step in the advance of optics over the last 50 years—initially in photography, where multielement lenses are common, and later in almost every type of optical system.

If the lens is used in air, then the simplest AR coating is a single layer of dielectric with an index equal to the square root of the lens index.[7] This reduces the index difference that the incident beam sees while adding a secondary reflection at the coating-to-lens interface. If the layer thickness is a fractional multiple of a wavelength ($\lambda/4$) then the primary and secondary reflections can, in principle, add out-of-phase ("interfere") to zero for that wavelength. Figure 4.6 shows performance data at normal incidence for a commonly used magnesium fluoride (MgF_2) AR coating ($n = 1.38$ at 546 nm) on a BK7 lens ($n = 1.5187$ at 546 nm), giving a net reflectance $R \approx 1.35$ percent at 0° incidence. The thin layers used to make a coating have very little absorption loss, so the low reflectivity shown in the figure translates into a direct increase in throughput.

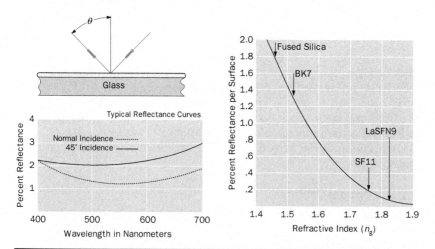

Figure 4.6 The addition of a quarter-wave-thick layer with the appropriate refractive index can reduce the reflectivity of a surface to less than 1 percent. (*Credit: CVI Melles Griot.*)

Optical coatings can also be used to make high-reflectivity (HR) optics such as dielectric mirrors, beam splitters, and edge filters that transmit only long or short wavelengths (Fig. 4.7). Band-pass filters, which reflect light on both sides of a given wavelength, are also useful. An example is the solar rejection filter (SRF) used to transmit a specific wavelength (such as 777.4 nm, to measure lightning) while *reflecting* the solar background and preventing the detector from being becoming saturated or damaged.

Dielectric mirrors and filters usually consist of multilayer coatings; SRFs sometimes use as many as 100 layers, making manufacture on large-area optics difficult. Optical systems engineers should know the basics of the design of such coatings—but only to the extent of understanding how to use them and whether or not what's available will meet the system requirements. In addition to available textbooks on the topic,[7-9] the manufacturers' catalogs from JDSU, CVI Melles-Griot, etc. are also excellent sources of information. Typical specifications for optical coatings are shown in Table 4.2, including the effects of angle of incidence (AOI), polarization, and change in filter temperature. The table entries indicate the wide range of issues that must be considered when a coating is needed for an optical system.

In addition to dielectrics, optical coatings can also be metals, which almost always are used to increase reflectivity. Mirrors, for example, are typically made with very thin layers of gold or aluminum. Silver is sometimes used, but it is difficult to deposit with any quality on a decent-sized optic.

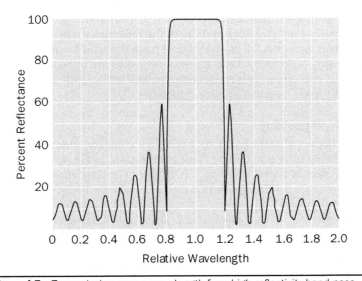

FIGURE 4.7 Transmission versus wavelength for a high-reflectivity band-pass coating. (*Credit: CVI Melles Griot.*)

Specification	Comments
Filter type	Broadband, notch, edge, etc.
Bandwidth ($\Delta\lambda$)	Specify value from peak (1%, 0.1%, etc.)
Center or edge wavelength (λ_c)	Tolerance and slope required
Transmission or reflection (%)	Over λ at 0° AOI and room temperature
Angle of incidence, AOI (°)	"Blue" shift with angle
Beam convergence (°)	"Blue" shift with larger NA
Polarization	Polarization splitting at nonzero AOI
Temperature range (ΔT)	"Red" shift with temperature
Scattering (dB)	Increases with number of layers
Durability	Scratch/dig, peel, salt spray, etc.
Wavefront error (waves RMS)	May be affected by coating stresses
Clear aperture (mm)	Aperture over which specs apply
Diameter (mm)	Physical diameter; ~ 5–7× thickness
Thickness (mm)	Limits diameter

TABLE 4.2 Typical specifications for an optical coating.

Figure 4.8 shows how reflectivity varies with wavelength for unprotected aluminum and protected gold at normal incidence. The curves indicate that aluminum reflects the best in the short-to-medium visible band whereas gold is best from the NIR out to the LWIR. A protective dielectric overcoat—oftentimes a single-layer MgF_2 coating—is also used to prevent the metal from being scratched or tarnished, even though it reduces the reflectivity a bit. Enhanced overcoats also protect the metal and have the additional benefit of increasing reflectivity. Both types of overcoats are options on metal mirrors offered by most manufacturers, and it's not difficult to guess which type of overcoat is more expensive!

4.1.3 Vignetting

It can happen that the entire cone of light incident on a lens is blocked from being fully transmitted. This is illustrated in Fig. 4.9, where the center sketch shows that only half the diverging cone has been collected by a collimating lens placed after a focus; the other half has been "vignetted."

Similarly, if the mechanical structure holding the optical instrument together gets in the way of the extreme rays that contribute to the image, then throughput will decrease. For a thin lens, for example, the chief ray defining the FOV goes through the center of the lens without loss. However, for the extreme rim rays (upper and lower marginal rays) defined by the f/# cone angle, the housing on

Radiometry 145

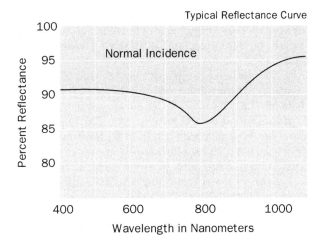

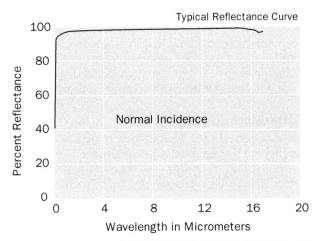

FIGURE 4.8 Reflectance variation with wavelength for the (a) unprotected aluminum and (b) protected gold coatings used for mirrors. (*Credit: CVI Melles Griot.*)

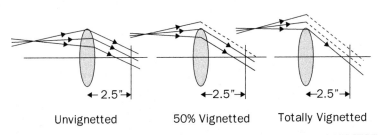

FIGURE 4.9 Vignetting can occur when lenses are not properly sized to transmit the entire beam at the given FOV. (*Credit: Warren J. Smith,* Modern Optical Engineering, *McGraw-Hill, www.mcgraw-hill.com.*)

the image side of the lens must not interfere with the extreme rays as they are imaged onto the upper edge of the detector. Such interference is also known as vignetting, and it must be avoided to maximize throughput.

Figure 4.10 shows an example of how to avoid vignetting for an f/1 lens with a 100-mm aperture and 100-mm focal length. The detector is 10 mm tall on each side, with the slope of the chief ray thus equal to 5 mm divided by 100 mm, or 0.05. The slope of the upper marginal ray on the image side can be calculated from Eq. 2.9 ($u' = u - y\phi = 0.05 - 50/100 = -0.45$); alternatively, the ray drops 45 mm over a distance of 100 mm, which also gives a slope of −0.45. The minimum mechanical radius of the surrounding structure thus varies linearly with distance along the optical axis—that is, it describes a cone.

If we want to estimate the potential contribution of both FOV and f/# on vignetting, an expression that is sometimes used to find the mechanical radius needed to avoid vignetting is $|y_c| + |y_m|$, where y_c is the chief-ray height and y_m is the marginal-ray height for an on-axis point at infinity. As shown in Fig. 4.10(b), the difference in this sum at any two points along the focal length, when divided by the length between them, equals the slope just calculated for the upper rim ray. The contribution of the FOV to the required mechanical aperture is the chief ray, and that resulting from the f/# is the marginal ray.

4.1.4 Throughput Budget

The purpose of Secs. 4.1.1–4.1.3 was to develop the background for understanding the throughput (or power) budget. This is a common optical engineering tool for summarizing an instrument's transmission. The budget consists of a list of each optical element,

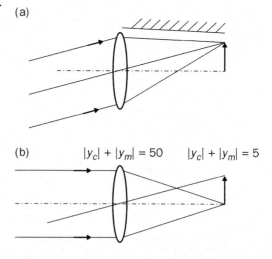

Figure 4.10 Ray trace of the upper chief ray, which shows the minimum cone required to avoid vignetting (not to scale).

and the fraction of light it transmits, while taking into account material properties, coatings, and vignetting. The total throughput is then the product of the transmission for each element. Example 4.2 describes a typical budget.

> **Example 4.2.** In this example we develop the throughput budget for a narrow-field, reflecting afocal telescope operating at a wavelength of 600 nm. Following the telescope are a beam splitter (to send a fraction of the light down a separate path), a solar rejection filter, and a lens for imaging the beam onto an FPA. There is no vignetting in the optical paths.
>
> The two telescope mirrors (M1 and M2) are coated with unprotected aluminum. Figure 4.8 shows that they each have a reflectivity at normal incidence of about 90 percent at 600 nm; this is referred to as the optical system's "transmission" even though it is physically due to reflection of the mirrors. The beam splitter (B/S) is designed to transmit half of the incoming light, and the vendor tells you the actual transmission is 48 percent at 632 nm with a fairly broadband response. A solar rejection filter just before the lens removes most of the solar spectrum as well as 10 percent of the light that's needed for the measurement. The lens (L1) has 1 percent AR coatings on both surfaces (S1 and S2) and an internal transmission (T1) of 96 percent; therefore, the lens throughput $T = (1 - R)^2 T_1 = 0.99 \times 0.99 \times 0.96 = 0.941$ at 600 nm.
>
> Table 4.3 summarizes these results and reveals that the total power on the image plane is only about 33 percent of the power incident on the first optic. In practice, the number of elements will likely be much larger, reducing the throughput to the point where there's not enough light on the FPA to collect images within a reasonable time (see Chap. 6). Insufficient throughput is a common problem, and trading the optical losses against a bigger aperture or a more sensitive detector will take up much more design effort than is suggested by this simple example.

A common instrument specification is a "rolled-up" summary of the throughput budget at different wavelengths. One approach is for

Element	Transmission
Mirror 1 (M1)	0.90
Mirror 2 (M2)	0.90
Beam splitter (B/S)	0.48
Solar rejection filter (SRF)	0.90
Lens 1 (L1)	0.94
Surface 1 (S1)	0.99
Internal transmission (T1)	0.96
Surface 2 (S2)	0.99
Throughput	0.33

TABLE 4.3 Throughput budget for a two-mirror afocal telescope, beam splitter, SRF, and imaging lens. The Fresnel losses for the FPA have not been included.

Wavelength	Throughput
400 nm	> 65%
500 nm	> 70%
600 nm	> 74%
700 nm	> 72%

TABLE 4.4 Typical throughput specifications for an instrument operating in the visible.

a customer to specify $T > 75$ percent (for example) over the entire visible band (400–700 nm). Also common is a throughput budget at three or more wavelengths, such as is shown in Table 4.4.

Some loss items are not included in the *instrument* throughput budget but still can be a limiting factor in the overall system goal of getting enough power to the detector. An example is atmospheric loss, which often controls how much power is available for laser radar and free-space laser communications systems. These losses are included in a *system* power budget so that the system designer can make trades between transmitter power, instrument throughput, detector sensitivity, and the like. How much power is considered "enough" will be covered in Chap. 6.

4.2 Irradiance

Although throughput is an important requirement for any optical system, an equally important issue is the detector size over which the transmitted power is distributed. Compact digital cameras, for example, have "shutter lag" in part because the pixels are extremely small; as a result, there is not much power on each pixel and so the shutter must stay open for a relatively long time in order to collect enough energy to yield a decent picture. Film-based cameras allowed some degree of control over this aspect of photography: "slower" films with smaller grains of silver halide required a longer exposure time, but the result was higher resolution from the smaller grains.

Some cameras also allow control of the exposure time by changing the aperture size (via the f/# or f-stop). Going in the other direction, too much power over too small an area can cause overheating and damage, as is clear to anyone who has used a magnifying glass to image the Sun. Figure 2.10 compares two f-stop designs with different radiometric properties: an f/1 and an f/2. The f/2 has a focal length that's twice that of the f/1; for an object at infinity with the same angular FOV, geometrical optics can be used to demonstrate that its image size is also twice that of the f/1. The total power on the image plane is the same since the entrance-pupil diameters are the same size and since the throughput for both lenses is (assumed to be) the same.

Yet because the area of the f/2 image is 4 times larger (each dimension is twice as big as the f/1), the power density [W/m^2] is 4 times smaller for the f/2. As a result, the exposure time required to collect enough energy for the f/2 is 4 times longer than that of the f/1, which explains the terminology by which smaller-f/# lenses are "faster." The concept of power distributed over an image or pixel area is known as *irradiance* and is measured in watts per square meter (W/m^2) of exposed area.

Although the f/# is a common metric for image irradiance, it is a limited concept in that it is a *relative* aperture (consisting of the ratio of two quantities) and so does not give absolute values of irradiance. For an absolute measurement, the throughput reduces the irradiance and a larger aperture collects more of the power density incident on the entrance pupil. The power (or flux) Φ on the focal plane is thus given as follows:

$$\Phi = TEA_p \quad [W] \tag{4.5}$$

Here the power depends on the throughput T of the optics and on the irradiance E [W/m^2] collected by the entrance pupil area A_p. The image irradiance on the focal plane is then Φ/A_i if the power is distributed uniformly over the image area A_i. As a result, the average irradiance in the image plane is equal to the incident irradiance multiplied by the ratio of aperture area to image area—and reduced by the instrument throughput.

In practice, the irradiance is not distributed uniformly over the image. Instead, pixels at the outer edge of the field have a lower irradiance than those at the center, even for a completely uniform source. There are two reasons for this. The first is that, as shown in Fig. 4.11, the distance $R = f/\cos\theta$ from the center of the exit pupil to the edge of the field is longer than the distance to the center of the field by a factor of $\cos(\theta)$; given the $1/R^2$ dependence of power density, this contribution to the irradiance nonuniformity is equal to $\cos^2(\theta)$.

The second contribution is that the exit pupil and detector planes are tilted at angle θ with respect to the field axis, which results in a smaller exit pupil area emitting toward the FPA as well as a smaller effective collection area for the outer pixels. For moderate to large f/#, the decrease in irradiance for each pixel again varies with $\cos(\theta)$. The net result of the longer distance and smaller areas is that the irradiance drops off proportionally to $\cos^4(\theta)$ for area sources. This dramatically affects the irradiance at the edges of systems with even moderate FOV; a half-angle of 30 degrees yields $\cos^4(0.52) = 0.56$. However, not all four $\cos\theta$ terms need be present in all lens forms; for instance, a typical design curve for a high-definition TV lens is shown in Fig. 4.12. It is the worst-case irradiance at the *edge* or corner of the field that controls the integration time for FPAs.

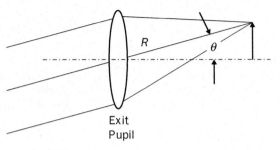

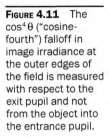

FIGURE 4.11 The cos⁴ θ ("cosine-fourth") falloff in image irradiance at the outer edges of the field is measured with respect to the exit pupil and not from the object into the entrance pupil.

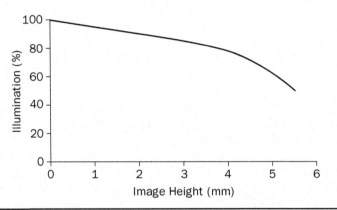

FIGURE 4.12 A typical design curve showing the drop in irradiance with chief-ray angle (image height) for a commercial high-definition TV lens. Data is for the Canon YJ20 × 8.5B KRS zoom lens.

In addition to spatial variations across an FPA, irradiance can also vary with wavelength; this leads to the concept of *spectral irradiance*. Light from the Sun, for example, has a wide range of wavelengths. Familiar to most are the coarse distinctions (of red, green, blue, etc.) that are seen in rainbows, and it's straightforward to measure how much irradiance from the Sun is in the "red" band, or in any other band that can be defined. Finer distinctions than "red" or "green" are usually made; it is common to measure the Sun's power in a small band that's only 1 nm wide, and the solar irradiance in that band is measured in watts per square meter per micron of bandwidth (W/m²·µm). This is known as the spectral irradiance, where the word "spectral" indicates that the measurement is per unit wavelength. Given the spectral irradiance E_λ (where the subscript λ indicates irradiance per unit wavelength[10]), the total power on the focal plane is obtained from

$$\Phi = TE_\lambda A_p \Delta \lambda \quad [\text{W}] \qquad (4.6)$$

Example 4.3 illustrates the distinction between irradiance and spectral irradiance while also showing the limitations of the use of Eq. (4.6).

Example 4.3. Is it possible to take pictures of the Sun with a digital SLR *without* destroying the camera? By pointing the camera directly at the Sun, the solar irradiance on the aperture is on the order of 1000 W/m^2. Of course, this value will be bigger on the top of a Colorado "14er" and smaller in the morning, but 1 kW per square meter is a good working number.[11] Will this "fry" the FPA detector used for digital imaging?

The collected solar power and the Sun's image area jointly determine the solar irradiance impinging on the FPA. According to the manufacturer's specifications, the camera has an aperture of 35 mm at f/1.4, for an area of 962 mm^2. When dealing with a bright source, it's best to first stop the aperture down to an f/64 (the smallest stop on the camera); the entrance pupil diameter is then 962 mm^2 × (1.4/64)^2 = 0.46 mm^2.

The solar power on the FPA is given by Eq. (4.5). For a throughput of 75 percent, the power $\Phi = TEA_p = 0.75 \times 1000$ W/m^2 × 0.46 × 10^{-6} m^2 = 0.345 × 10^{-3} watts. This not a large amount of power, but even a small amount of power in a small spot does not have sufficient surface area to conduct heat efficiently. Thermal damage is thus determined by the amount of heat in a unit area—in other words, it is the irradiance that is of concern, not the total power.

The image irradiance is found from the angular size of the Sun, which is known to be approximately 0.5 degrees (8.75 mrad) and can be translated into an image size if we know the camera's effective focal length. This parameter is given in the spec sheets for digital cameras as 50 mm for normal shots (neither telephoto nor wide-angle); alternatively, it can be obtained from the aperture and f/#. In either case, an EFL of 50 mm multiplied by 8.75 mrad gives a solar image that's 0.4375 mm in diameter (or an image area of only 0.15 mm^2). The image irradiance is then 0.345 × 10^{-3} watts divided by 0.15 × 10^{-6} m^2, or E = 2300 W/m^2 distributed over all the pixels that are lit up by the Sun's image.

If the FPA has square pixels that are 10 microns on a side, then this image covers about 44 pixels in diameter or about 1520 pixels for the Sun's image. The irradiance on a pixel is thus 2300 W/m^2 divided by 1520 pixels, which is approximately 1.5 W/m^2 per pixel (1.5 × 10^{-10} W/μm^2) if we assume a uniform solar irradiance. This is not enough to destroy the pixels but may well cause saturation, which washes out detail in the photo (see Chap. 6).

Saturation can be reduced by adding an optical filter to the camera. If the filter bandwidth is "small" enough, Eq. (4.6) and knowledge of the solar spectrum can be used to calculate the irradiance per pixel. Many filters used in photography are color filters, which let through fairly wide bands of, for example, "yellow," "red," or "green." Because Eq. (4.6) assumes that spectral irradiance is relatively constant over the filter bandwidth, the equation doesn't work with these filters for broadband sunlight, and an integration or discrete summation must be used instead (see Chap. 5). However, it is useful when applied to narrowband filters for which the irradiance doesn't change much with wavelength.[12]

4.3 Etendue, Radiance, and Intensity

A closer look at how much power is collected from an area source such as the sky reveals that the foregoing analysis of throughput and irradiance is incomplete. Equation (4.5) and Eq. (4.6), for example, show that a large collection aperture puts more power on the image plane. Intuitively, however, collecting light from a larger FOV must also collect more power than that from a smaller field. Clearly the image power cannot depend only on the aperture size, as Eq. (4.5) implies.

Since the field increases with detector size [see Eq. (2.3)], the collected power depends on the size of both the aperture and the detector. Specifically, the detector area is important because the total power in the image plane is increasing both in aperture area and in detector area. For instance, compact digital cameras exhibit a shutter lag in part because their pixels are extremely small; hence there is not much power on each pixel and so the shutter must stay open for a relatively long time in order to collect enough energy to give a decent picture.

Equation (2.3) also shows that the field angle depends on the effective focal length; therefore, the detector area is useful for comparing different designs only if the EFL is the same. To circumvent this restriction, a factor known as a *solid angle* is used instead. The FOV in one dimension varies with d/f, so the square of the FOV is used to take both dimensions into account. This is consistent with the use of the detector area (d^2 for square pixels), which—when combined with the $1/f^2$ term—leads to the concept of the solid angle $\Omega \approx \text{FOV}^2 = \theta^2 = (d/f)^2$.

A solid angle sweeps through space in three dimensions and not just in the two-dimensional arc of a planar angle. As shown in Fig. 4.13, the sweep can be conical, rectangular, square, or any other shape imaginable. Physically, it connects the surface area A of a sphere with the square of the sphere's radius R^2 such that the area

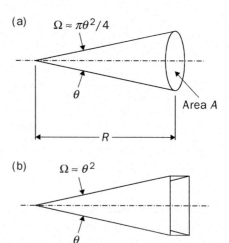

FIGURE 4.13 Projected solid angle Ω depends on the distance to an arbitrarily shaped surface, which in optical systems is typically a disk or a square.

(a) $\Omega \approx \pi\theta^2/4$

Area A

(b) $\Omega \approx \theta^2$

$A = R^2\Omega$. This is the three-dimensional analogue of the idea that a planar angle connects the arc length S of a circle with the radius R (where $S = R\theta$). With nonspherical surfaces, the projected area onto a sphere is used instead—in the same way a "squiggly" arc is projected onto a circle to determine the angle subtended in two dimensions—and creating what is known as a *projected* solid angle.

Solid angles are measured in units of square radians (rad^2); this unit is given the name *steradian* (sr), a three-dimensional version of the flatlander's two-dimensional radian. For circular optics (sources, lenses, mirrors, detectors, etc.), the solid angle has the shape of a cone and is given by

$$\Omega = 2\pi[1-\cos(\alpha)] \approx \pi\sin^2(\alpha) \approx \pi\alpha^2 \quad [\text{sr}] \tag{4.7}$$

where α is the half-angle FOV as used in Chap. 2. The range of α is 0 to 180 degrees; a full sphere ($\alpha = 180°$) subtends a solid angle of 4π steradians, and a hemisphere ($\alpha = 90°$) subtends 2π steradians.

The approximations on the right-hand side of this equation, which are applicable to paraxial $\alpha \approx \pi/6$ rad (30 degrees) or less, make clear the calculation of solid angle for the FOV: a circular detector area ($=\pi d^2/4$) divided by the square of the distance to the detector ($=f^2$) gives $\Omega = \pi d^2/4f^2 = \pi\alpha^2$ [using Eq. (2.3), where $2\alpha \approx d/f$]. Similarly, for square pixels we have $\Omega \approx 4\alpha^2$; for rectangular pixels, $\Omega \approx 4\alpha_x\alpha_y$ for a field angle that's different in the x and y directions. To gain a sense for the numbers, a 0.1-rad (5.73°) full-field angle for a square FPA corresponds to a solid angle $\Omega = (2\alpha)^2 = (0.1 \text{ rad})^2 = 0.01$ sr.

For small solid angles, the approximations also imply the use of the projected surface area rather than the actual surface area. For example, Fig. 4.13(a) shows a cone with a half-angle $\alpha = \pi/20$ rad (9°); the spherical cap on the end of this cone has an actual surface area of $2\pi R^2[1 - \cos(\alpha)]$, giving a solid angle $\Omega = A_{surf}/R^2 = 2\pi[1 - \cos(\alpha)] = 0.0774$ steradians [as follows also from the exact form of Eq. (4.7)]. The projected surface area—based on the shadow that the cap throws onto a perpendicular plane—results in a projected solid angle $\Omega_p = \pi\alpha^2 = \pi(\pi/20)^2 = 0.0775$ sr, so the approximation is pretty good for this small cone angle.

A cone with a half-angle $\alpha = \pi/2$ rad (90 degrees), on the other hand, is a hemisphere with a solid angle $\Omega = A_{surf}/R^2 = 2\pi R^2/R^2 = 2\pi$ sr. The projected area of the hemisphere is πR^2, which gives a projected solid angle of $\Omega_p = \pi$ steradians; this incorrect result illustrates that the approximations in Eq. (4.7) are not valid for such large angles.

Returning to the main story, both aperture and field of view determine the amount of light that creates an image. Aperture is measured as the entrance pupil area; field is measured as the solid angle within which light is collected from both field dimensions. Combining the concepts of aperture and solid angle leads to the important radiometric ideas of etendue and radiance.

4.3.1 Etendue

Whereas f/# is a common measure of radiometric performance, etendue is an equally important (but often neglected) concept.[13] Combining the effects of both aperture (f/#) and FOV, *etendue* is calculated as the product of aperture area and the projected solid angle of the field.[14] Based on the previous section, we can say that etendue is a component of the total power collected by the telescope and transferred to the image plane. It does not depend on how much light is incident on the optical instrument; rather, it is a geometric property of an optical system that allows one to compare different designs with different apertures, detectors, and focal lengths. Etendue is also known as the $A\Omega$-product, which is measured in units of area [m^2] times solid angle (steradian) or m^2·sr. It is also known as "throughput", but optical engineers typically use the word throughput for optical transmission and almost never for etendue.

By comparing two different imager designs, Fig. 4.14 illustrates the difficulty of relying solely on f/# as a measure of radiometric performance. Figure 4.14(a) shows a thin-lens f/1.5 imager and an FPA pixel size that captures a single-pixel IFOV of 0.1 degrees. A second design, shown in Fig. 4.14(b), uses the same aperture but now the pixels are twice as big and are located twice as far away from the lens. Hence the device captures flux from the same angular size of the scene (IFOV) but has twice the f/#.

Because the f/# in Fig. 4.14(b) is twice that in Fig. 4.14(a), a common mistake is to assume that the pixels in the second design collect only

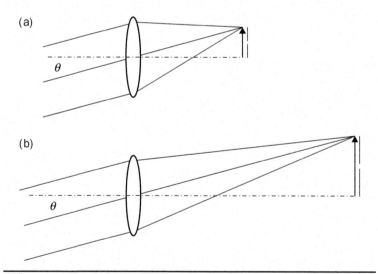

FIGURE 4.14 A larger f/# does not necessarily imply that less power will be collected by the optical system. The etendue is the same in both *(a)* and *(b)*, which results in the same amount of power incident on the FPA.

one-quarter the power of the first. This assumption is wrong; although the irradiance on the larger pixels is smaller, the larger pixel area ensures that both designs place the same amount of power on the detector. Physically, both designs are collecting light from the same IFOV with the same-size entrance pupil, so the power on the bigger pixels must be the same as on the smaller pixels. Quantitatively, the product of entrance pupil area and the solid angle of the field— that is, the etendue— is the same for both designs. Example 4.4 gives some typical numbers associated with two different designs for a space-based imager.

Example 4.4. A space-based imager measures lightning flashes over a 10-km × 10-km ground sample distance (GSD) from geosynchronous orbit (GEO) at 36,000 km above the Earth. (The GSD is the size of the Earth's surface that images onto one pixel on the FPA.) Starting with square pixels each 15 μm on a side and equating the field angle on the object and image sides of the entrance pupil, the EFL = (36,000 km × 15 × 10^{-6} m)/(10 km) = 54 mm. A separate calculation by a lens designer shows that at least 100 mm of aperture (EPD) is required and that the needed f/# = EFL/EPD = 0.54 is almost impossible to achieve.

The etendue of this system is also calculated based on the area of the aperture and the IFOV of the pixel: $A\Omega_p = [\pi(0.1^2)/4][(10/36000)^2] = 6.06 \times 10^{-10}$ m²·sr. When calculating etendue, it is almost always easier to use the pixel IFOV rather than the FOV of the entire FPA, because there may be gaps between pixels that reduce the etendue.

To minimize redesign costs, the same FPA is used and the focal length is doubled to get to an f/1.08 system. In doing so, however, the power on each pixel has been reduced by the square of the f/#-ratio (a factor of 4), because the 10-km image is now spread out over four pixels (a result of keeping the same pixel size). The etendue has also decreased by a factor of 4 because the IFOV is halved by using the same pixel size at twice the focal length [$\Omega \approx \alpha^2 \approx (d/f)^2$]. It turns out that the detector is not sensitive to this smaller power level, so another approach is needed for increasing the f/#.

If the focal length is doubled but now a different FPA with twice the pixel size is used, then the IFOV is the same and the etendue does not change. The f/# again doubles to f/1.08, which is needed to make this design more manufacturable. But by not being careful with the use of the concept of f/#, it is easy to believe that doubling up on the f/# decreases the power on a pixel by a factor of 4.

Instead, the collected power per unit area has decreased and, to compensate, the pixel area has been increased—keeping the total power on a pixel the same. So even though the f/# has increased, the etendue and collected power remain constant as a result of increasing the FPA pixel size. For those who are familiar with photographic film, this is identical to using faster film with a larger grain size to increase the etendue. The design freedom allowed by relatively unrestricted choice of pixel size is a major advantage of electronic FPAs over film.

Why not double up again and use an f/2.16 system or higher? This is a reasonable systems engineering trade of lower cost against the increase in size and weight of a bigger f/# and FPA. Yet to *increase* the etendue requires either a larger IFOV (using larger pixels or a smaller focal length) or a larger aperture, and each option has its own set of trade-offs. Limitations on aperture are absolute size and aberration correction; limitations on increasing the solid angle via larger pixels include not only absolute size of the entire FPA and aberrations but also detector noise (see Chap. 6).

The pixel size in Example 4.4 (15 μm) is much larger than the diffraction-limited spot size for the f/1 design (B = 2.44λ × f/# = 2.44 μm for λ = 1 μm). If the image quality is improved by reducing each pixel to the size of the Airy disk for a diffraction-limited spot, then the etendue of an extended (nonpoint) source is also reduced by a factor of $(15/2.44)^2 \approx 38$. Diffraction-limited image quality, then, is not necessary for large etendue and radiometric efficiency; in practice, image quality and radiometric efficiency often trade against each other. This is a common theme in radiometer design, where the pixel is intentionally chosen to be something larger than the diffraction-limited blur to obtain the best radiometer, not the best imager. Not surprisingly, the design of an imaging radiometer involves a delicate balance between the two. The differences between the two types of instruments will be explored in more detail in Chap. 6.

4.3.2 Radiance

Etendue is helpful for understanding the radiometric performance of an optical instrument, but the concept needs to be extended somewhat to describe the instrument's images. As we've seen, a bigger etendue puts more power in the image, making visible images look brighter. A more powerful light bulb, however, can also make the image brighter, and *radiance* is the notion that captures the brightness of images and sources both.

Looking first at sources, we remark that a small source gives off a brighter signal than a larger one emitting the same power. In this context, the notion of radiance probably evolved because a smaller source has less surface area to dissipate heat. The smaller source therefore radiates at a higher temperature (see Chap. 5)—allowing it to glow red, orange, yellow, or even white—while the large source with the same output power may still appear cool. The same is true for images, where using a magnifying glass to focus the Sun (for example) into a small spot heats a surface more than a larger spot would; our eyes tell us that the hotter spot is brighter and more "radiant," even though the power in a larger spot is the same.

Source power that is also directed into a smaller solid angle (e.g., with a lens) is more concentrated than when directed into a larger angle, and again it looks brighter (Fig. 4.15). For the same output power, a smaller area and solid angle thus gives a brighter source, leading to the definition of source radiance as the power divided by its projected area (giving irradiance) and divided again by the solid angle into which it radiates. Radiance is measured in units of W/m^2·sr; like irradiance, it can also be measured in small wavelength

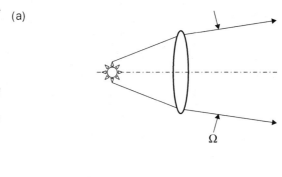
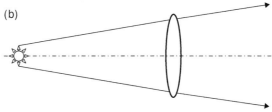

FIGURE 4.15 For a given level of power collected by a lens, the source in (b), with the smaller solid angle Ω, is brighter. Because of the larger distance to the lens, this also requires the source power in (b) to be greater than that in (a) (see Chap. 5).

bands. It is then called *spectral radiance* and measured in units of W/m²·sr·μm.

The larger the radiance L of the source and the greater its angular size (projected solid angle Ω_p) as seen from the entrance pupil of the telescope, the more power that is imaged onto the focal plane. For light that is either emitted by or scattered off the object, the power on the FPA is given by

$$\Phi = TLA_p\Omega_p \quad [W] \qquad (4.8)$$

where the telescope's transmission T and aperture A_p also control the image power. Equation (4.8) has a spectral counterpart; in this case, the spectral radiance [W/m²·sr·μm] is used to calculate the total power on the focal plane:

$$\Phi = TL_\lambda A_p\Omega_p\Delta\lambda \quad [W] \qquad (4.9)$$

As before, image *irradiance* is found by dividing the power in Eq. (4.8) or (4.9) by the area of the image. Image *radiance* is found by dividing the image irradiance by the solid angle of the cone illuminating the image, based on the f/# of the lens.[15] This is a bit tricky conceptually, because a faster image cone with a larger solid angle should produce a brighter image. However, in dividing the irradiance by the larger solid angle, the math shows that the image should be dimmer, not brighter.

Resolving this is straightforward once you realize that the image area and image cone are related. This was seen previously in Fig. 4.14, when f/1 and f/2 lenses were compared. Reexamining the figure makes it clear that, as the solid angle of the lens becomes bigger (moving in from an f/2 to an f/1), the image area becomes proportionally smaller. This means that A and Ω are connected, and the math disagrees with our physical reasoning only when they are viewed as being independent. The connection between A and Ω is established through the conservation of etendue, which we discuss in Sec. 4.4.

One of the more common calculations with radiance involves reflections off particles or a diffuse surface. The soft, radiant glow of desert sunsets, for example, is due to lightwaves being scattered by dust particles in the air. Unlike a mirror, which reflects light in a specific direction, a diffuse surface can scatter light in all directions. When light hits a diffuse surface straight on, some of the light is reflected straight back toward the source and some goes off to the sides. A painted wall or ground glass, for example, does not reflect like a mirror into a specific angle but instead reflects light into a range of angles (Fig. 4.16). From the perspective of solid angle, the light is scattered into a hemisphere of 2π steradians. Working through the math, this is the same as scattering into a projected solid angle of π steradians onto a flat surface parallel to the diffuse surface.

The polar scattering profile shown on the left side of Fig. 4.17 is due to a so-called Lambertian surface. These surfaces have a number of important properties, the first of which is that the power scattered into a small solid angle decreases with $\cos(\theta - \theta_s)$ with a maximum at the specular angle θ_s and no power scattered into 90 degrees

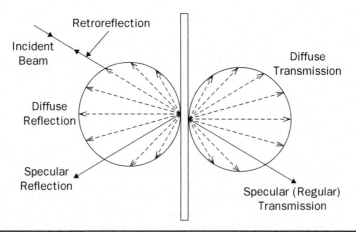

Figure 4.16 In addition to the specular laws of reflection and refraction, light scatters from rough surfaces in a diffuse manner with specific angular distributions. (*Credit:* Handbook of Optics, *McGraw-Hill, www.mcgraw-hill.com.*)

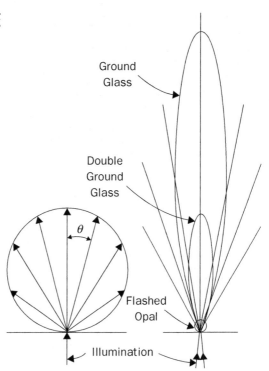

FIGURE 4.17 Scattering from a Lambertian surface *(left)* and a directional surface *(right)*. The intensity [W/sr] seen from the Lambertian surface varies with the cosine of the off-specular angle (off vertical, in this figure); it peaks in the direction of specular angle and drops to zero at 90 degrees. *(Credit: Warren J. Smith, Modern Optical Engineering, McGraw-Hill, www.mcgraw-hill.com.)*

off-specular. In the form of an equation that bears Lambert's name, the intensity $I = I_o \cos(\theta - \theta_s)$, where I_o is the intensity (in units of W/sr; see Sec. 4.3.3) measured at the specular angle.

Although the scattered power becomes smaller at larger angles, the radiance is the same as it is at the specular angle—in other words, radiance does not vary with angle. This is seen in the solar reflection off the Moon, whose appearance is that of a disk of constant brightness (radiance) rather than a sphere whose brightness falls off at the edges. The reason is that the larger area seen by the observer at the edges of the Moon reflects more power but is also angled so that more scatters away from the observer's FOV, thus keeping constant the power-to-area ratio that is scattered into a small solid angle.

For flat Lambertian surfaces, the irradiance scattered into a small solid angle (i.e., the radiance) is also the same in all directions. The radiance of the reflected light [W/m²·sr] is then the incident irradiance [W/m²] reduced by the surface reflectivity and then divided by the π-steradian projected solid angle into which the incident irradiance is equally scattered. The solid angle would be 2π steradians if the scattering were into a sphere; our assumption of π steradians is based on scattering onto a flat plane parallel to the object such that $L = \rho E / \pi$.

Coatings that are Lambertian at relatively small angles of incidence include Spectralon, Martin Black anodize, and Aeroglaze Z306. Most optical components (lenses and mirrors) are *not* Lambertian—and scatter more power into a smaller range of angles than a surface that is Lambertian (see right side of Fig. 4.17). In addition to off-specular angles, other factors that affect how much light is scattered include how clean the optic is, as well as how rough the surface is in comparison with the optical wavelength. Almost no surfaces are truly Lambertian, but many are treated as if they were. The following example illustrates a typical application of the concept.

Example 4.5. A telescope on a GEO satellite is pointed toward Earth and is used to measure lightning in the clouds. Sunlight reflected from the clouds makes it difficult to measure the lightning, so it's necessary to estimate how much background power from the Sun ends up on the focal plane and then compare it with the power from the lightning signal. The ratio of the two powers is an indicator of signal quality (as distinct from image quality).

In using the solar spectral irradiance to find the solar spectral radiance, we assume that the clouds are Lambertian reflectors at all wavelengths so that the solar irradiance is reflected into a projected solid angle of π steradians. The solar spectral radiance incident on the entrance pupil of the telescope is then $L_\lambda = \rho E_\lambda / \pi$ W/m²·sr·µm regardless of distance or viewing angle. Lightning has a lot of energy at a wavelength of 777.4 nm; a solar spectral irradiance of E_λ = 1159 W/m²·µm at 780 nm is given in Vol. 1, Table 3.2 of the *Infrared and Electro-Optic Systems Handbook*.[11] Using a value of ρ = 1 for the cloud reflectivity yields a solar spectral radiance of $L_\lambda \cong$ 370 W/m²·sr·µm, which is a constant at any angle of reflectance.

The solar power on the focal plane depends on the radiance reflected off the clouds and also on the etendue and spectral filtering of the telescope. The size of the satellite limits the telescope to a 100-mm aperture; with this aperture, a spectral filter with a 1-nm bandwidth over the FOV is difficult but not impossible. Including the transmission through the filter, the overall throughput of the telescope is T = 0.5. Equation (4.9) shows that the IFOV is also required; using the same GSD as was used in Example 4.4, α = (10 km)/(36,000 km) = 2.78 × 10⁻⁴ rad.

After substituting the IFOV into Eq. (4.9), we find that the reflected solar background imaged onto each of the focal plane's pixels is

$$\Phi_s = TL_\lambda A_p \Omega \Delta \lambda$$
$$= (0.5)(370 \text{ W/m}^2 \cdot \text{sr} \cdot \mu\text{m})(0.00785 \text{ m}^2)(2.78 \times 10^{-4} \text{ rad})^2(0.001 \text{ µm})$$
$$\approx 1.0 \times 10^{-10} \quad [\text{W}]$$

The lightning, whose radiance at 777 nm is 0.01175 W/m²·sr,[16] images onto each pixel with a power

$$\Phi = (0.5)(0.01175 \text{ W/m}^2 \cdot \text{sr})(0.00785 \text{ m}^2)(2.78 \times 10^{-4} \text{ rad})^2$$
$$\approx 3.6 \times 10^{-12} \quad [\text{W}]$$

or approximately 30 times *less* than the solar power. To extract the lightning signal from the solar background will thus take some work, the details of which are given in Chap. 6.

4.3.3 Intensity

Everything said about radiometry up to this point applies only to extended objects with a finite field size. Within the field, however, there may be point sources such as stars, and our understanding of radiometry must be extended to take these into account.

As shown in Chap. 3, a point source images to a diffraction-, aberration-, or atmospheric-limited blur. Real objects of finite size can image to such a blur if they are far enough away and if their angular size is smaller than the blur (aka the point spread function) created by the optics. In such cases, the spot size has no connection with the actual size of the source, which is too small to measure. But if the source's image size is not known, then neither is its irradiance. It is always possible, of course, to divide the source power by the area of the blur and so derive a number for irradiance on the detector, but there's no connection between that number and irradiance of the source. For this reason, the notions of source radiance and irradiance do not apply to point sources; the concept used instead is *intensity*.

A point source is considered intense if it puts out a lot of power into a steradian. Because the source area cannot be known, the intensity I is measured in units of watts per steradian (W/sr) into which the source emits or reflects. The power on the detector then depends on the source intensity and the solid angle intercepted by the instrument's entrance pupil:

$$\Phi = TI\Omega \quad [W] \tag{4.10}$$

Here the solid angle Ω from the source to the optics is approximately equal to the entrance pupil area divided by R^2 (the square of the distance to the source). Spectral intensity I_λ is also used, though by now it should be clear how to calculate the power on the detector in this case ($\Phi = TI_\lambda \Omega \Delta\lambda$).

Point sources are sometimes specified in terms of power rather than intensity. If the distance to the source is known then the inverse square law determines its irradiance at the entrance pupil; Eq. (4.5) can then be used to calculate the power on the detector. A source that emits uniformly into a sphere with an intensity of 1 W/sr, for example, has a power of 4π watts. At a distance R away, the irradiance from the source is 4π watts divided by $4\pi R^2$ m², which yields $E = 1/R^2$ W/m². Multiplying by the entrance pupil area in Eq. (4.5) then gives the same result as Eq. (4.10). Unfortunately, the distance to the source is not always known; in that case, intensity is the more useful quantity for describing the source.

Because they are point sources, the rules for imaging stars are a bit different than those for extended objects such as mountain scenes. To get the most power from an extended scene, a large aperture combined with a short focal length (and thus large solid angle) gives the highest irradiance on the focal plane. A large aperture is also required for a star, again for the purpose of collecting more light. But instead of a small f/#, the design uses a *long* focal length to create a large f/#. This collects the background from a smaller solid angle (decreasing its irradiance), but does not affect the measured irradiance of the star (as long as its blur size is still smaller than a single pixel and keeps most of the energy on the pixel). As a result, stars increase their contrast against the background sky and, in principle, can even be seen during daylight with a large-f/# telescope.

4.4 Conservation Laws

So far we have established that, depending on the type of source, the power incident on the FPA may depend on the source irradiance [Eq. (4.5)], radiance [Eq. (4.8)], intensity [Eq. (4.10)], or any of the spectral variants of these three quantities.

A unit analysis makes it easier to keep all the concepts and vocabulary straight; see Table 4.5. Unfortunately, the committees that define radiometric terms use words to mean something other than what engineers have commonly used in the past. *Flux*, for example, is well known to physicists as any quantity (e.g., electric field) passing through a unit surface area. However, the radiometric folks have decided that "flux" is the word that should be used for power. That redefinition is ignored in this book, and throughout we simply use "power" (although the radiometric symbol Φ is used in the equations).

Common Name	Radiometric Name	Units
Energy, E	Radiant energy, Q	joules [J]
Power, P	Flux, Φ	watts [W]
Intensity, I	Irradiance, E	W/m^2
	Exitance, M	
	Intensity, I	W/sr
Brightness, B	Radiance, L	$W/m^2 \cdot sr$
Spectral brightness, B_λ	Spectral radiance, L_λ	$W/m^2 \cdot sr \cdot \mu m$
$A\Omega$-product	Etendue, G	$m^2 \cdot sr$

TABLE 4.5 Common names and radiometric names for various physical quantities.

"Intensity" is another word whose meaning has been re-defined from its common usage. In this case, it has actually been *reassigned* from W/m² to W/sr. In a profession where crisp language is required to avoid errors, this is literally inviting an engineering disaster.

Radiometry also seems to be the only field that changes the name of a concept depending on whether it's coming or going. Irradiance, for example, is formally valid only for the number of watts per square meter *incident on* a surface. But when talking instead about the number of W/m² *coming from* a surface, then the recommended word is exitance (and the symbol also changes from E to M). These sorts of pedantic games do not help the learning process, but they do need to be understood because both sets of names in Table 4.5 will be used throughout one's career.

It's also common to hear phrases such as "watts per square" or "watts per square per stare" to describe irradiance (W/m²) and radiance (W/m²·sr), respectively. Spectral radiance even gets its own word ("flicks") when it is given in units of W/cm²·sr·µm (note the change to cm² for the area in the denominator).

From the list in Table 4.5, there are three conserved quantities: power, etendue, and radiance. Power conservation is something expected from the laws of thermodynamics; etendue is a bit trickier and is explained in Sec. 4.4.2. Radiance is just the ratio of power to etendue, so it must also be a conserved quantity.

4.4.1 Power

Conservation of power is the most intuitive of the conservation laws. Power may change form from optical to electrical to heat to mechanical motion, but as the late Jim Palmer taught in his classes at the University of Arizona: "Everybody has to be someplace." When power sources such as the Sun are included, the total power in a system is still conserved, though in this case the power entering the system must balance that leaving if the system is in thermal equilibrium with its surroundings (i.e., if its temperature is not changing).

An important distinction is that *optical* power is never conserved. That is, some light is reflected, some is transmitted, and some is absorbed to end up as heat (thermal power). What is transmitted to the detector is the incident power minus the losses (reflection and absorption). The process is summarized by Eq. (4.1), where conservation of optical power requires that there be no absorption ($A = 0$), or $T + R = 1$. Yet because the second law of thermodynamics guarantees that there will always be some absorption, it follows that optical power is never conserved. Vignetting is another example of how not to conserve power; in this case, low throughput is the inevitable consequence of poor design. In Secs. 4.4.2 and 4.4.3 to follow, we show that the optical quantities of etendue and radiance can be conserved—but only if we ignore absorption, reflections, and scattering and if the instrument is specifically designed to conserve them.

4.4.2 Etendue

The idea that etendue is conserved for extended (nonpoint) objects has acquired a nearly Kafkaesque quality of defeatism: "Sorry, etendue is conserved and there's nothing we can do to change it. The telescope design stays as it is." But we have already shown that the system designer can control etendue with a bigger aperture or a bigger field (by using a bigger detector for the same EFL or using a smaller EFL with the same detector). Clearly these options have costs, such as possible loss of image quality (associated with the bigger field), increases in size and weight (by adding more lenses in an attempt to keep diffraction-limited image quality), and so forth. None of these limitations are fundamental, however, so in what sense is etendue conserved?

Once an optical design is decided upon, then the etendue *can* be—but isn't necessarily—conserved at different planes inside the instrument. This is handy for calculations and also is conceptually useful when making system-level evaluations of feasibility. Designers not familiar with radiometry, for example, often try to invent an optical system where the image is brighter than the object. A back-of-the-envelope calculation of the etendue at the entrance pupil and image plane shows that this is not possible without injecting power into the image from elsewhere.

Specifically, at an aperture such as an entrance pupil, the etendue is calculated from the area of the pupil and the solid angle of the field [Fig. 4.18(a)]. At images, however, $A\Omega$ is more readily calculated from the area of the image and the solid angle of the cone of light determined by the exit pupil that creates the image [Fig. 4.18(b)]. If the telescope is designed correctly then, according to the conservation of etendue, the etendue at any pupil is the same as that at any image or at any plane between them.

This argument can be understood physically based on ideas introduced in Chap. 2. There it was shown that all of the power from the object that ends up in the image must first pass through the exit pupil. The exit pupil, then, can be thought of as a source with a known angular size when seen from the image. The image, in turn, can be thought of as an "aperture" that collects light over its area. The product of the exit pupil's solid angle and the area of the image is then physically the same as the product of the object's solid angle (that of the FOV) and the entrance pupil area.

That etendue is conserved can be shown mathematically by comparing its value at the object and entrance pupil with its value at the exit pupil and image planes. At the entrance pupil [Fig. 4.18(a)]

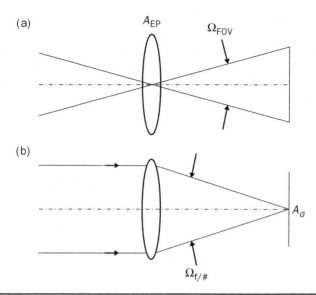

FIGURE 4.18 Etendue can be calculated using either (a) the entrance pupil area and detector solid angle or (b) the detector area and the solid angle of the cone angle illuminating the image. The detector FOV may be for entire FPA (as shown) or for a pixel IFOV, in which case the pixel area is used.

the etendue equals $A_{EP}\Omega_{FOV}$, where A_{EP} is the entrance pupil area and Ω_{FOV} is the solid angle of the object. At the image plane [Fig. 4.18(b)] the etendue equals $A_d\Omega_{f/\#}$, where A_d is the detector image area and $\Omega_{f/\#}$ is the solid angle of the exit pupil as seen from the image plane. Now the image area $A_d \approx (\theta f)^2 \approx \Omega_{FOV} f^2$ and, with $\Omega_{f/\#} \approx A_{EP}/f^2$, the etendue at the image plane is equal to the etendue at the exit pupil (within the limits of the approximations used).

Instead of the physical quantities of area and solid angle for etendue, some books use the one-dimensional equivalents of radius and angle. At a pupil, for example, the chief-ray angle u_c determines the FOV solid angle and the marginal-ray height y_m determines the aperture area. Therefore, the product of marginal-ray height and chief-ray angle ($u_c y_m$) is conserved. Similarly, the product of the chief-ray height at an image plane (which determines the image height) and the marginal-ray angle (which determines the solid cone angle illuminating the image) are conserved. More generally, the difference between these products is conserved: $u_c y_m - u_m y_c =$ constant, which is known as the *optical invariant* or the *Lagrange invariant*.[20]

When is the etendue or optical invariance not conserved? In addition to material absorption, most often this occurs when the optical designer does not take radiometry into account and thus makes an aperture or a detector too small to conserve them. This was illustrated in Sec. 4.1.3, where we observed that a too-small aperture results in a loss of power due to vignetting. The smallest etendue in the optical path is what controls the power that finally makes its way to the detector; in this sense, etendue is minimized rather than conserved. Reflections and scattering also ensure that etendue is not conserved—as we should expect for processes that transfer light into a range of solid angles that do not contribute to the image.

Example 4.6. Chapter 2 showed that an afocal telescope has an internal chief-ray angle (in image space) that is bigger than the external chief-ray angle (in object space) by a factor equal to the telescope magnification. This example takes a look at conserved etendue for such afocal systems.

The afocal telescope described in Example 2.9 has an EPD of 50 mm and a 7× magnification, giving an exit pupil diameter of 7.1 mm. The external FOV is 8 degrees, and the internal chief-ray angle at the exit pupil is 7 times greater (i.e., 56 degrees), a relatively large field for the eyepiece design to correct for aberrations. The etendue at the entrance pupil is $G = A_{EP}\Omega_{FOV} = 0.00196$ m² × $(0.1396$ rad$)^2 = 3.8 \times 10^{-5}$ m²·sr; at the exit pupil, it is $A_{XP}\Omega_{XP} = 0.00004$ m² × $(0.977$ rad$)^2 = 3.8 \times 10^{-5}$ m²·sr. Thus etendue is conserved with respect to the entrance and exit pupils.

Looking next at the intermediate image at the field stop, the etendue is based on the image area and the cone angle that illuminates it. For a square FPA, we have $A_i \approx (\theta f)^2 = (0.1396$ rad × 0.112 m$)^2 = 2.45 \times 10^{-4}$ m² and $\Omega_{f/\#} \approx A_{EP}/f^2 = 0.00196$ m²/(0.112 m)² = 0.156 sr; the etendue at the intermediate image is $A_i\Omega_p = 2.45 \times 10^{-4}$ m² × 0.156 sr $= 3.8 \times 10^{-5}$ m²·sr. Hence the etendue between entrance pupil and intermediate image is also conserved, as expected for a good design.

How can etendue end up not being conserved? Common causes are vignetting and stops that are too small. If a stop smaller than 7.1 mm were placed at the exit pupil, for example, then the area A_{XP} would be reduced—as would the etendue at that pupil. Even though the entrance pupil area and FOV have not changed, the etendue for the entire telescope would not be conserved because only the smallest etendue can make it through the system. The vignetting associated with making any of the optics (e.g., the eyepiece) smaller than necessary to accommodate the full FOV also reduces the etendue, as do material absorption, reflections, and scattering off the optical surfaces.

4.4.3 Radiance

Like the power associated with a wavefront, *radiance* propagates through a system. At the same time, it is also a geometric concept like etendue. Radiance is thus a combination of the two and is also conserved for extended objects. This is a bit difficult to accept at first: Is the radiance of the Sun measured from Mercury or Venus the same as it is measured from Earth? Is a telescope's magnified image of the Moon really no brighter than the Moon itself? Part of the difficulty here is that people often say "bright" when they mean "powerful"; the other part of the difficulty is conceptual.

We can use the Sun as an example to clean up the conceptual part. In propagating from the Sun to Earth, solar power spreads out over a larger and larger area. Just outside the atmosphere, the Sun has an irradiance of about 1370 W/m² and an angular diameter θ of 0.53 degrees (or a solid angle of 6.7×10^{-5} sr). The solar radiance at the Earth is then simply 1370 W/m² divided by 6.7×10^{-5} sr or about 2×10^7 W/m²·sr.

If the optical system is moved closer to the Sun by a factor of 1/2, then the irradiance increases by a factor of 4 (owing to $1/R^2$); at the same time, the Sun's solid angle also increases by a factor of 4 (owing to α^2). The net result is that the irradiance per solid angle is the same and so the radiance of the source is conserved. The radiance is the same if the system is close to the source or far away—at least within the limits of the paraxial approximations used.

Still, that statement is not correct under all conditions. Again using the Sun as an example, the solid angle becomes smaller when imaging through water (or any material with refractive index $n > 1$). The reason for the smaller angle is Snell's law, according to which the Sun has a smaller angular size. Because this size is smaller by a factor of n for each dimension, it follows that the solid angle is smaller by n^2. The Sun's radiance is thus larger by the same factor, so it's not really radiance that's conserved but, instead, a quantity known as *basic radiance*, or radiance divided by n^2. Similarly, it is not etendue per se but rather "basic" etendue ($n^2 A\Omega$) that is conserved in such situations.

4.5 Stray Light

Although most of this chapter has been devoted to understanding how power from "the" source ends up on the focal plane, this is not the full story of radiometry. In addition to light from the source of interest, there is always light coming from other sources besides the ones being measured. In obtaining images of a dim star, for example, light from the Moon or nearby cities has a number of ways of ending up on the focal plane. Such sources are known as "stray light," and they reduce the contrast of the star's image against its background. Similarly, the power measured by a radiometer may measure the star's power plus the lunar background, leading to an inaccurate measurement.

Stray light can come either from in-field or out-of-field sources. An obvious example is when the Moon is within the instrument's field and some moonlight finds its way onto the pixel where a star is located, rather than the pixels where the image of the Moon is. This is not predicted by geometrical optics but can be explained by a form of diffraction known as scattering. Another example is when moonlight is outside the instrument's FOV but can still makes its way through the entrance pupil, bounce around the walls of the instrument, and end up on the focal plane. This may or may not be predicted by geometrical optics, and this section reviews the design options for dealing with both in-field and out-of-field stray light.

4.5.1 Sunshades, Baffles, and Stops

Common sources of stray light are objects outside the instrument's field of view. These objects include the Sun, the Moon, and any other sources that are emitting or reflecting light. Because such light is outside the field determined by the lens EFL and the detector size, it is not *directly* incident on the detector. Nonetheless, light from these objects can still scatter as it transmits through the lenses, bounce off the interior walls of the mechanical structure holding the instrument together, and make its way onto the detector through indirect paths. Figure 4.19 illustrates how this can happen.

One way to keep this from happening is to block the out-of-field light from reaching the lens—that is, preventing it from ever making its way to the inside walls. Figure 4.20 shows the relevant geometry. The generic term for such a light blocker is a "sunshade," even if the source being blocked is not the Sun.

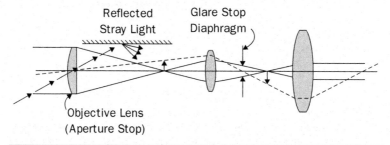

FIGURE 4.19 One source of stray light is an out-of-field source reflected off the walls of the mechanical structure that holds the optical system together. (*Credit: Warren J. Smith,* Modern Optical Engineering, *McGraw-Hill, www.mcgraw-hill.com.*)

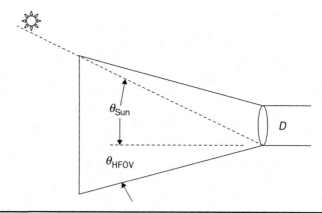

FIGURE 4.20 Sunshades prevent out-of-field light from illuminating the optics. The angle over which the shade blocks the source depends on the shade length, the system FOV, and the aperture diameter D.

Unless the sunshade extends all the way to the object, however, there is always the chance that some light will make its way into the instrument as stray light. A sunshade that long is not practical in most cases, so it is also common to coat or paint interior walls of the sunshade and the instrument with an absorbing layer that minimizes reflectivity. For visible and IR wavelengths, black anodize (such as Martin Black) or black paints are often used. These coatings must be applied with care, since most materials will continue to reflect fairly well at steep angles.

Figure 4.21 shows reflectance data for a black coating typically used in the MWIR and LWIR; the *Handbook of Optics* has a more complete collection of data for a variety of coatings over a range of wavelengths.[17] Although some of these coatings may not look black to the human eye, they are still "black" (i.e., absorbing) in nonvisible wavelengths such as UV or IR.

A variation on the sunshade theme is to use vanes (or "baffles") inside the sunshade or the instrument itself; these keep any light that does reflect and scatter off interior walls from making its way into the instrument's FOV. As shown in Fig. 4.22, the basic idea is to block the scattered light with the vane. If the vanes are designed properly then each one will also bounce some light backward toward the other vanes, and each bounce reduces the amount of light that can make its way into the aperture stop. Details concerning how deep to make the baffles—and how far apart they should be spaced—can be found in Wyatt[19] and Smith.[20]

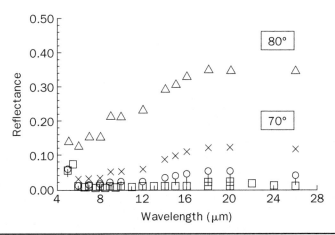

FIGURE 4.21 Infrared reflectivity of Martin Black coating for various incident angles. The coating remains relatively black and exhibits low reflectivity up to angles of about 60 degrees. (*Reprinted with permission from Persky.*[18] *Copyright 1999, American Institute of Physics.*)

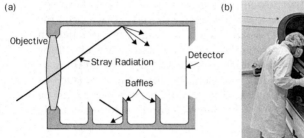

FIGURE 4.22 Stray radiation can be prevented from reaching the detector by *(a)* baffles distributed along interior walls or *(b)* an external sunshade. (*Credits: (a) Warren J. Smith, Modern Optical Engineering, McGraw-Hill, www.mcgraw-hill.com; (b) NASA, www.kepler.nasa.gov.*)

Baffles can also be used in the interior of the instrument, and they are often as simple as threads, grooves, or cones on the inside diameter (Fig. 4.23). Black paints or anodize are also used to lower the reflectivity of baffles. The disadvantage of using baffles is that the vanes introduce diffracted light, which must also be kept away from the detector.

Stops are very useful at keeping both diffracted and scattered light off the detector. A field stop at the intermediate image of an afocal Keplerian telescope, for example, restricts light from outside the field from entering the eyepiece. There are many optical designs where this is not always possible or practical—for instance, an imaging radiometer for which the field stop is the detector itself.

In these cases, a glare (or Lyot) stop may prove useful (Fig. 4.19). This is a stop located at an image of the aperture stop, such that that scattered and diffracted light that appears to be coming from outside the aperture will be imaged onto the stop's outer rim, preventing it from reaching the detector.[21] If these stops are not sized correctly then they will become the limiting apertures for the system etendue. Because a glare stop requires a lens after the aperture stop to effect the imaging, it is not always a realistic design option.

Example 4.7. Sunshades can easily become longer and heavier than expected. Recall from Fig. 4.20 that the length of the shade depends on the diameter of the aperture and the angle where light from the edge of the Sun just begins to hit the far edge of the stop. The geometry shows that $\tan \theta_{Sun} = D/L + \tan \theta_{HFOV}$ in Fig. 4.20, which for small angles is well approximated by $\theta_{Sun} \approx D/L + \theta_{HFOV}$.

It's straightforward to plot the length as a function of solar exclusion angle θ_{Sun}. This is shown in Fig. 4.24 for a telescope with a 100-mm aperture. For small angles, the length varies inversely with angle for small HFOV, with a 1-degree exclusion angle requiring a sunshade that's 6.4 meters long for an HFOV = 0.1 degrees. This is a large, massive structure to be hanging on the end of most instruments, though cutting back to a 10-degree exclusion angle would still

Radiometry 171

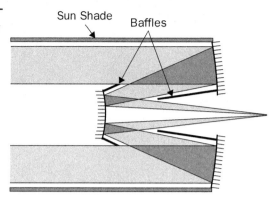

FIGURE 4.23 Both a sunshade and baffles/vanes can be used to prevent direct and scattered light from being transmitted into the system. (*Credit: Warren J. Smith, Modern Optical Engineering, McGraw-Hill, www.mcgraw-hill.com.*)

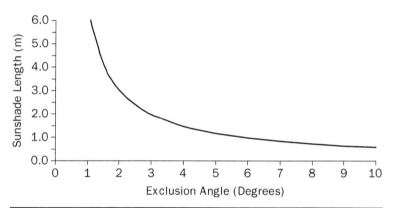

FIGURE 4.24 Sunshade length required to shade a 100-mm aperture and a 0.1-degree HFOV as a function of the exclusion angle of the source (θ_{Sun} in Fig. 4.20), ranging here from 1 to 10 degrees.

require a sunshade that's about 0.57 meters long. With a larger aperture, such as the meter-class telescopes common on NASA missions, a 5.7-meter sunshade is required for the same 10-degree angle.

The systems engineering decision of whether to use 10 degrees or some other number for the exclusion angle depends not only on the size and weight of the shade but also on how much light the optics are expected to scatter into the field of view (see Sec. 4.5.3).

4.5.2 Ghosting

Another source of stray light is multiple reflections of a bright source (Fig. 4.5). This occurs when there are multiple optics in the design and so multiple reflections within or between the elements end up heading back toward the detector (Fig. 4.25). This phenomenon is

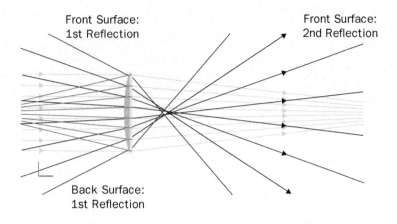

FIGURE 4.25 Second and higher-order reflections from the front surface of a lens can be reflected back in to the instrument, creating ghost images or diffuse background illumination. (*Credit: Photon Engineering LLC.*)

occasionally seen in camera photos, where the picture includes a mysterious "ghost" or "flaring" of the Sun even though it was nowhere in the field of view.

It is difficult to remove ghost images completely, they can be minimized by following a few design rules. The first is to use antireflection coatings on all optical surfaces. We saw in Sec. 4.1.2 that an uncoated surface with an index of 1.5 reflects about 4 percent of the incident light, of which the second surface reflects 4 percent back toward the detector. With a bright source, this reflection can easily be picked up by the detector as a ghost. Depending on the source brightness, an AR coating that reduces the inherent Fresnel reflections of each optic is an effective and usually inexpensive option for reducing this source of stray light.

If one of the optical elements is a flat then there is a second option, which is shown in Fig. 4.26. The idea is to tilt the optic by a small angle; this doesn't change its power but it does redirect the reflections away from the detector. It's also possible to change the curvature of individual lenses and mirrors to make sure that the reflections are not reimaged. Sometimes index-matching fluids are also used to reduce ghosting; however, making sure that the fluid is contained and does not contaminate other parts of the instrument adds quite a bit of cost to the design.

4.5.3 Scattering

Even with sunshades, baffles, stops, and coatings, it is still possible for light outside the FOV to hit the first optic in the design and then scatter into the field. This is rather amazing because it entails a completely different mechanism than light passing through the lens and bouncing

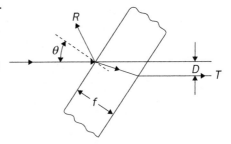

FIGURE 4.26 An angled flat in the optical path can prevent back-reflections from reaching the detector. (*Adapted from Warren J. Smith, Modern Optical Engineering, McGraw-Hill, www.mcgraw-hill.com.*)

off the walls of the instrument. The complicating factor added in this section is that light incident on an optic from outside the FOV can also end up on the detector via optical scattering. Such scattering explains why a laser beam lights up a room: light scatters off small particles of dust in the air, sending the photons in all directions.

In the case of such elements as lenses, mirrors, and beamsplitters, scattering also occurs not only when the optics are "dusty" but also in response to rough surfaces, scratches, dig marks, and so forth. For example, if the surface of a window has that dusty, rugged "Western" look, then light doesn't just bounce off or go through the window at the anticipated specular angle. Some of it also comes off the surface at "crazy" (nonspecular) angles that don't follow the laws of reflection and refraction, and the resulting diffuse light is known as *scattering*. Chapter 3 showed how diffraction sends light in directions not predicted by the laws of reflection or refraction; scattering is a type of diffraction, although it is sometimes referred to as a separate phenomenon.

Scattering affects the performance of most optical systems, including imagers and radiometers. An image that needs a good modulation transfer function, for example, may not meet its requirements if stray light from outside the field is scattered into the image; such "veiling glare" reduces the contrast between light and dark and thereby makes extremely difficult the discovery of extrasolar planets against a bright-star background. Similarly, if a radiometer needs to make highly accurate measurements of an object's irradiance, then any scattered light that's not coming directly from the object will be misread as if it were.

Figure 4.17 shows how light scatters off a Lambertian surface. The polar plot illustrates how the intensity varies with the cosine of the off-specular angle: peaking in the direction of specular reflection and dropping to zero at 90 degrees. As was shown in Sec. 4.3.2, a unique property of Lambertian scattering is that the radiance is constant because the off-axis drop in intensity is exactly compensated by the smaller projected emitting area $[L = I/A = (I_o \cos \theta)/(A_o \cos \theta) = I_o/A_o =$ constant$]$. The reflected irradiance is therefore scattered equally in all directions such that $L = \rho E/\pi$.

The sections that follow demonstrate calculating how much light scatters from both Lambertian and non-Lambertian surfaces at a given angle, surface roughness, wavelength, and level of cleanliness (absence of dust and particles).

Total Integrated Scatter

The first criterion for "how much" light is the total power reflected as scatter into all nonspecular angles. Known as total integrated scatter (TIS), this metric is the fraction of reflected light that is not specular, and computed as the difference between how much power reflects into the specular angle and the total power that reflects off the surface. This difference is then normalized by dividing by the total reflected power; hence the TIS doesn't depend on the initial power, and optics can be compared under different conditions:[22,23]

$$\text{TIS} = \frac{P_o - P}{P_o} = 1 - \exp\left\{-\left(\frac{2\pi \times \Delta n \times \sigma}{\lambda}\right)^2\right\} \approx \left(\frac{2\pi \times \Delta n \times \sigma}{\lambda}\right)^2 \quad (4.11)$$

Here P is the power reflected into the specular angle and P_o is the total reflected power, so the total scatter does not include the specular power (TIS = 0 for $P = P_o$).

The last expression for TIS is not self-evident but does have an important physical interpretation. The Δn term shows that the TIS increases as the index difference between the optic and its surroundings increases, behavior that is consistent with Fresnel reflections [note that $\Delta n = 1 - (-1) = 2$ for mirrors]. Equation (4.11) also shows that TIS depends on the surface roughness as compared with the optical wavelength λ; thus, scatter is low when σ/λ is small.

As illustrated in Fig. 4.27, the surface roughness is the RMS depth σ of the surface irregularities due to polishing (the effective depth decreases in proportion to cos θ). Roughness is measured in Angstroms (Å); typical numbers are 100 to 200Å for an off-the-shelf optic and 10 to 20Å for highly polished optics. Values as small as 5Å

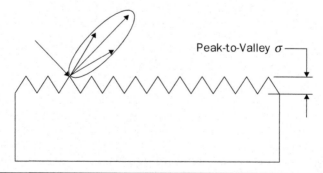

Figure 4.27 As summarized by Eq. (4.11), the total integrated scatter (TIS) is determined by the surface roughness σ and the wavelength λ.

are possible for super-polished surfaces, which are extremely expensive and should be used only when absolutely necessary. Depending on stray light requirements, the depth in terms of the number of optical waves must be small (< 1/100) to avoid significant scatter. For example, a mirror with $\sigma = 50$Å at $\lambda = 1$ µm has a TIS ≈ 0.4 percent (–24 dB) at normal incidence. An ultraviolet instrument designed for shorter wavelengths will exhibit even more scatter unless a finer polish is specified.

Although TIS is a useful starting point for the study of scattering, it is often necessary to know how the scatter varies with angle—especially near the specular angle. For example, it is often necessary to calculate the light scattered into an instrument's FOV from an out-of-field object such as the Sun, a task for which the TIS is not suited. Furthermore, Eq. (4.11) does not include the effects of scattering due to cleanliness factors. The next section introduces a new concept which circumvents these limitations.

Bidirectional Scatter Distribution Function

In Sec. 4.3.2 we saw how a diffuse surface can scatter light into angles not predicted by the laws of reflection and refraction. Here we quantify how much light can be scattered as well as the variables that affect it. A Lambertian surface, for example, has a radiance that's the same in all directions, and Example 4.5 shows that the scattered radiance equals the reflected spectral irradiance (ρE_λ) divided by a π-steradian projected solid angle. However, very few surfaces are Lambertian, so in this section we calculate the scattered radiance for non-Lambertian surfaces.

The calculation starts with measurements of scattered radiance for a given irradiance incident on a surface. For Lambertian surfaces, the value is the same for any off-axis scatter angle; but for surfaces that are not Lambertian, the value depends on the angle from specular and on the angle of incidence.[24] A plot of how this radiance-to-irradiance ratio (in units of 1/sr) depends on the off-axis angle is the *bidirectional scatter distribution function*, or BSDF. Figure 4.28 shows typical BSDF curves for Lambertian and non-Lambertian surfaces.

Since the BSDF represents the scattered radiance per unit irradiance, it follows that multiplying BSDF by spectral irradiance (BSDF × E_λ) gives the spectral radiance L_λ. We can now incorporate the math from Eq. (4.9) to find the power (Φ_s) scattered into the FOV and onto the FPA. Writing everything out yields:

$$\Phi_s = \text{BSDF} \times E_\lambda A_p \Omega_p \Delta\lambda \quad [\text{W}] \quad (4.12)$$

Here the reflectivity $\rho(\lambda)$ is implied in Eq. (4.12) through the BSDF for a reflecting surface; an additional assumption is that the BSDF is approximately constant over the solid angle Ω of a pixel or the FPA. The scattering surface thus acts as an optical source with a constant

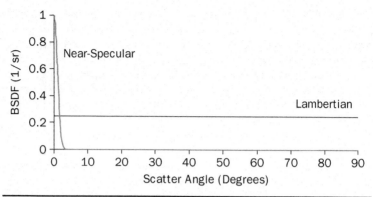

FIGURE 4.28 The BSDF quantifies the scattered radiance per incident irradiance; this value is constant for Lambertian surfaces, which scatter in all directions to give a constant radiance, but it peaks near the specular reflection angle for clean, highly polished surfaces. At a given wavelength, the constant for a Lambertian surface is BSDF = $L/E = [\rho(\lambda)E/\pi]/E = \rho(\lambda)/\pi$.

radiance for a Lambertian surface (for which the BSDF is constant); for a non-Lambertian surface, it behaves as a source with a radiance that varies with wavelength and with the incident and scatter angles.

In physical terms, the BSDF depends not only on the surface roughness of the optic but also on how clean it is. A rough surface (compared with the wavelength) scatters more light, as does a dirty optic with lots of dust particles that cause more scatter. Both are extremely important effects and are often underestimated with a cavalier, "cowboy" mentality. Figure 4.29 plots the distribution function versus the off-specular scatter angle at four different levels of surface cleanliness. Example 4.8 shows how these curves can be used with Eq. (4.12) to calculate the scattered power.

Example 4.8. Laser-based instruments such as laser radar systems and fluorescence microscopes are difficult to design because the same telescope that's used to transmit a relatively large amount of power is also used to collect nanowatts of reflected signal. As a result, it's critical to control stray light, including reflections (from the telescope mirrors) that are scattered backward toward a highly sensitive detector.

To reduce scatter, an off-axis reflective Keplerian (Mersenne) telescope can be used to take advantage of: (1) the drop-off in BSDF with angle; (2) an image of the aperture stop (the primary mirror) created by the secondary, where a glare stop can be placed; and (3) a real intermediate image to locate a field stop [Fig. 4.30(a)]. A common alternative for reducing scatter is the three-mirror anastigmat, which gives higher-quality, low-astigmatism images; however, this method depends on the afocal magnification (maximum $M \approx 10\times$) and allowable aberrations over the field.[19] This example uses a two-mirror Mersenne telescope to illustrate the stray-light problem. The nominal specs are a 50-mm entrance pupil, a 0.2-mrad field of illumination equal to the FOV in object space, and a 5× afocal magnification.

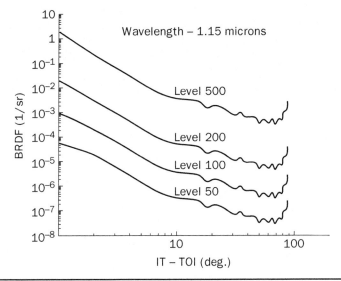

FIGURE 4.29 Typical dependence of the BSDF—in the case of mirrors, known as the bidirectional *reflectance* distribution function (BRDF)—on the off-specular scatter angle defined by IT − TOI for different cleanliness levels. The smaller numbers represent cleaner surfaces with fewer (and smaller) particles that can scatter light. (*Credit: Spyak and Wolfe.*[25])

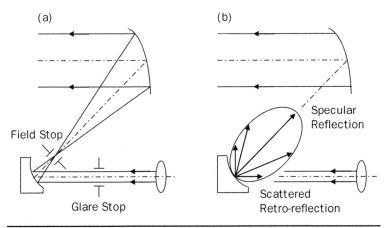

FIGURE 4.30 The reflective Keplerian (or Mersenne) telescope allows for the use of a field stop and glare stop; the off-axis design also reduces the retro-reflected components of the scattered light.

Figure 4.30(b) shows how the transmitted light can be reflected backward into the detector FOV. The transmitted laser beam scatters off the secondary mirror at a 45-degree angle back toward the primary. Because of scatter, some of the beam also reflects back to the detector. If too much light works its way back in this fashion, then the scatter will swamp out the weak signal returning from the sample.

Equation (4.12) estimates how much power is scattered backward from the secondary mirror. Using Fig. 4.29 for the BSDF of the secondary, we have BSDF $\approx 10^{-3}$/sr at an off-specular angle of 45 degrees for a Level 500 (CL500) mirror; this is assumed to be constant over the small FOV of the collecting optic. Other numbers needed for the equation are the solid angle of the collecting optics in image space (internal FOV = M × external FOV = 5 × 0.2 mrad, giving $\Omega = 7.85 \times 10^{-7}$ sr), the area of the beam on the scattering optic [$A \approx \pi(0.05 \text{ m}/5)^2/4 = 7.85 \times 10^{-5} \text{ m}^2$], and the irradiance of the laser on the scattering optic ($E = E_\lambda \times \Delta\lambda \approx 1300$ W/m^2 for a 1-mW laser; see Chap. 5). Substituting into Eq. (4.12), we find that the scattered power $\Phi_s \approx 1300$ W/m^2 × 10^{-3}/sr × 7.85×10^{-5} m^2 × 7.85×10^{-7} sr = 8×10^{-11} W. Other surfaces, including the primary mirror, will also contribute to additional back-scatter on the FPA. The calculation in this example is relatively straightforward in that the BSDF is approximately constant over the small FOV—as required by Eq. (4.12).

If the laser power is scaled up to 1 W, then the scattered power also increases by a factor of 1000 to 0.08 mW, an easily detectable quantity. Is this "too much" power, requiring a cleaner, more finely polished mirror? To answer this question, it's necessary to compare the scattered power with the return signal and with the detector sensitivity, topics that will be covered in Chap. 6.

Although it's comforting to have an equation into which we can plug numbers, the scattered power predicted by Eq. (4.12) is the least accurate of all the equations in this book. The reason is that theoretical estimates of BSDFs can be off by as much as a factor of 10 or so, and no one's explanation as to why is any better than anyone else's. As a result, unless the BSDF is measured for the actual optic that will be used under the same cleanliness conditions for the final system, the result obtained from Eq. (4.12) is accurate at best to an order of magnitude.

Despite their potential inaccuracies, the equations in this section do capture the physical trends that reveal how to reduce stray light and scattering. The design options to be considered for reducing stray light are summarized as follows:

- Sunshades and baffles
- Paints or anodize on interior walls, low-reflectivity "black" for the design wavelengths
- Field and glare stops
- Off-axis optics to reduce the value of the BSDF [cf. Eq. (4.12)]
- A small IFOV [Ω in Eq. (4.12)] so that stray light can be filtered spatially
- A narrowband optical filter so that stray light can be filtered spectrally [affecting $\Delta\lambda$ in Eq. (4.12)]

- Smoothly polished optics to reduce the surface roughness [affecting BSDF in Eq. (4.12)]
- Clean optics to reduce the scatter from dust particles [affecting BSDF in Eq. (4.12)]

In all cases, back-of-the-envelope and order-of-magnitude calculations using the equations developed in this chapter are the first tools for evaluating these options. Then, for a more detailed analysis, one can employ stray-light software packages such as ASAP, FRED, or ZEMAX. These design options may not all be feasible and in any case must be traded against other aspects of the system design and against each other. For example, reducing the aperture would reduce stray light but would also reduce the signal to be measured. This trade-off between signal and background noise (from scattering and other sources) will be examined in more detail in Chap. 6.

Problems

4.1 Why does spherical wavefront divergence result in an inverse square ($1/R2$) law to describe the power (or energy) density incident on a surface?

4.2 A solar cell is typically made from a semiconductor, such as silicon, that has a refractive index of $n \approx 3.5$ over the solar band. What fraction of incident light reaches the cell to generate electricity? In principle, these cells can be antireflection coated to reduce Fresnel loss and improve efficiency. What are the practical limitations on such an improvement? *Hint:* Compare the cost of AR-coated solar cells with the conventional methods of obtaining electricity.

4.3 An expression that is sometimes used to find the minimum mechanical radius to avoid vignetting is $|y_c| + |y_m|$, or the sum of the absolute values of the chief- and marginal-ray heights. For the f/1 lens and 10-mm detector in Sec. 4.1.3, what is the slope of this sum on the image side of the lens? How does this compare with the slope found using $u' = u - y\Phi$? How does it compare with the slope found in Sec. 4.1.3?

4.4 What is the sum of the absolute values of the chief- and marginal-ray heights on the object side of a thin lens for an object at infinity? What's the physical meaning of the sum in this case?

4.5 When used to create an image of the Sun, what is it that a "magnifying" glass is magnifying?

4.6 In Sec. 4.2, the irradiance for an f/1 lens is 4 times greater than an f/2 for an object at infinity. Does this conclusion change for objects that are not at infinity? Place the object at 2 times the focal length for your first set of calculations; for your second set, place it at 4 times the focal length.

4.7 The t-number (t/#) is occasionally used instead of the f/# to describe relative irradiance on an image plane; it is defined as the f/# divided by the square root of the throughput ($t/\# = f/\# \div \sqrt{T}$). Why divide by the throughput to quantify relative irradiance? Why use the square-root of the throughput?

4.8 In Sec. 4.2, the net result of a longer distance and smaller projected area is that the off-axis irradiance drops off proportionally to $\cos^4(\theta)$ for area sources. What if the object is a point source? *Hint:* Does the emission area of a point source change as it is viewed off-axis?

4.9 Equation (4.7) is based in part on the following approximation: $\Omega = 2\pi[1 - \cos(\alpha)] \approx \pi \sin^2(\alpha)$. Plot the approximate form and the exact form of the solid angle versus α for $\alpha = 0$ to 180 degrees.

4.10 How big a magnifying glass would be needed to ignite a piece of paper using a full Moon as the source? *Hint:* Assume the Moon is Lambertian with an average solar reflectivity of 17 percent over all wavelengths.

4.11 Are the words "radiance" and "irradiance" related in the same way as the words "relevant" and "irrelevant"?

4.12 Is intensity conserved? What about spectral radiance? Why isn't irradiance conserved?

4.13 For a given $f/\#$, do smaller pixels collect more or less light? Can the $f/\#$ be changed to increase the amount of light collected? If the pixel size is halved so that it has one-quarter the area, how much would the $f/\#$ need to change in order to maintain the same amount of light collected?

4.14 For a given aperture, does a smaller IFOV collect more or less light? Can the aperture be changed to increase the amount of light collected? If the IFOV is halved, then how much would the aperture need to change in order to maintain the same amount of light collected?

4.15 Would an AR coating of many layers be expected to have more scatter than one of only a few layers?

4.16 A full Moon is located 3 degrees away from a star that is being imaged with a SWIR-wavelength camera that features an aperture of 50 mm, a HFOV of 2 degrees, and a pixel IFOV of 1 μrad. How much light from the Moon is scattered into the FOV of the star? Assume that the surface roughness of the imaging lens is 20Å, its cleanliness level is CL500, and the star is imaged onto 1 (on-axis) pixel.

4.17 Example 4.8 calculated the scattered power for a Mersenne design with a 45-degree angle. Will more or less power be scattered for a design with a 30-degree off-axis angle? Using Fig. 4.29 for the BSDF, plot the scattered power as a function of design angle for a range of 10 to 60 degrees.

4.18 Design a telescope that can image the second-brightest star in the Northern Hemisphere at noon on a sunny day. You will need to look up the intensity of that star (not the Sun) as well as the spectral radiance of the sky. Also assume that the detector requires 1 nW of optical power to produce a reasonable signal. How big and how long does the telescope need to be? How narrow a spectral filter must it have? Are the numbers found for EPD, focal length, and filter band-pass the only answers, or can these variables be traded against each other to come up with different answers?

Notes and References

1. Jim Breckinridge and Chris Lindensmith, "The Astronomical Search for Origins," *Optics and Photonics News*, February 2005, pp. 24–29.
2. Barrie W. Jones, *Life in the Solar System and Beyond*, Springer-Verlag, www.springeronline.com, 2004.
3. In creating an image of the Sun, what's being increased ("magnified") by such a lens is the irradiance, not the brightness (see Sec. 4.2).
4. Eugene Hecht, *Optics*, 4th ed., Addison-Wesley (www.aw.com), 2001.
5. The plot for the external transmission T includes the Fresnel reflection losses for two identical surfaces; the transmission is therefore referred to as "index limited."
6. Fused silica is also known as fused quartz. Both have the same basic composition (silicon dioxide, or SiO_2) but undergo different processing from crystalline quartz, whose birefringence makes it suitable for polarization components. Cleartran™ is a version of ZnS that transmits with low loss in the visible.
7. H. Angus MacLeod, *Thin-Film Optical Filters*, 3rd ed., Institute of Physics (www.iop.org), 2001.
8. James D. Rancourt, *Optical Thin Films: User Handbook*, SPIE Press (www.spie.org), 1996.
9. Philip W. Baumeister, *Optical Coating Technology*, SPIE Press (www.spie.org), 2004.
10. The term E_λ indicates a quantity per unit wavelength (e.g., $W/m^2 \cdot \mu m$), and $E(\lambda)$ is W/m^2 as a function of wavelength; a quantity such as $E_\lambda(\lambda)$ is both.
11. D. Kryskowski and G. H. Suits, "Natural Sources," in G. J. Zissia (Ed.), *Infrared and Electro-Optic Systems Handbook*, SPIE Press (www.spie.org), 1993, chap. 3.
12. The mathematical explanation is that E_λ is approximately constant over the filter bandwidth $\Delta\lambda$; hence it can be pulled out of an integral to yield E_λ times $\Delta\lambda$.
13. Etendue is a word of French origin.
14. W. H. Steel, "Luminosity, Throughput, or Etendue?" *Applied Optics*, 13(4), 1974, pp. 704–705.
15. It was shown in Chap. 2 that all of the power from the object that ends up in the image must pass through the exit pupil. As a result, the exit pupil can be thought of as a source with a paraxial angular size $\theta \approx D_{xp}/L_{xp}$ when seen from the image.
16. H. J. Christian, R. J. Blakeslee, and S. J. Goodman, "The Detection of Lightning from Geostationary Orbit," *Journal of Geophysical Research*, September 1989, pp. 13329–13337.

17. S. M. Pompea and R. P. Breault, "Black Surfaces for Optical Systems," in M. Bass, E. W. Van Stryland, D. R. Williams, and W. L. Wolfe (Eds.), *Handbook of Optics*, 2nd ed., vol. 2, McGraw-Hill (www.mcgraw-hill.com), 1995, chap. 37.
18. M. J. Persky, "Review of Black Surfaces for Space-Borne Infrared Systems," *Review of Scientific Instruments*, 70(5), 1999, pp. 2193–2217.
19. Clair L. Wyatt, *Electro-Optical System Design*, McGraw-Hill (www.mcgraw-hill.com), 1991.
20. Warren J. Smith, *Modern Optical Engineering*, 4th ed., McGraw-Hill (www.mcgraw-hill.com), 2008.
21. Robert E. Fischer, B. Tadic-Galeb, and Paul Yoder, *Optical System Design*, 2nd ed., McGraw-Hill (www.mcgraw-hill.com), 2008.
22. Harold E. Bennett, "Optical Microroughness and Light Scattering in Mirrors," in *Encyclopedia of Optical Engineering*, Marcel Dekker, 2003, pp. 1730–1737.
23. John C. Stover, *Optical Scattering: Measurement and Analysis*, 2nd ed., SPIE Press (www.spie.org), 1995.
24. Ed Friedman and John L. Miller, *Photonics Rules of Thumb*, 2nd ed., McGraw-Hill (www.mcgraw-hill.com), 2004, chaps. 5, 13–14, 16–17.
25. Paul R. Spyak and William L. Wolfe, "Scatter from Particulate-Contaminated Mirrors. Part 4: Properties of Scatter from Dust for Visible to Far-Infrared Wavelengths," *Optical Engineering*, 31(8), 1992, pp. 1775–1784.

CHAPTER 5
Optical Sources

The nighttime view of the planet shows large areas of land lit up by Thomas Edison's invention. With most of the land mass in the Northern Hemisphere, Fig. 5.1 shows a scene as captivating as the images captured by Hubble and the other outward-looking space telescopes. The photo also reveals the presence of a natural source—namely, fire emitting in the visible and NIR spectra. The IR emissions from such thermal sources, a result of an object's temperature and transparency, can also reveal many properties of a scene not visible to the human eye (Fig. 5.2).

What for some is a necessity is for others a nuisance. Astronomers using ground-based telescopes put a great amount of effort into *avoiding* human sources of light. For example, streetlights in the cities of Tucson and Flagstaff (both in Arizona) are specifically selected to minimize emission of the wavelengths at which astronomers observe. As was shown in Chap. 4, unwanted sources—including normal streetlights—can be detected as stray light even when the telescopes are not pointed directly at them, thereby "washing out" the stars. The unwanted light could be filtered out, but doing so leads to throughput loss that limits visibility of the most distant stars.

In addition to streetlights, conventional types of human-made light sources are everywhere. Incandescent bulbs are commonly used at home, for example, while the more efficient fluorescent lamps buzz away at the office. Light-emitting diode (LED) lamps are quietly becoming more common, with applications ranging from reading lamps to solar-powered outdoor nightlights. Photography is also a lot more difficult when the Sun has gone down for the day, in which case flashbulbs make the job quite a bit easier for close-ups.

Beyond the use of sources for general lighting and illumination,[1] liquid-crystal display (LCD) and plasma flat-panel TVs allow viewing of the Discovery Channel and LED TVs are used for marketing or outdoor events requiring *very* large screens. As a result, what's happening in Vegas can be rebroadcast to the rest of the world via large LED displays. Video projectors are yet another method of creating large images; these have also been developed with LEDs as the source, yielding an improvement over the short lifetime, large size, and inefficiencies of arc lamps.

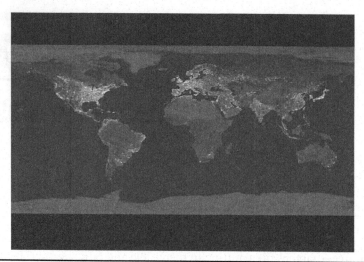

Figure 5.1 Composite image showing the planetary distribution of nighttime lights emitting visible and near-IR wavelengths. (*Credit: NASA, www.nasa.gov.*) (*See also color insert.*)

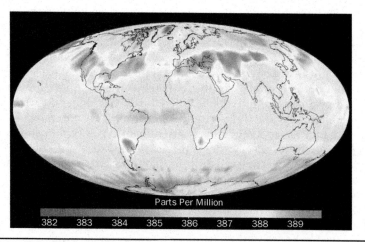

Figure 5.2 Composite image showing the planetary distribution of carbon dioxide as inferred from the atmospheric emission of MWIR and LWIR wavelengths. (*Credit: NASA Jet Propulsion Lab, http://airs.jpl.nasa.gov.*) (*See also color insert.*)

Although natural sources of light such as the Sun and Moon are rarely seen by many who go to Vegas, they play an important role for the rest of us. At one time the reflected sunlight from the harvest Moon gave farmers more time to gather their crops; now it serves mostly as a guide on nighttime hikes. Of course, the Sun and other stars create their own light, and their color is based not on reflection but on their temperature. Red stars, for example, are much cooler than yellow-white stars such as the Sun. Even room-temperature objects that seem only to be reflecting light are also emitting LWIR radiation that's not visible to the human eye, and this phenomenon can actually complicate the design of scientific and industrial instruments.

These instruments include optical systems such as imagers, radiometers, and spectrometers, where the mechanical structure and even the optics themselves are emitting large quantities of MWIR and LWIR radiation. Clearly, this will make the problem of measuring IR radiation from an external source that much more difficult. These infrared "sources" are everywhere in the form of electromagnetic radiation that may be a useful signal or may simply be background noise. Solutions to this problem include cooling the optics and choosing optics that have low emissivity; these concepts are examined in Sec. 5.1.

Finally, there are a number of nonconventional applications for natural light sources. Ultraviolet (UV) light is used as a water purifier and germicidal agent for pools, reducing the use of poisonous chlorine in the backyard. Using the Sun itself to generate electricity is still considered unusual, though it's becoming more common as a way to avoid the greenhouse gases associated with traditional power generation.

In this chapter, the various types of optical sources are reviewed. They are discussed in Sec. 5.1 with the goal of providing some basic familiarity with different source technologies. There are thousands of human-made sources of light radiation from which to choose, but the most common are categorized into five types: blackbodies, tungsten-halogens, arc lamps, light-emitting diodes, and lasers. This chapter discusses each of these types as well as how to choose among them (Sec. 5.4); more specialized sources—such as high-intensity discharge (HID), plasma displays, and electroluminescent displays (ELDs)—are not reviewed. An important part of using sources is understanding not only the available types but also how to use them when designing an optical system (Sec. 5.2) and what performance can be expected from them (Sec. 5.3).

There is much interest in developing new light sources that conserve energy, with high-brightness LEDs (HB-LEDs) and compact fluorescent lamps (CFLs) the leading contenders at this point. Companies such as General Electric, Osram Sylvania, Philips, and Cree are spending a large amount of research money on the development of high-efficiency lamps. In response to the U.S. Federal Energy bill of 2007, for instance, Philips introduced an incandescent bulb that is 30 percent more efficient than conventional bulbs. Even in the realm of conventional commercial products, optical source design is a surprisingly sophisticated endeavor. For example, GE Research Labs has generated thousands of patents and products over the years that target higher efficiency, lower cost, more brightness, smaller size, or any other options that system designers look for in a source, regardless of type.

5.1 Source Types

There are several different ways to classify sources, but the most useful from an engineering perspective is in terms of available products—an approach necessitated by the important task of source selection. In contrast, the conventional academic perspective is to categorize sources in terms of thermal versus quantum (or continuum versus line). That approach is certainly valid, but it's of little use to a systems engineer developing optical instruments.

What the designer does find useful is classifying light sources by available products: blackbodies and tungsten-halogen lamps (both thermal sources), arc lamps (mostly thermal but partly quantum), light-emitting diodes (quantum), and lasers (also quantum). All sources create light from electron motion. Thermal sources generate broad-spectrum light from oscillating electrons that have been heated and thus have a wide range of energies, whereas quantum sources generate light from electrons that give off electromagnetic energy in a more focused, somewhat digital or "quantum" manner corresponding to a much smaller range of electron energies.

The intent of this section is not to be exhaustive but rather to provide an overview of the most important engineering properties of the most common types of sources (Table 5.1). More systematic reviews of sources can be found in the *Handbook of Optics*.[2]

Wavelengths	Abbreviation	Source Type
0.2–0.4 μm	UV	Deuterium, Xe arc, Hg arc
0.4–0.7 μm	VIS	Halogen, LED, Hg arc, Xe arc
0.7–1 μm	NIR	Halogen, LED, Xe arc
1–3 μm	SWIR	Halogen, LED, Xe arc
3–5 μm	MWIR	Blackbody
8–12 μm	LWIR	Blackbody
12–30 μm	VLWIR	Blackbody

TABLE 5.1 Source types commonly used to obtain optical power within specific wavelength bands.

5.1.1 Blackbodies

Things not normally thought of as sources of photons are, in fact, everywhere. All room-temperature objects continuously emit radiation at a peak wavelength of about 10 µm—a property that allows thermal cameras to work. Even sources that are not generating internal heat are emitting long-wavelength infrared radiation, which is invisible to the human eye but obvious to a detector that is sensitive to those wavelengths.

The reason for this radiation is that anything whose temperature is greater than absolute zero—which is everything—emits electromagnetic radiation when its electrons oscillate. As temperatures increase, the higher-energy electrons oscillate faster and emit radiation of higher frequency and shorter wavelength; such sources are classified as thermal. At high enough temperatures, the radiation includes wavelengths that are "optical." At about 240 K, the radiation peaks at an LWIR wavelength of approximately 12 µm, and there is also a significant amount of energy at shorter wavelengths.

The infrared spectrum is often observed as heat; in fact, IR radiation was discovered by Newton when the thermometer he placed outside a visible rainbow (that he had created with sunlight and a prism) became hotter. This eventually led to the idea of electromagnetic radiation as a form of heat, and the reflective jackets used by firefighters are actually an optical method of protecting skin and clothing from overexposure to the intense radiation produced by a hot fire.

Thermal sources are also called *blackbodies*. This may seem a bit strange given that one can clearly see that the Sun is yellow-white and that campfires are red, orange, and yellow (and sometimes even have a touch of green). The explanation behind this strange terminology is that objects that *absorb* the visible spectrum appear black to the human eye. We shall see that an object that absorbs well also emits efficiently, so a good absorber in the visible is an efficient thermal source when it is hot enough to emit in the visible. Some materials absorb infrared wavelengths, and these make good IR thermal sources; these sources may also absorb well in the visible, making them appear black to the human eye.

Manufactured versions of thermal sources provide some degree of control over emitted power and wavelengths. Incandescent and tungsten-halogen bulbs are both thermal sources, as are the low-temperature calibration sources (e.g., laboratory blackbodies) used when stability and uniformity are important. In this section we review the basic physics of blackbodies, including the total emitted (radiated) power, spectral distribution (exitance), and peak wavelength in the spectrum (Wien's law), as well as absorption and emission (emissivity). Also discussed are some of the engineering issues involved in designing blackbody sources.

Radiated Power

Everything gives off energy in the form of electromagnetic waves. How quickly the energy is emitted—that is, the radiated power Φ—depends on the object's temperature and surface area as follows:[3]

$$\Phi = \varepsilon \sigma A T^4 \quad [\text{W}] \tag{5.1}$$

Here T is the absolute temperature[4] in degrees Kelvin (K) and A is the emitting surface area. The constant σ is the Stefan–Boltzmann constant, whose numerical value is 5.67×10^{-8} W/m²·K⁴. The other "constant" in this equation (the emissivity ε) actually depends on the surface properties. For now, we assume that emissivity $\varepsilon = 1$ over all wavelengths—the definition of a "blackbody" source. More typical emissivity values are between 0 and 1; these values characterize a "graybody" source.

An example of how much power is emitted as radiation will prove instructive. One that is often used is that of the human body at room temperature; its approximate dimensions are 2 meters by 0.6 meters by 0.3 meters, giving a surface area $A \approx 4$ m². Substituting into Eq. (5.1), we calculate the emitted power as $\Phi = 5.67 \times 10^{-8}$ W/m²·K⁴ × 4 m² × (300)⁴ K⁴ ≈ 1.84 kW at room temperature if human skin is a blackbody. This is a stunning amount of power, leading to the question: Why can't this power be used for, say, baking brownies? The reason is that almost everything else on the planet is also at 300 K and it's a temperature *difference* that's required for useful things such as baking.

Spectral Exitance

As shown in Chap. 4, the total power emitted from a source is not the whole story. Power that is radiated as electromagnetic waves from a thermal source is emitted over a wide range of wavelengths. Depending on the temperature, some of these wavelengths are optical (as defined in Chap. 1). In perhaps the most useful physics of the 20th century, Maxwell Planck showed that the wavelength distribution of emitted power is best described by a spectral irradiance:[3]

$$M_\lambda = \frac{2\pi\varepsilon hc^2}{\lambda^5} \frac{1}{\exp\{E_p/kT\} - 1} \quad [\text{W/m}^2 \cdot \text{m}] \tag{5.2}$$

(the new symbols will be identified in paragraphs that follow). Because this spectral irradiance is emitted, its name has been changed by the radiometry committees to "spectral exitance" and denoted M_λ (see Table 4.4). Even so, many manufacturers continue to use the word irradiance in their catalogs, as the units of power per unit area per meter (W/m²·m) or power per unit area per micron (W/m²·μm) have the same physical meaning.

Equation (5.2) is plotted in Fig. 5.3 for various blackbody temperatures; the plot shows that blackbodies emit over a wide band (or spectrum) of wavelengths and that the spectrum becomes broader as the temperature increases. At the same time, the peak exitance shifts to shorter and shorter wavelengths. For example, objects at room temperature (300 K) emit their peak exitance at a wavelength of approximately 10 µm whereas those at 1000 K have peak emission at a wavelength of about 3 µm.

Room-temperature objects do not emit at a visible wavelength, so they are seen with the hues and colors of reflected light. However, a red-hot coal (with a temperature of about 3000 K) has a peak emission at approximately 1 µm. Because of the width of the emission spectrum, the coal also emits a fair amount in the long-wavelength part of the visible; this gives the coal its red and orange colors (despite the seeming implication of the terms "blackbody" and "graybody").

Planck discovered Eq. (5.2) by making what was, at the time, an off-the-wall assumption: that electromagnetic waves carry energy in digital "clumps" known as *photons*. The photon energy in Eq. (5.2) is $E_p = h\nu = hc/\lambda$, which indicates that a higher-frequency wave (i.e., one with a shorter wavelength λ) has more energy because it has more oscillations per second. Planck was rewarded with an eponymous constant for his work: the Planck constant $h = 6.626 \times 10^{-34}$ J·sec. Typical values of the photon energy for visible wavelengths are then $hc/\lambda = 6.626 \times 10^{-34}$ J·sec multiplied by 3×10^8 m/sec divided by 0.5×10^{-6} m $\approx 4 \times 10^{-19}$ joules, or about 2.5 electron volts (eV).

Boltzmann, too, had a constant (k) named after him; the average thermal energy of an oscillating electron is proportional to kT, where $k = 1.381 \times 10^{-23}$ J/K. At room temperature, this energy is 1.381×10^{-23} J/K multiplied by 300 K = 4.14×10^{-21} J. Equation (5.2) shows that the term comparing the photon energy with the thermal energy

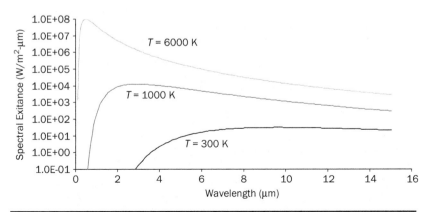

Figure 5.3 Blackbody curves from Eq. (5.2) in units of W/m²·µm for $\varepsilon(\lambda) = 1$; the graph includes room temperature (300 K) and the approximate surface temperature of the Sun (6000 K).

[1/(exp$\{E_p/kT\}$ – 1)] increases with wavelength and that the $1/\lambda^5$ term decreases with wavelength; the resulting trade-off determines the skewed bell shape of the blackbody spectrum.

Figure 5.3 is an extremely common plot, but several other forms are also used. One example is a graph of exitance in terms of number of photons per second. These values are found by dividing the exitance (J/sec·m²) by the photon energy (hc/λ joules per photon) to yield the number of photons per second per unit area (photons/sec·m²). The net result is that the $2\pi\epsilon hc^2/\lambda^5$ term in Eq. (5.2) becomes $2\pi\epsilon hc/\lambda^4$, so that spectral exitance has been converted to spectral photon exitance (photons/sec·m²·μm).

It is assumed that the intensity of a blackbody has no directional preference (i.e., that a blackbody is Lambertian), so the spectral radiance at any wavelength can also be found by dividing Eq. (5.2) by π steradians; this gives $L_\lambda = M_\lambda/\pi$. The same calculations apply for photons, yielding the spectral photon radiance [photons/sec·m²·sr·μm].

In all cases, the total exitance is found by integrating the spectral exitance over the wavelengths of interest. Integrating Eq. (5.2) over all wavelengths results in Eq. (5.1); a more common use of the concept is to find the total irradiance transmitted through an optical filter. For example, a filter that transmits over the MWIR band blocks most of the source exitance outside that band. Integrating Eq. (5.2) over the range of 3–5 μm reveals how much is transmitted, a figure that can then be put into the instrument power budget. Example 5.2 (in Sec. 5.1.2) goes through the details.

Wien's Law

For a given temperature, the peak exitance in Fig. 5.3 is due to the balance between the denominator terms [$1/\lambda^5$ and $1/(\exp\{E_p/kT\} - 1)$] in Eq. 5.2. The peak moves to shorter wavelengths as the object grows hotter, balancing the increase in thermal energy with the emission of higher-energy photons. There is also a nonobvious algebraic connection between the source temperature and peak exitance wavelength for blackbodies:[3]

$$\lambda_{max} T = 2898 \; \mu m \cdot K \approx 3000 \; \mu m \cdot K \tag{5.3}$$

This expression is known as Wien's displacement law, and its effects can be seen in the peaks of the curves in Fig. 5.3. Values computed using Wien's law are shown in Table 5.2 for various thermal sources.

Emissivity

The most electromagnetic power a blackbody source can radiate is given by Eq. (5.1) with emissivity $\epsilon = 1$ over all wavelengths. However, almost all sources have a wavelength-dependent emissivity $\epsilon(\lambda) < 1$, and thus less total irradiance than a true blackbody at the same

Thermal Source	Temperature [K]	λ_{max} [µm]
Sun	5900	0.5
Tungsten-halogen	3000	1
Blackbody source	1000	3
Room temperature	300	10

TABLE 5.2 Approximate values of the peak exitance wavelength for various thermal sources.

temperature. In addition to its implications for source efficiency, emissivity has a number of other important properties.

The first of these properties affects the conservation of energy given in Eq. (4.1). That equation is incomplete because it doesn't take sources into account. The optical power measured from a hot plate, for example, is greater than the reflected component R, which is zero for an ideal blackbody. A blackbody neither reflects nor transmits, and Kirchhoff was the first to realize that what can be "seen" depends only on what the object emits. If the blackbody isn't heating up or cooling down—in other words, if it is in thermal equilibrium—then any radiation that's absorbed must be reemitted. Therefore,

$$\varepsilon(\lambda) = \alpha(\lambda) \tag{5.4}$$

at a specific wavelength λ. When the emissivity $\varepsilon(\lambda)$ and absorption $\alpha(\lambda)$ are constant over all wavelengths, thermal emitters are known as blackbodies if $\varepsilon(\lambda) = 1$ and as graybodies if $\varepsilon(\lambda) < 1$. More common are spectral emitters for which $\varepsilon(\lambda)$ equals $\alpha(\lambda)$ over specific wavelength bands, such as the visible or MWIR, but not across all wavelengths.[3]

Together, Eqs. (4.1) and (5.4) define some important terms. The word "black" is a description of visible wavelength appearance, but it also applies to objects with high emissivity at wavelengths outside the visible. The term implies the object absorbs everything over a band of wavelengths—which is why it is called "black"—and thus also has high emissivity. As a result, an object can be red, green, or blue in the visible yet still be "black" (high absorption and high emissivity) in the UV or IR.

At the other extreme is the word "shiny", a description in the visible spectrum of a good reflector. The spectral form of Eq. (4.1) shows that $\varepsilon(\lambda) + \tau(\lambda) + \rho(\lambda) = 1$; thus, high reflectivity is associated with low emissivity and low absorption [since $\rho(\lambda) = 1 - \varepsilon(\lambda)$ for $\tau(\lambda) = 0$]. As with the blackbody concept, an object can be bright and shiny in the visible but look completely different at other wavelengths.

An efficient blackbody source, then, is one that has a high emissivity at the wavelength for which the source is designed. In practice, emissivity depends in part on surface properties and is therefore affected by coatings, paints, and surface roughness. Figure 5.4 plots the emissivity of a surface coating that's black in the MWIR band, which would make this an excellent choice for a blackbody operating at 600–1000 K [see Eq. (5.3)]. A lower emissivity still gets the job done, but in that case a higher temperature is required to obtain the same spectral radiance as the blackbody.

To obtain an emissivity higher than that available from coatings and paints, source geometry can be employed. As shown in Fig. 5.5, an open cavity absorbs more light than a flat plate because there are more reflections off the cavity's interior surface and each reflection is

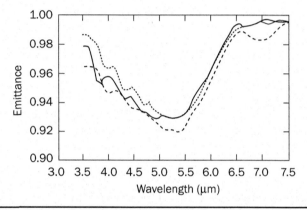

Figure 5.4 The spectral emissivity of Martin Black coating at an incident angle of 0 degrees indicates good absorption and emissivity in the MWIR band. (*Credit: A. L. Shumway et al.,* Proceedings of the SPIE, *vol. 2864.*)

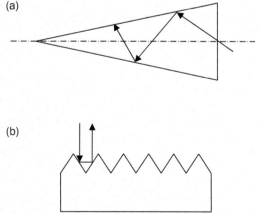

Figure 5.5 The absorption from multiple reflections explains why the effective emissivity of a cavity [cones as in (a), corrugations as in (b), spheres, etc.] is greater than that of a flat surface. Absorption is a function of the cavity's surface area compared with the area of the aperture that lets light in.

partially absorbed. The higher the ratio of interior surface area to aperture area, the greater the total absorption.

Equation (5.4) then shows that the effective emissivity is also higher—so much so that even a stack of shiny razor blades can have an effective emissivity of about 0.99.[5] At the extreme, the highest emissivity would result from an open sphere with lots of interior surface area to absorb light and a small aperture to let light out. Recent research has taken advantage of this concept (on a microscale level) even more efficiently than Martin Black, which (as shown in Fig. 5.6) exhibits surface roughness and micropatterning that resembles small cavities.

Blackbody sources using these principles are available as commercial products from a number of vendors, including Newport, Electro-Optic Industries, Santa Barbara Infrared, CI Systems, and Mikron. The most typical use of these sources is for laboratory calibration of other sources and detectors. The largest possible emissivity is not critical for most blackbody sources, but knowing exactly what the emissivity is allows them to be used for calibration. Emissivities for open-cone designs can be as high as 0.997 to 0.998 over the 3–12-µm range. Stability and spatial uniformity over the aperture is also important, and the open cavity is a critical part of such homogenized sources.

In addition to optical sources, there are other components of an optical system for which emissivity plays an important design role.

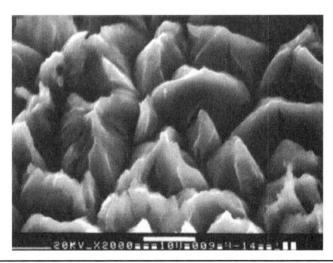

FIGURE 5.6 Electron micrograph of the surface of highly absorbing Martin Black. The horizontal white bar at the bottom of the photo is 10 µm in length. (*Credit: Reprinted with permission from M. J. Persky, "Review of Black Surfaces for Space-Borne Infrared Systems,"* Review of Scientific Instruments, *70(5), 1999, pp. 2193–2217. Copyright 1999, American Institute of Physics.*)

For example, mirrors have relatively low emissivity because they have high reflectance and low absorption. As a result, mirrors are sometimes preferred over lenses for MWIR and LWIR systems, where thermal emission of the optics results in background noise. Components such as lenses, mirrors, and baffles—even though they are not sources in the conventional sense—are just as effective at generating photons as the lamps and LEDs reviewed in the following sections.

One way to reduce thermal emissions from certain components is to recognize that, even though the spectral emissivity equals the spectral absorption at any given wavelength, nothing says they can't be different at different wavelengths. Coatings and paints, for example, can be a poor emitters/absorbers at one wavelength (e.g., at 0.5 µm) yet good emitters/absorbers at a different wavelength (e.g., at 10 µm).[6] This is a common approach used to control the temperature of systems exposed to solar radiation, where a coating with a low absorption α_s of solar light in the visible (e.g., white paint) also has a high emissivity ε_{LWIR} in the LWIR wavelengths (corresponding to a 300-K temperature for the object). Example 5.1 illustrates another way in which the same concept is used for the design of an IR imager.

Example 5.1. As will be shown in Chap. 6, it is common to cool a detector in order to reduce its inherent noise and improve its sensitivity. Unfortunately, if the rest of the instrument is warmer than the detector—as it often is—then blackbody radiation from the higher-temperature instrument becomes an additional source of noise. If it is not reduced, this so-called background noise can limit the performance of even a cooled detector.

For an instrument at room temperature (300 K), Eq. (5.3) shows that the peak emission wavelength of this 300-K background is about 10 µm; a plot of Eq. (5.2) indicates that there is also a significant amount of power at shorter wavelengths such as the MWIR band (see Fig. 5.3).

The amount of background noise is determined not only by the instrument's temperature but also by the detector's FOV. This is not the same as the *system* FOV, which is determined by the detector size and the focal length of the optics. Since the detector can "see" the inside of the instrument, the FOV of the detector by itself is the hemisphere of the instrument's interior toward which the detector points. Unless something is added to the system, blackbody background from the interior hits the detector from the entire solid angle.

What's usually added is a shield that keeps the background off the detector, much as the Sun is shielded on a hot day by a tree or an umbrella. As shown in Fig. 5.7, optic shields are typically a baffled cone or tube located close to the detector. The entire FOV can't be shielded, of course, for then no signal could be collected from the object. However, unlike a simple umbrella, the shield must be cooled down (to approximately the same temperature as the detector) so that it emits little blackbody radiation of its own. Such "umbrellas" are also known as cold shields.

Placing a cold stop at an exit pupil location will not affect the instrument's FOV but will exclude background radiation; the cold stop is thus the IR equivalent of a glare or Lyot stop. In practice, some light can make its way through the stop's aperture and reflect off the shield's interior walls onto the detector, since

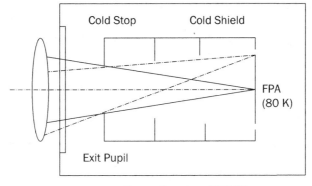

Figure 5.7 A cold shield reduces the radiation incident on a cooled FPA from hotter sources. A cold stop must also be placed at the exit pupil to prevent out-of-field thermal radiation from entering the system.

the shield's placement and the mechanical tolerances on its diameter make it difficult to obtain 100 percent exclusion. A cold-stop efficiency of 98 percent is typical,[7] and much lower efficiency characterizes systems that do not place the cold stop at an exit pupil.

This inefficiency allows background radiation to make its way through the stop's aperture and reflect off of the interior surface. As a result, a shield that's reflective (low emissivity) on the inside is *not* better than one that's black (high emissivity). Depending on the cold-shield efficiency, a "shiny" interior generally increases reflected background flux more than a low-temperature cold shield emitting with an absorbing, high-ε surface whose flux Φ varies as εT^4.

However, it is best to leave the shield's exterior surface reflective, so that radiation falling on the shield is not absorbed but rather reflected out the dewar or transferred to the cooler that's keeping the detector and shield at their low temperatures. Low temperatures for components such as cold shields are not "free," and any heat load which makes them more difficult to maintain should be avoided (see Chap. 7).

Even with an efficient cold stop, background radiation is still emitted from the optics that are used to image the object. Two methods are employed to keep this background to a minimum for IR systems. The first is to use reflective optics wherever possible, since they have a lower emissivity than refractive lenses. The second is to use a cold filter, which removes much of the background radiation and, depending on its temperature and emissivity, emits little radiation of its own.

5.1.2 Tungsten-Halogen Lamps

Unlike commercial blackbody sources, which are used mainly in laboratories, tungsten-halogen sources have a number of scientific and industrial applications. One of the most common is automobile headlamps; other applications include heating semiconductor wafers with IR radiation as part of their manufacturing process, alignment aids, and scientific instruments (e.g., spectrometers) that require broadband sources.

The basic principle of the tungsten-halogen lamp is that of a thermal emitter, where electric current resistively heats a tungsten filament to about 2500–3200 K in order to obtain mostly visible, NIR, and SWIR radiation (Fig. 5.8). Tungsten is used because of its high melting point, but no effort is made to increase the metal's emissivity to anything near that of a blackbody; instead, tungsten's natural emissivity ($\varepsilon \approx 0.35$ at $\lambda = 1$ µm) is used. Heating a low-emissivity metal does not give the best emission efficiency, but does result in a relatively low-cost source of lighting.[8]

The common incandescent bulb also uses a tungsten filament, but it has a much shorter lifetime than the tungsten-halogen. The reason is that tungsten *evaporates* from the incandescent filament as it heats up. The tungsten-halogen lamp addresses this problem by filling the bulb with a halogen gas (typically iodine) that recycles some of the evaporated tungsten back onto the filament (Fig. 5.9).

The process is not perfect, of course, so the lamp's lifetime depends on how hot the filament is and on how efficiently the halogen recycles. Recycling efficiency is not within the optical system designer's control, but temperature can be controlled by reducing current to the lamp to less than its rated value. As shown in Sec. 5.3, this "derating" can result in significantly longer lifetime[9] but at the expense of reduced output power.

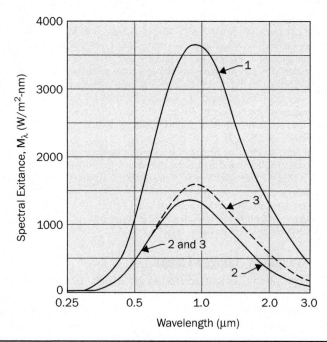

Figure 5.8 Emission spectra of a tungsten-halogen lamp (curve 2). Curve 1 is for an ideal blackbody ($\varepsilon = 1$) and curve 3 is for an ideal graybody (here, $\varepsilon = 0.45$); all curves are at 3100 K. (*Credit: Newport Corp.; used by permission.*)

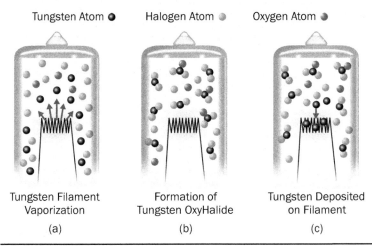

FIGURE 5.9 Operating principle of the tungsten halogen lamp: evaporated tungsten is redeposited on to the filament by the halogen gas. (*Credit: Carl Zeiss, Inc.*) (*See also color insert.*)

A quartz bulb is used to transmit the visible and IR radiation, so these lamps are also known as quartz-halogens; the names "halogen," "quartz-iodine," and "tungsten-iodine"—indeed, almost any combination of the words quartz, tungsten, halogen, and iodine—are also found on the market. The quartz is resistant to thermal fracture, which is a significant advantage when bulb temperatures are changing rapidly.

The electrical input power of common tungsten-halogens ranges anywhere from 10 to 2000 watts. Generating light from the resistive heating used to increase the filament's temperature is an efficient process: about 80 percent of the electrical power is converted to optical output,[8] and the remaining 20 percent heats the gas or bulb (whose thermal emission is not usually counted as lamp output). Tungsten-halogen lamps are not blackbodies, but neither are they graybodies. Instead, they are thermal sources known as *spectral emitters* with an emissivity that depends on wavelength. Figure 5.10 shows how emissivity varies with wavelength for tungsten; Fig. 5.8 shows the effects on light output, spectral shape, and spectral peak.

Because tungsten-halogen lamps are often used as visible sources, emission efficiency is not a sufficient description of their performance. What's also needed is a measure of how sensitive the human eye is to the visible light (0.4–0.7 µm) being emitted. The amount of visible light is not measured in watts but rather in *lumens* (lm). This is the photometric equivalent of the watt, and is roughly equal to 1/683 watts of source power emitted at $\lambda = 0.555$ µm, the wavelength to which the eye is most sensitive under daylight conditions (Fig. 5.11). Because the eye's sensitivity drops off as the wavelength moves away from 0.555 µm, in such cases there are fewer

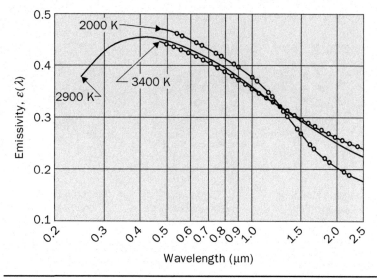

FIGURE 5.10 The tungsten filaments in tungsten-halogen lamps are neither blackbodies nor graybodies; rather, they are spectral emitters whose emissivity depends on wavelength. (*Credit: Newport Corp.; used by permission.*)

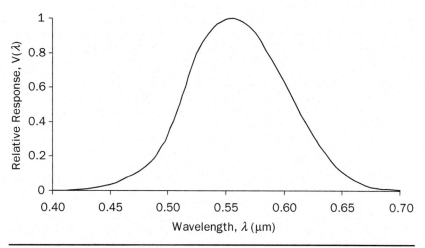

FIGURE 5.11 Relative daytime response of the human eye. The graph illustrates why the 0.4–0.7-micron band is defined as visible light (the nighttime response is shifted slightly to shorter wavelengths).[10]

lumens available from a watt of source power. The net lumens emitted from a source depends on the overlap of the source spectrum with the eye's sensitivity.[10] This effect is captured by the notion of *luminous efficiency* (aka *efficacy*), which is calculated as the ratio of lumens emitted to watts of electrical input power (units of lm/W, also written as LPW). Typical luminous efficiencies for tungsten-halogen lamps are on the order of 20–25 lm/W, as calculated in Example 5.2.

Example 5.2. In this example, Eq. (5.2) is used to calculate how much power a tungsten-halogen lamp emits in the visible as well as its luminous efficiency. The tungsten-halogen lamp emits a reddish-orange glow, which Eq. (5.3) shows corresponds to a peak wavelength of about 1 μm for a filament temperature of 3000 K. Since the visible spectrum runs from 0.4 to 0.7 microns, this is clearly not the most efficient source for visible lighting.

Most of the power that goes into heating the filament doesn't end up as visible light but instead becomes NIR and SWIR photons. One measure of efficiency, then, is the fraction of power that goes into creating visible photons. From Eq. (5.1) we know the total power emitted, so Eq. (5.2) can be integrated over the visible band to find the visible power. The ratio of these two quantities is the visible output efficiency.

To obtain this efficiency, we must integrate Eq. (5.2) numerically. As shown in Fig. 5.12, the basic idea is to break the integration down into "small" steps—that is, small enough that the spectral exitance M_λ is approximately linear over the wavelength difference. In that case, the integral becomes a sum:

$$M = \int M_\lambda \, d\lambda \approx \sum M_{\lambda,\text{avg}} \Delta \lambda = \sum [0.5(M_{\lambda_1} + M_{\lambda_2})(\lambda_2 - \lambda_1)] \quad (5.5)$$

where better accuracy is obtained by using the average exitance $[M_{\lambda,\text{avg}} = 0.5(M_{\lambda_1} + M_{\lambda_2})]$ over the wavelength span $\lambda_2 - \lambda_1$ rather than the exitance value at the beginning or end of the step.

A spreadsheet is especially useful for numerical integration; Table 5.3 shows its use in integrating Eq. (5.2). In the left column are the wavelengths integrated over in steps of 0.01 μm (10 nm). The next column is the value of M_λ at that wavelength as calculated using Eq. (5.2). The third column ("Radiometric") is the running total of the sum given in Eq. (5.5), starting from zero at $\lambda = 0.4$ μm and reaching a value of 2.17×10^5 W/m² at $\lambda = 0.7$ μm for a graybody emissivity

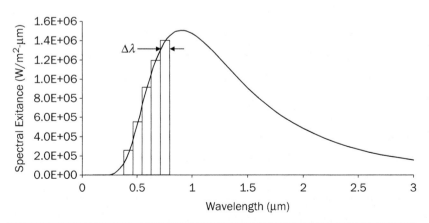

Figure 5.12 The total exitance M from a source is found by numerically integrating the spectral exitance M_λ in Eq. (5.5) in small, discrete steps of width $\Delta\lambda$.

Wavelength [μm]	M_λ [W/m²·μm]	Radiometric $\int M_\lambda\, d\lambda$ [W/m²]	$V(\lambda)$	Photometric $\int M_\lambda V(\lambda)\, d\lambda$ [W/m²]
0.40	1.674E+05	0.0	0.0004	0.0
0.41	1.946E+05	1.810E+03	0.0012	1.448E+00
0.42	2.240E+05	3.903E+03	0.0040	6.891E+00
0.43	2.555E+05	6.301E+03	0.0116	2.559E+01
0.44	2.889E+05	9.023E+03	0.023	7.268E+01
0.45	3.240E+05	1.209E+04	0.038	1.662E+02
0.46	3.608E+05	1.551E+04	0.060	3.339E+02
0.47	3.989E+05	1.931E+04	0.091	6.207E+02
0.48	4.383E+05	2.350E+04	0.139	1.102E+03
0.49	4.787E+05	2.808E+04	0.208	1.898E+03
0.50	5.199E+05	3.307E+04	0.323	3.223E+03
0.51	5.618E+05	3.848E+04	0.503	5.457E+03
0.52	6.040E+05	4.431E+04	0.710	8.992E+03
0.53	6.466E+05	5.057E+04	0.862	1.391E+04
0.54	6.891E+05	5.724E+04	0.954	1.997E+04
0.55	7.315E+05	6.435E+04	0.995	2.689E+04
0.56	7.737E+05	7.187E+04	0.995	3.440E+04
0.57	8.153E+05	7.982E+04	0.952	4.214E+04
0.58	8.564E+05	8.818E+04	0.870	4.975E+04
0.59	8.968E+05	9.694E+04	0.757	5.688E+04
0.60	9.363E+05	1.061E+05	0.631	6.324E+04
0.61	9.748E+05	1.157E+05	0.503	6.866E+04
0.62	1.012E+06	1.256E+05	0.381	7.305E+04
0.63	1.049E+06	1.359E+05	0.265	7.638E+04
0.64	1.084E+06	1.466E+05	0.175	7.872E+04
0.65	1.117E+06	1.576E+05	0.107	8.028E+04
0.66	1.150E+06	1.689E+05	0.061	8.123E+04
0.67	1.181E+06	1.806E+05	0.032	8.177E+04
0.68	1.211E+06	1.925E+05	0.017	8.206E+04
0.69	1.239E+06	2.048E+05	0.0082	8.222E+04
0.70	1.265E+06	2.173E+05	0.0041	8.229E+04

Table 5.3 Numerical integration of Eq. (5.5) in steps of $\Delta\lambda = 0.01$ μm for a graybody radiator at 3200 K with $\varepsilon = 0.35$; the effects of the human eye's relative response $V(\lambda)$ are also shown ("E+05" is shorthand for "× 10^5").

$\varepsilon = 0.35$ and filament temperature $T = 3200$ K. The total exitance is 2.08×10^6 W/m² [= Φ/A in Eq. (5.1)], yielding a visible emission efficiency of only about 10 percent.

The remaining 90 percent of the radiation emitted from a tungsten-halogen lamp is mostly in the NIR, SWIR, and longer wavelengths. This energy is eventually absorbed by the surroundings as heat, which is why these sources are sometimes sold as "quartz heaters," not optical sources. When used as a source, it's possible to remove this heat from the system by using an optical filter that transmits the visible but absorbs or reflects the longer wavelengths. That's not an efficient use of energy, and there are better ways to generate visible photons without the heat (see Sec. 5.1.4).

How much power ends up in the visible does not completely capture the response of the human eye. We have already mentioned that the eye is most sensitive at 0.555 µm during the day, so any light not at this wavelength will not produce as strong a response. The eye's overall sensitivity to a source thus depends on how well the source's spectrum matches that of the eye, and for this metric luminous efficiency is used.

For example, a 1-W source (emitted optical power) whose spectrum is a small band centered at 0.555 µm emits 683 lumens of photometric power and has a luminous efficiency of 683 lm/W. If we use a different 1-W source whose narrow spectrum is centered at 0.61 µm—where the eye is only half as sensitive as it is at 0.555 µm—then the luminous efficiency is 0.5×683 lm/1 W = 341.5 lm/W. Using both of these sources at the same time yields a total output of 683 + 341.5 ≈ 1025 lumens, for which 2 watts of power are used; hence the luminous efficiency for this combination source is (683 lm + 341.5 lm)/2 W ≈ 512 lm/W.

For an arbitrary source distribution such as that from a blackbody, the integration is performed numerically. Equation (5.5) is changed slightly, because the spectral response V(λ) of the eye in daylight is included:

$$M = \sum [0.5(M_{\lambda_1} + M_{\lambda_2})] \times 0.5 \,[V(\lambda_1) + V(\lambda_2)](\lambda_2 - \lambda_1)]$$

The average value $V(\lambda)_{avg} = 0.5[V(\lambda_1) + V(\lambda_2)]$ is also used over the wavelength span, along with the tabulated values for $V(\lambda)$ given in Table 5.3.

Using a spreadsheet to find the summation, we find that $M = 8.23 \times 10^4$ W/m² for the photometric exitance emitted by the tungsten halogen within the visible range over which the eye is sensitive (see the last column in Table 5.3). This result is then multiplied by 683 lm/W to give 5.6×10^7 lm/m². If we divide through by the total exitance $M = \Phi/A = 2.08 \times 10^6$ W/m² from Eq. (5.1), then the luminous efficiency is about 27 lumens per watt of optical output. Measured values for tungsten-halogens are closer to 20–25 lumens per watt of electrical input, indicating that there are only small inaccuracies in the assumption of a 3200-K graybody with emissivity $\varepsilon = 0.35$ over all wavelengths.

In short: tungsten-halogen lamps are inexpensive, high-power sources for the visible to SWIR bands. Manufacturers include General Electric, Osram Sylvania, Mitsubishi, and Ushio; their Web sites have the detailed design data needed for using one of these sources. Some vendors, such as Newport and CVI Melles Griot, sell lamps in housings with a cooling fan and collimating optics (Sec. 5.2). Of course, tungsten-halogens are only one of many source options; another common option is the arc lamp, which is described in the next section.

5.1.3 Arc Lamps

Arc lamps are probably more common than tungsten-halogens. Their range of applications includes the fluorescent lamps seen in *The Office*, neon signs announcing that the coffeehouse is open, white-light sources for video projectors, flashbulbs for cameras, and the high-intensity discharge lamps recently introduced as more efficient replacements for tungsten-halogen auto headlights. The common theme behind these applications is the need for a bright source (in the UV-to-visible range) that does not require a megawatt of electricity to operate. These are applications for which the arc lamp excels.

The basic principle behind these lamps is the creation of an arc that emits light, just as lightning paints arcs across the sky during summer storms. The arc is discharged through a gas that emits light when excited; the most common gases are xenon, mercury, and a combination of these two. The discharge occurs between electrodes (Fig. 5.13) that carry large amounts of current and create the arc. The electrodes are tungsten, and a pointed cathode is heated to help eject electrons toward the anode. Here, too, the bulb is quartz; it transmits the UV-to-visible wavelengths and protects against thermal shock.

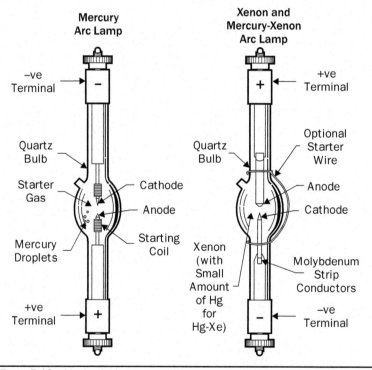

Figure 5.13 Mechanical structure of the mercury, xenon, and mercury-xenon arc lamps commonly used in optical systems. (*Credit: Newport Corp.; used by permission.*)

One reason for using these sources is that, except for lasers, high-pressure lamps with short arcs are among the brightest manufactured light sources available. Since the output power of an arc lamp and a tungsten-halogen lamp is about the same (~ 50–5000 W for the arc lamp) and since the solid angle into which the different lamps emit is also about the same, it follows that the arc lamp is brighter because it emits from a smaller area (and thus has more radiance). Applications such as movie and video projectors rely on this brightness.

Another reason for using arc lamps is the range of wavelengths they emit. As shown in Fig. 5.14,[11] the output spectrum for a xenon lamp has little longwave IR emission but lots of power in the UV, visible, and NIR. The spectrum also has a number of unusual features when compared with a thermal source. Because the gases in the bulb are hot, there is a thermal component to the spectrum that accounts for approximately 85 percent of the total output power.[12] Another component, which is particularly strong for mercury arc lamps, consists of large peaks in the spectrum. These peaks cannot be explained as thermal emission; rather, they result from quantum emissions associated with the electronic transitions of the emitting gas. Such peaks may or may not be a problem, depending on the application.

A disadvantage of these lamps is that the arc tends to move around, which is a problem in a system where it's expected that the image of the source won't "wander." Their output power also tends to fluctuate with time, as the arc continues to crackle even after warm up. Tungsten-halogens do not have these problems, so they are preferred when a quiet source is needed that doesn't change much spatially or over time.

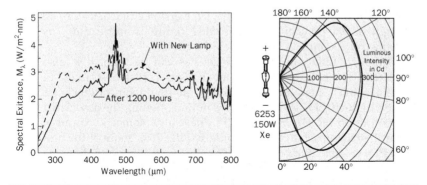

Figure 5.14 Output spectrum and spatial distribution emitted from a xenon arc lamp; note the asymmetry in the spatial distribution. (*Credit: Newport Corp.; used by permission.*)

It is possible to create photons without the inefficiency of broadband thermal emission—in other words, generating only the spectral peaks. The next section shows how this can be done with a light-emitting diode.

5.1.4 Light-Emitting Diodes

Whereas tungsten-halogens and arc lamps are based on relatively old technologies, light-emitting diodes (LEDs) are a fairly new approach to generating photons. There are a number of companies developing high-brightness LEDs, whose high ecological rating stems from the efficiency with which the photons can be generated. Compact fluorescent lamps (CFLs) output more lumens than LEDs, but there is currently little use for CFLs in optical systems design. In addition to the possibility of replacing incandescent bulbs with solid-state LEDs for general lighting, LEDs have a number of applications that are useful in optical systems. Examples include displays (ranging from cell phones to flat-panel TVs to stadium video screens), camera flashlamps, high-efficiency sources for space-based instruments—even replacements for arc lamps in video projectors, resulting in portable, hand-held projectors.

LEDs are efficient because they can generate light close to a certain wavelength without wasting a lot of energy while heating up a tungsten filament or creating an arc between electrodes. Tungsten-halogens and arc lamps have a relatively *broad* output spectrum, and much electrical energy goes into creating wavelengths that may not be needed. The LED, in contrast, uses a voltage difference to bring electrons to a higher-energy state and then obtains photons within a *narrow* band of wavelengths when the electrons lose this energy [Fig. 5.15(a)]. Without going into the details, a diode made from a semiconductor p-n junction is an efficient way to help electrons lose their excess energy.[13] As a result, LEDs are very good at converting electrical energy into optical energy. Conversion efficiencies can be as high as 95 percent—although this figure is misleading, as explained in Example 5.3 (later in this section).

Although LED conversion efficiencies can be high, it is also important to look at the output power available. Games of "specmanship" often find companies quoting their products as having high efficiencies but perhaps not mentioning the associated low output power. For example: CFLs have a photometric output of about 1000 lumens; however, it would require 10 or more of the best HB-LEDs to equal that power even if both source types had the same luminous efficiency (typically about 70 lm/W).

In terms of radiometric power, LED output is also relatively small; it ranges from less than 1 mW up to about 5 W, which is quite a bit less than what's available from most tungsten-halogen or arc lamps. Observe that LED power is specified as the optical *output* power, whereas most other sources are specified in terms of electrical

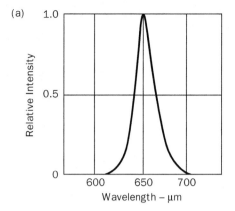
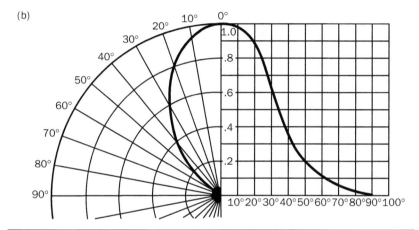

FIGURE 5.15 *(a)* Output spectrum and *(b)* spatial distribution emitted from a conventional lensed LED. (*Credit: Avago Technologies, www.avagotech.com.*)

input power. The concept of *wall-plug efficiency* (see Sec. 5.3.6) is used to connect these two metrics for a fair comparison.

In addition to being highly efficient, LEDs also are available in a wide range of wavelengths. Colors run from blue to green to yellow to orange to red and on through the NIR and SWIR, typically up to about 1600 nm and with longer wavelengths also available. There is a surprising number of LED "geeks" who are keen on knowing all the different semiconductor materials used to create these various colors. It's not important for a systems engineer to be that knowledgeable, but some familiarity is useful.

No single LED material can give the entire range of wavelengths needed for all systems, and what differs among semiconductor materials is their *bandgap energy*. This is the energy required to put an electron into a relatively stable, high-energy state. Because electrical potentials are normally used to change the electron's energy, bandgap energy is measured in electron volts (eV). When the electron is encouraged (by a p-n junction) to fall back to a lower-energy state, the energy of the emitted photon is approximately the same as the electron energy (which is also the same as the bandgap energy). The resulting photon energy $E_p = hc/\lambda = 1.24$ eV·µm/λ (with λ measured in microns); this expression connects a material property (bandgap energy) with the wavelength of the emitted light. For instance, green LEDs emitting at a wavelength of 0.5 µm are made with a semiconductor whose bandgap energy is approximately (1.24 eV·µm)/(0.5 µm) = 2.48 eV.

Different semiconductors are needed to obtain the range of bandgap energies required for the LED's many potential applications. For NIR wavelengths, fabrication starts with gallium and arsenic mixed together with aluminum to create aluminum gallium arsenide (AlGaAs). The wavelength range of 750–900 nm is obtained by varying the relative proportions of these three elements, which changes the bandgap energy of their combination.[13] If indium and phosphorus are instead added to the GaAs, then InGaAsP LEDs (with longer wavelengths) are obtained (Table 5.4).

Now omitting the arsenic and adding indium and phosphorus in various proportions to the gallium yields shorter-wavelength InGaP LEDs, down to about 690 nm. High-brightness LEDs emitting in the visible are made by adding aluminum to the InGaP (to produce AlInGaP), which is used to create red, orange, and yellow wavelengths.[14] Blue diodes are now possible by adding phosphorus to GaAs, and new materials—including gallium nitride (GaN) and zinc selenide (ZnSe)—are being explored. The best UV-to-blue LEDs are currently made from indium mixed with GaN (or InGaN).

LED material	Peak wavelengths
InGaN, GaN	300–405 nm
AlInGaN, GaP	440–550 nm
AlInGaP, GaAsP	570–645 nm
InGaP	600–690 nm
AlGaAs	750–900 nm
InGaAsP	910–1600 nm

TABLE 5.4 Available wavelengths for various LED sources.[13,17] The semiconductor materials are indium gallium nitride (InGaN), aluminum gallium indium phosphide (AlGaInP), indium gallium phosphide (InGaP), aluminum gallium arsenide (AlGaAs), and indium gallium arsenic phosphide (InGaAsP). High-brightness LEDs are made from AlInGaN and AlInGaP (sometimes written as AlGaInP).

Different-wavelength LEDs can be combined to produce colors that aren't possible otherwise. White light from red, green, and blue (RGB) in the correct proportions is the most common example of combining wavelengths. This concept has been applied for years to television screens and is now used with LEDs for outdoor stadium displays and flat-panel TVs. White light can also be obtained by using a blue LED with a phosphor which emits yellow, red, green, and orange in various proportions to look more or less white. This is less efficient than directly generating RGB from three individual LEDs, but it's often more convenient to use one LED than three. Note that there's also quite a bit of theory behind what color proportions are needed to produce various shades of white light (e.g., warm white, neutral white, cool white); see Schubert for more details.[15]

In addition to the available wavelengths, two additional features that make LEDs useful are their low operating voltage (~ 2–5 V) and extremely long lifetimes (~ 50,000 hours). This makes them good choices for critical (high-reliability) applications ranging from space-based instruments and automobile brake lights and also for consumer toys that require small power supplies. Furthermore, LEDs are readily available from major manufacturers, including Osram, Philips Lumileds, and Nichia. On the negative side, an LED's most annoying features are its spatial distribution and tolerance on pointing angle, as well as its intensity and wavelength variation with temperature. These first two parameters are determined by the inexpensive plastic lens attached to the semiconductor chip. Removing these lenses has unfortunate consequences, as Example 5.3 illustrates.

Example 5.3. Although LEDs may be efficient at converting electrons into photons, they are notoriously inefficient at *emitting* those photons. The culprit is total internal reflection (TIR). Absorption and Fresnel losses also reduce the emission efficiency, but TIR is the biggest loss mechanism. Using an LED with 100 percent conversion efficiency, this example shows how little of the light that's generated actually makes its way out of a conventional LED.

Figure 5.16(a) shows a typical geometry for a common, surface-emitting LED. Photons are created at the junction of the p layer (p-GaN) and n layer (n-GaN), where electrons lose energy after being brought to higher-energy states by a voltage applied across the junction. The rays associated with the wavefronts of these photons hit the top surface of the LED, where they are either transmitted or reflected.

For rays which hit the surface at an angle less than the TIR angle, Fresnel losses reflect about 30 percent of the photons back into the LED; and for rays that hit the surface at an angle greater than or equal to the TIR angle, *all* the photons are reflected back into the LED. Thus, the total amount of power that makes it out depends on the TIR solid angle as compared with the solid angle for all photons emitted from the junction.

For emission into air, the external TIR (or "critical") angle is given by $\theta_{TIR} = \sin^{-1}(1/n)$, where n is the index of the semiconductor used for the p-n junction.[16] For $n \approx 3.4$, $\theta_{TIR} \approx 0.3$ rad (17.1 degrees); hence the TIR solid angle

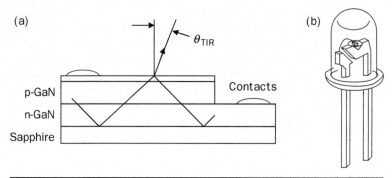

FIGURE 5.16 *(a)* Semiconductor structure and *(b)* packaging of a conventional LED die immersed in a plastic lens. The LED in panel *(a)* is about 0.1 mm wide; the lens diameter in panel *(b)* is approximately 5 mm. (*Credit: (b)* The Handbook of Optics, *McGraw-Hill, www.mcgraw-hill.com.*)

from Eq. (4.7) is $\Omega_{TIR} = 2\pi[1 - \cos(0.3)] \approx 0.28$ steradians. This is only 2.2 percent of the 4π-steradian solid angle into which the Lambertian p-n junction emits, and the result is an extremely low emission efficiency. The actual efficiency is 30 percent less (multiplied by 0.70) owing to Fresnel reflections at the surface for photons that do escape [see Eq. (4.2)]. Overall, then, only 1.5 percent of the photons generated at the p-n junction actually make their way out of the conventional LED as usable optical power!

A number of methods have been tried to increase this extraction efficiency, including developing lower-index semiconductors (it's difficult to find any at the wavelengths needed), applying AR coatings (even single-layer coatings are relatively expensive and the TIR losses remain large, improving to only 2.2 percent efficiency at best), or using a lens to reduce the index difference between the semiconductor and air. To keep costs low, the latter are often inexpensive plastic lenses, which has long been the preferred method to increase LED output for the consumer market [Fig. 5.16(b)].

To see how much better output can be obtained, note that common plastic lenses have an index of $n = 1.5$, giving $\theta_{TIR} = \sin^{-1}(1.5/3.4) \approx 26.2$ degrees. The solid angle into which the LED emits is thus 0.64 steradians, an improvement by a factor of about 2.3 over the unlensed LED. Fresnel losses are also reduced because reflectivity at the surface is now lower: $R = [(3.4 - 1.5)/(3.4 + 1.5)]^2 \approx 15$ percent. Fresnel losses (of about 4 percent) at the lens-to-air interface reduce the transmission somewhat, and the final result is a factor of about $2.3 \times (0.3/0.15) \approx$ 4 times better efficiency using a lens. So instead of 1.5 percent the lensed LED has about 6 percent emission efficiency, which is still extremely small. The lens also has the added benefit of providing some degree of control over the angular size of the emitted beam, as in Fig. 5.15(b). It's also possible to buy LEDs with a lens which has been roughened into a diffuser, thus scattering the light into a *wider* range of angles.

While putting lenses onto LEDs is not typically an issue an optical systems engineer is concerned with, Example 5.3 is a good illustration of the light-extraction issues faced in many areas of optical engineering. For LEDs, current research is directed toward chemically roughening or micro-patterning the semiconductor surface, giving more photons an average angle of incidence less than the TIR angle, increasing the emission efficiency by a factor of 2 to 3 times over that of conventional LEDs.[17]

5.1.5 Lasers

A close relative of the LED is the semiconductor laser, and there are many other types of lasers as well. Independent of type, these sources were originally viewed as an interesting curiosity with little commercial potential. However, we know now that they are useful in many applications: CD and DVD players, bar-code scanners, laser radar, fiber-optic communications, and directed energy—to name just a few. You should proceed with caution when considering the use of a laser in an optical system. The unique properties of lasers present numerous difficult engineering issues (such as speckle). These issues and many of their solutions are well known,[18,19] but there is not room enough in this book to cover them. Hence these sources should not be selected based solely on the limited information given here.

What makes lasers unique is a combination of properties not found in any other source. Perhaps the most obvious of these is their spectral purity, which is evident in the rich colors that they emit. For instance, a red laser pointer has a deep, red hue—a result of the extremely narrow output spectrum inherent in these sources. Also obvious to the eye is an emitted beam that stays more or less the same size over a long distance. As shown in Sec. 5.2.1, a point source combined with a telescope can have the same diffraction-limited divergence angle as a laser, but much power is wasted in making it so. In contrast, a laser emits all of its power directly into this narrow angle—a result of near-diffraction-limited wavefront error ("spatial coherence") across the entire beam diameter.

A source which emits all of its power into a small, diffraction-limited solid angle is extremely bright. This is the third unique property of lasers, and it makes them much brighter than arc lamps that have many times more power. At the same time, the laser power is emitted into a narrow spectrum (< 0.1 nm). The result is an extremely high spectral brightness, much greater than that of tungsten-halogen lamps (with their broadband blackbody emission) and even the Sun.

Despite these advantages, using lasers in optical systems can lead to unexpected difficulties such as speckle, fringes, thermal lensing, and optical power damage. Before deciding to use these sources, consult Hecht[18] and Hobbs[19] for more details on how to get around these problems.

5.2 Systems Design

Our goal in this book is not learning how to design sources but rather understanding how to use them in optical systems. Part of this is knowing the capabilities and limitations of the sources themselves, but an equally important aspect is incorporating them intelligently into a system. When the sources and optics are combined, typical functions for sources include illumination, collimation, and source coupling, and the resulting performance depends on whether the source is a point of light or has an illuminating area. This section takes a more detailed look at the design issues involved with two types of source-plus-optics systems.

5.2.1 Field of Illumination

Target illumination is perhaps the most common application of the source-plus-optics system. The idea is to efficiently collect light from a source and to place it with utmost care on a target or field of view, which must be illuminated. A camera flash, for example, must place as much light as possible on the scene being shot and not waste any light on things that won't end up in the image. Likewise, automobile headlamps are not useful if they're illuminating the sky instead of helping us move on down the road.

To avoid wasting light and electrical power, then, the source must be directed to illuminate the field that's being imaged, and nothing more. Unfortunately, most sources emit into a different angle than the image FOV, so it's necessary to augment the illumination system with optics that redirect the source's diverging beam. The most common system for doing this is a *collimator*. As shown in Fig. 5.17, it usually consists of an extended source along with collimating optics.

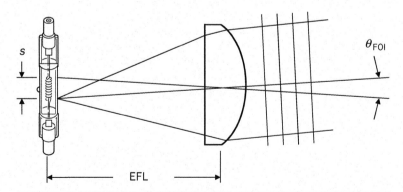

FIGURE 5.17 If a source is placed at the effective focal length of an optical system, then the wavefronts in any direction within the field of illumination (FOI) are planar and the system is known as a collimator; note that the FOI angle $\theta_{FOI} \approx s/EFL$. (*Credit: Newport Corp. for the lamp drawing.*)

Optical Sources

Placing the source at the effective focal length of the optics results in the diverging wavefronts from the source being converted into planar wavefronts, which define the source as being collimated. From the perspective of imaging, planar wavefronts image to a spherical wavefronts centered at the focal length to create an image. Collimation is the reverse process in that it starts with spherical wavefronts emitted by a source and ends up with planar ones to illuminate a field. The idea is simple, but one must pay attention to many engineering details.

Collimators are useful for a number of reasons. One is that the source need not be at infinity in order to produce planar wavefronts for other parts of the system (such as a beam splitter, which introduces aberrations when used with noncollimated light). A second reason is that it's often necessary that a source image *not* be placed on a target in order to allow for the possibility that the target may be changing distance from the source. Finally, collimated wavefronts allow the source irradiance to be independent of the distance from the collimator—up to a "crossover" point, where the wavefronts from each point on the source do not completely overlap, as determined by θ_{FOI} and the collimator diameter.[8] This constant-irradiance condition occurs over a smaller FOI than the source emission itself, but it increases the irradiance (though not the brightness) of this source-plus-optics system.

A common misconception about collimator use is supposing that the source can only illuminate on-axis. The reasoning is that a collimator illuminating a field of on-axis and off-axis points must have a diverging beam, and a diverging beam cannot be collimated. Yet this conclusion is wrong, because there is no connection between divergence angle and collimation. Regardless of the angle from which a collimator is looked at, the wavefronts coming out are planar, and it is planar wavefronts that define collimation—not their divergence angle or the size of the field that's being illuminated.

Not surprisingly, the field being illuminated is known as the *field of illumination* (FOI), which is the source analogue of the detector's FOV. Chapter 2 showed that the FOV depends on the detector size and focal length of the imaging optics; for sources, the FOI depends on the source size and focal length of the collimating optics.

To put some numbers on the FOI, a collimated source is identical to imaging an object at infinity *in reverse*. As a result, the ray that goes straight through the center of the lens is useful for finding the FOI into which a collimator can project light. This is illustrated in Fig. 5.17, which shows the center rays and the angle they make with the source. Straightforward geometry reveals that, for an extended source, the divergence angle depends on the size s of the source and the effective focal length f of the lens:

$$\text{FOI} = 2\tan^{-1}\left(\frac{s}{2f}\right) \approx \frac{s}{f} \quad [\text{rad}] \tag{5.6}$$

Here the FOI is approximately equal to s/f for "small" angles. We must emphasize the idea of a "full-FOI," since the factor of 2 is often left out before the arctangent to give only a half-FOI (or HFOI, also known as the semi-FOI).

It was shown in Chap. 3 that imaging a point source is different from imaging an extended source, and the FOI defined by a collimator follows the same rules. As the source size s in Eq. (5.6) is made smaller and smaller, geometrical optics shows that the FOI likewise decreases and goes to zero if $s = 0$. Diffraction prevents that from happening, however, and there is a size for which the FOI due to diffraction is greater than that predicted by Eq. (5.6). This occurs when the source size equals the diffraction-limited image size ($s \leq 2.44\lambda \times f/\#$), assuming the collimator is used in reverse to image an object at infinity to obtain a blur size. The FOI for a point source—that is, any source whose angular size is less than or equal to the angular blur β obtainable with the optics—is thus given by

$$\text{FOI} = \frac{2.44\lambda}{D} \quad [\text{rad}] \tag{5.7}$$

Placing the 2.44 "prefactor" in front of λ/D puts 84 percent of the source energy in the FOI for a diffraction-limited optical system with zero wavefront error (WFE). The remainder of the energy is spread out into larger field angles, and Table 3.9 shows how much additional energy can be expected at larger angles.

The distinction between small and large sources also affects the radiometry of the collimator. Specifically, a collimator with a large FOI emits into a large solid angle, with the radiance unchanged in comparison with a smaller-FOI collimator that collects light farther away from the same source. Yet the use of higher-power sources does not necessarily increase collimator radiance. Example 5.4 shows why.

Example 5.4. Using a higher-power source does not necessarily result in a higher-radiance collimator. This is not obvious from the physics of collimation, since it is was shown in Chap. 4 that more power results in more radiance. However, more power oftentimes requires more source area (which *reduces* radiance), and it is this engineering limitation of source design that controls collimator radiance.

To understand why, we illustrate here the design a collimator using a tungsten-halogen lamp (Newport #6336) with an output power of 600 watts and filament dimensions of 4 mm in diameter × 13.5 mm in length.[11] Because the source dimensions are different in the x and y dimensions, the FOI is also different. Using f/2 collimating optics with a 100-mm EFL, Eq. (5.6) shows that $\text{FOI}_x \approx (4 \text{ mm})/(100 \text{ mm}) = 0.04$ rad whereas $\text{FOI}_y \approx (13.5 \text{ mm})/(100 \text{ mm}) = 0.135$ rad. The illuminated field is therefore elliptical, and the associated solid angle $\Omega = \pi(\text{FOI}_x)(\text{FOI}_y)/4 = \pi(0.04 \text{ rad})(0.135 \text{ rad})/4 = 0.00424$ steradians.

The lamp radiance is given by $P/A_s\Omega_{s'}$ where $P = 600$ W, A_s is the source area ($\pi \times 4$ mm $\times 13.5$ mm), and Ω_s is the solid angle into which the source emits (assumed to be 4π steradians). However, the collimator radiance is smaller because the collimating lens collects light from only a fraction of this solid angle. As a result, the power emitted by the collimator is $\Phi = TP\Omega_L/\Omega_s$, where $\Omega_L = \pi(D/f)^2/4$ is the solid angle of the lens as viewed from the source; thus, $\Phi = 0.75 \times 0.9 \times 600\text{ W} \times (\pi/16)/4\pi = 6.3$ W if we assume an electrical-to-optical conversion efficiency of 75 percent and an optical transmission of 90 percent.

The collimator brightness is then calculated from the power Φ and the etendue of the optics. It was shown in Chap. 4 that the etendue can be found either from the product of source area and f-cone angle or from the product of entrance pupil area and FOI solid angle. Using the FOI, we calculate the etendue as $A\Omega = \pi(0.05\text{ m})^2/4 \times 0.00424\text{ sr} = 8.33 \times 10^{-6}$ m²·sr. The resulting collimator brightness is then $L = \Phi/A\Omega = (6.3\text{ W})/(8.33 \times 10^{-6}\text{ m}^2\cdot\text{sr}) \approx 7.56 \times 10^5$ W/m²·sr.

Since this value may be too low for the rest of the optical system, we consider using a more powerful, 1000-watt source (Newport #6317), which has filament dimensions of 5 mm in diameter × 18 mm in length and the same filament temperature as the 600-W source. The illuminated solid angle is now $\Omega = \pi(\text{FOI}_x)(\text{FOI}_y)/4 = \pi(0.05\text{ rad})(0.18\text{ rad})/4 = 0.00707$ steradians, and the resulting etendue is $A\Omega = \pi(0.05\text{ m})^2/4 \times 0.00707\text{ sr} = 1.39 \times 10^{-5}$ m²·sr; this value is greater than that of the 600-W source by a factor of about 1.7. The collimator brightness with this lamp is $L = \Phi/A\Omega = (10.5\text{ W})/(1.39 \times 10^{-5}\text{ m}^2\cdot\text{sr}) \approx 7.55 \times 10^5$ W/m²·sr, or almost exactly the same brightness as we found for the 600-W source!

Physically, the reason for this nonintuitive result is that a higher-power source of the same type (tungsten halogen, in this case) requires a bigger filament. This is collimated into a larger FOI, which reduces the collimator's brightness even though the source itself has more power. Without question, the more powerful source will put more power in the FOI [see Eq. (4.8)]. Yet the field being illuminated is also larger, so the power in a given solid angle is approximately the same because the source irradiance (~ 8.3×10^6 W/m²) is nearly the same for both the 600-W and 1000-W lamps.

The brightness can be increased by using a smaller f/# for the optics. Collimators often employ fast optics for this reason—on the order of f/1 and faster using off-the-shelf aspheres. (The limitations on this technique are described in the paragraphs that follow.) Also, a reflector can be added behind the source to collect more power than can be obtained by using a lens alone. It may also be possible to improve the design by switching to a different type of source (e.g., an arc lamp).

Neither point sources nor extended sources are guaranteed to have diffraction-limited wavefront quality across the collimator aperture. Therefore, some effort must be put into designing the optics so that they have low wavefront error in applications that require the largest possible collimator radiance. It was shown in Chap. 3 that low wavefront quality led to spot sizes much larger than diffraction limited. Collimator design is exactly analogous: low wavefront quality leads to beam divergence angles that are greater than predicted by diffraction. This means that a design with poor WFE requires more source power to obtain a given level of radiance. The design trade-off between source power and optical quality is almost always resolved by investing design effort into reducing WFE.

Chromatic aberration with refractive collimators is a common problem with broadband sources. As we saw in Chap. 3, shorter wavelengths typically refract with a higher index than do longer ones, and Eq. (2.1) shows that the focal length of a thin lens is shorter for shorter wavelengths. If we design a collimator for the middle of the wavelength band, then the shorter wavelengths will be at a distance greater than the short-λ focal length away from the lens and thus will converge to a real image somewhere on the other side of the lens. At the same time, the longer wavelengths will be at a distance less than the long-λ focal length away from the lens and so will diverge to a virtual image. Achromats can be used, but they convert light to heat via absorption and are difficult to design for broad wavelength bands; hence mirrors are commonly used instead.

The temperature of the optics can also be a limiting factor in collimator design. In Example 5.4, only 10 W or so from a 1000-W source manage to make it through the collimating optics. The remaining 990 W are either absorbed or bouncing around somewhere in the system, and some of this may end up back on the lenses. Even 10 percent of this is a huge amount of power, and can easily cook the optics until they are hotter than a summer day in Tucson. The most obvious result is thermal fracture, and even though there may be enough heat transfer to avoid this, excess WFE remains a serious problem in engineering source-plus-optics systems with high-power sources. These thermal design issues will be looked at more closely in Chap. 7. There are also a number of other details involved in choosing an optical source, many of which will be reviewed in Sec. 5.3.

5.2.2 Source Coupling

Just as there are imaging applications that obtain an image from an object at infinity, there are also those that image finite conjugates. Similarly, there are illumination applications that require a collimated source and others that require finite conjugates. For example, coupling a large source into a smaller aperture is not efficient with a collimator: because the illuminated field is larger than the aperture, a large fraction of the collimated beam cannot be coupled.

A more efficient way to couple the source into the aperture is by reimaging the source directly onto the aperture with a coupling lens. This is shown in Fig. 5.18, where the source and aperture are at finite conjugates; the source image illuminates anything to the right of the

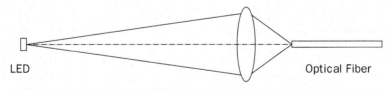

FIGURE 5.18 Finite-conjugate coupling of an LED into an optical fiber; the requirements for such an optical system are illustrated in Example 5.5.

aperture as if the source were at the aperture itself. This is most useful when the image needs to be smaller than the source or when the mechanical housing for the source is too big to fit directly in the aperture.

However, for efficient coupling it is not enough simply to match the image size to the aperture size. Optical fiber, for example, relies on the principle of total internal reflection, and any ray angle greater than the TIR angle won't be coupled into the fiber. A typical numerical aperture (NA) for standard telecommunications fiber with a 10-μm aperture at a 1550-nm source wavelength is about 0.1, which Eq. (2.6) shows is equivalent to a coupling optic with a working f/# of f/5. Even with an image size that matches the 10-μm aperture, faster lenses (with lower f/#) will not couple more light into the fiber.

Not surprisingly, the combination of image size and f/# is the critical factor when coupling sources. This was illustrated in Chap. 4 via the concept of etendue, which must be conserved for efficient coupling. Again using telecommunications fiber as an example, the etendue of the lens-plus-fiber combination is $A\Omega = 0.25\pi(10 \times 10^{-6} \text{ m})^2 \times \pi/[4(f/\#)^2] = 2.5 \times 10^{-12}$ m²·sr, an incredibly small value.

The consequence of a required etendue on the image side of the coupling lens is that there is no advantage for the etendue on the object side to be larger, and that puts a design restriction on the source area and object-side f/#. For a fiber, this is impossible to match efficiently with anything but a laser, which demonstrates the value of etendue calculations in selecting a source. Of course, one can always use a source with more power than is needed, but (say) a 1-W source will couple only a microwatt of power into the rest of the optical system. Example 5.5 illustrates the difficulties of matching of both aperture and field (or area and solid angle) to minimize losses when coupling an LED into optical fiber.

Example 5.5. Optical systems can often be made simpler and more robust by coupling the source into an optical fiber that transmits the light to other parts of the system. One drawback of this approach is the difficultly of coupling most sources into fiber without a huge loss of power. Optical fiber coupling is most often used in laser systems, though LEDs and arc lamps are also employed. In this example, an LED is used as a source to calculate the expected power loss.

The coupling geometry is shown in Fig. 5.18, where a lens is used to image the LED onto the face of the fiber. The fiber is standard telecommunications fiber, which has a 10-μm circular aperture for a 1500-nm source wavelength and an NA ≈ 0.1 (or a working f/# = 1/2NA = 5). Before jumping into the calculation implied by Fig. 5.18, we should ask this question: Why not just butt the LED right up against the fiber and thereby avoid the expense of using a lens?

The source is a 1 mm × 1 mm high-brightness LED, which we assume does not have a lens and thus is a Lambertian emitter. If the LED is butted against the fiber, then there will be losses associated with the mismatch in area and solid angle (i.e., the etendue). Specifically: the emitting area of the LED is 10^{-6} m² but the collecting area of the fiber is only 7.9×10^{-11} m², so at most the fiber will be able to collect only about 1 watt in 10^4 owing solely to area considerations.

However, there is also loss associated with the mismatch between the LED's emitted solid angle (2π steradians over which a Lambertian source emits) and the smaller angle collected by the fiber: $\Omega \approx \pi/[4(f/\#)^2] = 0.0314$ steradians. Thus, based on solid angle, only 1 watt in 200 (= $0.0314/2\pi$) is collected. The net effect of the etendue mismatch between source and fiber is therefore approximately 200×10^4, with only about 0.5 microwatts from each watt of LED source power making its way into the fiber if butt coupling is used.

Can a coupling lens do any better? To answer this question, it's best to start with the etendue required and then see what that implies for the coupling optics. For the area, the 1-mm^2 LED must be reimaged down to a 10-µm diameter; this requires a finite magnification $|m| = s_i/s_o = (0.01 \text{ mm})/(1 \text{ mm}) = 1/100$. For the solid angle, if we ignore the difference between square and circular geometries then an f/5 image cone with $|m| = 1/100$ implies an f/500 object cone.[20] The resulting solid angle intercepted by the coupling optics is $\Omega \approx \pi/[4(500)^2] = 3.14 \times 10^{-6}$ steradians; because the LED is emitting into 2π steradians, the result is a collection efficiency of only 1 part in 2×10^6. The lens thus couples the *same* power as does butting the LED against the fiber—before we consider that it is not possible to create an image size of 10 µm for a Lambertian source with an f/5 optic.[20]

Surely it's possible to do better with the lens. Tracking down the reason for the poor collection efficiency, it's clear that matching the size and solid angle of the image to those of the fiber will require a small object-side f/#. This entails using a smaller source so that the magnification is smaller and the object-side f/# is bigger. Standard LEDs with 0.1 mm × 0.1 mm area are common, and for these we have $|m| = 1/10$ with an f/50 object cone.

The solid angle collected by the coupling lens then becomes $(50/5)^2 = 100$ times greater than with the case of the larger LED, which would seem to imply a 100× improvement in collection efficiency. Unfortunately, the total power coupled into the fiber hasn't been increased because it was necessary to use an LED with 100 times less emission area (and therefore 100 times less output power).

In sum: for a source-plus-lens etendue that is larger than the fiber, it is not possible do any better than butt coupling because the etendue of the fiber is the smallest in the chain of the coupling system. Coupling lenses can be useful when the geometry does not allow butt coupling. But to collect more power from the source, it's necessary to use either a higher-power source or a fiber with a larger etendue (i.e., one with a larger area and/or collection angle). For this purpose, multimode fibers[21] or light pipes are available.

5.3 Source Specifications

In Secs. 5.1 and 5.2, five different types of sources and their common systems applications were reviewed. It was seen that tungsten-halogen lamps, for example, are relatively low-cost, high-power sources for visible, NIR, and SWIR photons. In this section, we take a more systematic look at the properties of optical sources by summarizing what must be specified for any type of source to be "designed in" to a system.

5.3.1 Output Power

The most basic specification of a source starts with this question: How many watts of optical power are available? Unfortunately, only LEDs and lasers specify output power (Fig. 5.19). Thermal

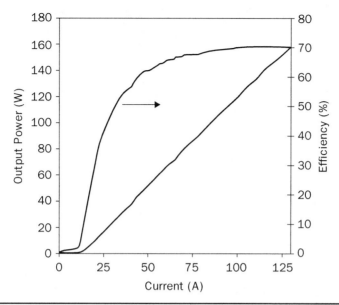

FIGURE 5.19 The output from a high-power diode laser varies linearly with drive current (up to its rated current). (*Credit: Paul Crump et al.*, Proceedings of the SPIE, *vol. 6456.*)

sources and arc lamps are specified in terms of the electrical input power (or power consumption). For example, a 100-W tungsten-halogen lamp uses 100 W of electrical power, and outputs approximately 80 W of optical power,[8] of which only 10 W is visible output. This information is not always available, but it may be found by using either the irradiance at a certain distance (Sec. 5.3.3) or the wall-plug efficiency (Sec. 5.3.6). Laboratory blackbodies are usually specified in terms of the emission temperature (e.g., "a 1000-K blackbody"), which allows the output power to be calculated from the surface area and Eq. (5.1).

5.3.2 Output Spectrum

The output power is usually specified as the total power over all wavelengths, even though there are many situations where it is necessary to know the power within a specific band of wavelengths. As described in Sec. 5.1.2, the visible band even has its own units of power (lumens) that take into account the relative response of the human eye. Output spectrum is important also because it affects what can be done with the source and the design of the optics-plus-source system, since chromatic aberrations may affect the optical design for a broadband source.

Two aspects of output spectrum must be considered for system design: the wavelength the source outputs at its peak emission, and the spectral width. The spectral width depends on the type of source. Thermal sources such as tungsten-halogens tend to be broadband (~ 2000 nm) whereas quantum sources such as LEDs are narrow (~ 50 nm). The spectrum may also depend on a number of other factors; these include source temperature (for blackbodies and tungsten-halogens) and drive current (for LEDs) as well as aging and use (Fig. 5.14). Care must also be taken to filter the wavelengths of the output spectrum that are not needed because otherwise they may end up on the detector as background that obscures the signal.

5.3.3 Spectral Irradiance

Spectral irradiance combines the output power and spectrum into a single specification. As shown in Fig. 5.14, vendors typically supply plots showing how the spectral irradiance varies with wavelength for a given source. The irradiance is often measured at a distance of 0.5 meters from the source, and it is given in units of $W/m^2 \cdot \mu m$ or $W/m^2 \cdot nm$.

The output power *collected* from a source is then found by first adjusting the spectral irradiance at 0.5 meters for the distance of the collection optics in the system. Because the intercepted solid angle is larger for a lens at a distance of (say) 100 mm, this lens collects approximately $(5/1)^2 = 25$ times more power (from a Lambertian source) than does the same lens at a distance of 500 mm. The spectral power [W/nm] is calculated as the product of this spectral irradiance and the entrance pupil area of the collecting optics; the collected power (W) is then found by integrating the spectral power over all wavelengths used in the system (or simply multiplying if the power is constant over the wavelength spectrum). The visible-spectrum output (lumens) can also be found by first converting the spectral irradiance to its visible equivalent (*lux*, with units of lm/m^2) using the method illustrated in Example 5.2.

Some vendors sell their sources pre-packaged in a housing, in which case they also supply a conversion factor for the lens area that takes Fresnel and transmission losses into account. In addition to a collection lens, some housings also include collection mirrors behind the source; this increases the output irradiance by 60 percent or so, and this value should also be available from vendors selling source-plus-optics systems.

For initial systems engineering calculations, vendor irradiance curves (see, e.g., Fig. 5.14) are acceptable but should not be relied upon too heavily for the actual hardware. During the design phase, the sources must be measured directly—especially if either the accuracy requirement or the power budget is not well established.

5.3.4 Spectral Radiance

Chapter 4 showed that one of the fundamental quantities of radiometry is brightness or radiance (units of W/m^2·sr). Although source *power* is not conserved as it is collected by a collimating lens, a brighter source ultimately puts more power in the FOI. It is also source radiance that determines how much power can be reimaged onto a target, and spectral radiance (units of W/m^2·sr·μm) is what determines the amount of power in a specific band of wavelengths.

A source that emits a lot of power from a small area into a small solid angle is the brightest possible design. Source radiance thus takes the following facts into account: (1) a source may have a relatively small emitting area (as found in short-arc lamps) or a large area (as in high-power tungsten-halogens); and (2) a source may emit in all directions (like lightbulb) or in a very specific direction (as with a laser or lensed LED). Even though the output power of a 100-W tungsten halogen is about 80 W and that of a helium-neon (HeNe) laser is 10,000 times less, the laser is much brighter (~ 10^{10} W/m^2·sr) because it emits into a smaller solid angle from a smaller area.

However, the laser's brightness is much *lower* if the radiance is measured at wavelengths that the laser doesn't emit. At wavelengths as small as 0.1 nm away from 632.8 nm, for example, the standard HeNe laser's output power is essentially zero, as is its brightness. The concept of spectral radiance [W/m^2·sr·μm] thus takes into account the possibility that sources can emit a relatively narrow band of wavelengths (as by an LED or laser) or a wide band (as by thermal sources such as blackbodies or tungsten-halogens). This concept also serves as a caution when selecting a source, since a laser's or LED's high level of spectral brightness is not useful unless it is emitted over the wavelengths needed for the application.

Radiance must often be derived or calculated from available data. The brightness of a 100-W tungsten-halogen lamp, for example, is calculated from the expression for brightness ($L = P/A\Omega$) along with vendor data. In the case of halogens, vendor data consists of: (1) electrical input power, which is used with the wall-plug efficiency (see Sec. 5.3.6) of 75 percent to give 75 W of output power; and (2) source dimensions—for example, 4.2 mm in length × 2.3 mm in diameter—from which the emitting area $A \approx 3 \times 10^{-5}$ m^2. An assumption must also be made regarding the solid angle Ω into which the source emits, which is taken to be 4π steradians (spherical) for a cylindrical Lambertian emitter. The source radiance is then 75 W/(3 ×10^{-5} m^2 × 4π sr) = 0.2 × 10^6 W/m^2·sr. This is an average radiance over the emitting area; the actual value may differ over specific regions owing to non-Lambertian emittance and source nonuniformities (see Sec. 5.3.9).

5.3.5 Source Geometry

One way to classify sources, then, is based on the size of the emitting area; system performance is also affected by the choice of a point source or an extended source. For example, we saw in Sec. 5.2.1 that the angular size of a collimator's field depends on the source size and lens EFL for an extended source [Eq. (5.6)] and that the lens diameter determines the FOI for a point source whose size is less than the Airy disk [Eq. (5.7)].

Most sources are extended sources, although short-arc lamps (with arc sizes less than 1 mm^2) are sometimes considered points, depending on the focal length of the lens. LEDs are even smaller, and they, too, can be used as point sources with the appropriate optics. Just as with the case of imagers, classifying a source as a "point" depends not only on the size of the source itself but also on its angular size, which in turn is determined by the source dimension and the focal length of the collimating lens.

As shown in Example 5.4, the extended sources used in collimators are not necessarily brighter just because they're bigger or output more power. This statement holds also for finite conjugates, where larger sources can actually waste energy if they are being imaged onto a small target area. As a result, a larger source will merely put more energy outside the target for a source image that is already larger than the target. Maximizing output radiance, not output power, is the reason that source designers try to pack a lot of power into a small area; for such applications, arc lamps and lasers are the preferred choice.

5.3.6 Efficiency

We have seen that source energy is wasted by a mismatch between source geometry and image size. In addition, the electrical energy that powers the source is often not converted efficiently into light. The term "efficiency" can be defined in many different ways. In this section we shall examine the most common: wall-plug, luminous, and spectral efficiencies.

Wall-Plug Efficiency

Anyone who has ever touched an incandescent bulb that's been burning for an hour knows that sources can get hot! Part of the reason for this is that some sources (e.g., blackbodies, incandescent bulbs, tungsten-halogens) must reach a certain temperature before they work properly. Another reason is that the electrical power that's required to run a source is not all converted into optical power, and the difference between the electrical and optical power ends up as thermal power (or heat).

The wall-plug efficiency η_{wp} is the ratio of optical output power to electrical input power ($\eta_{wp} = \Phi_{opt}/P_{el}$); it is so named because sources are often connected to a "wall-plug" outlet to obtain electrical power.[22] Thus, a 100-W tungsten-halogen lamp with a wall-plug efficiency of 80 percent uses 100 watts of electrical power, but outputs only 80 watts of optical power. The difference in power is heat, which is either removed or increases the temperature of the source (gas and bulb).

Unfortunately, wall-plug efficiency can be a surprisingly difficult number to obtain from a vendor. Even LED manufacturers don't always state this efficiency, although it can be calculated from the given output power and the product of operating current and voltage. A spec sheet for a common AlGaAs LED, for instance, lists an output power of 40 mW, for which 2.5 V and 200 mA are required. The wall-plug efficiency of the LED is thus 40 mW/(2.5 V × 200 mA) = 0.08 (or 8 percent).

Operating conditions dramatically affect wall-plug efficiencies. Higher temperatures reduce the efficiency of LEDs (Fig. 5.20). However, under higher temperatures both blackbodies and tungsten-halogens require less electrical energy to heat up the emitting area to the operating temperature, which improves wall-plug efficiency. Vendors can usually supply detailed information on request.

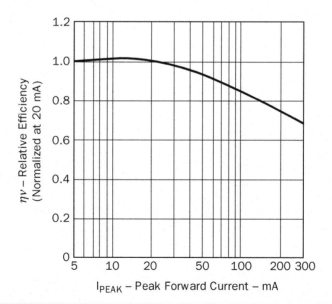

FIGURE 5.20 Relative efficiency versus operating current for a low-power LED; the plot illustrates the effect of power dissipation (and thus of temperature) on output power. The temperature increase also shifts the output spectrum to longer wavelengths (not shown). (*Credit: Avago Technologies, www.avagotech.com.*)

Luminous and Spectral Efficiency

Optical sources have been much in the news lately as source manufacturers seek to develop more efficient bulbs and lamps that save money and reduce carbon emissions. For example, LEDs are now being developed with the goal of making them more efficient than compact fluorescents, which in turn are more efficient than incandescent bulbs.

What the popular press has generally not made clear is that the "efficiency" being talked about is actually *luminous* efficiency (efficacy). This is a metric for how much visible power is emitted in the 400–700-nm band for a given electrical input power, and it is measured in units of lumens per watt. Any light that is created outside the visible band is wasted and therefore reduces efficacy. For example, tungsten-halogen lamps with wall-plug efficiencies near 80 percent have luminous efficiencies of only about 20 lm/W; the reason is that they create a lot of NIR and SWIR photons, which are outside the visible band (see Example 5.2).

In contrast, LEDs can be tailored to create nothing but visible photons, and they also have high luminous efficiencies (~ 70 lm/W) despite their low output power and low wall-plug efficiency. Arc lamps are unusual in that their luminous efficiency actually increases at higher power, increasing from about 20 lm/W for a 150-W xenon lamp to about 30 lm/W for a 450-W lamp.[12] Presumably, the increase in power shifts the output spectrum more toward the visible in response to a higher xenon gas temperature. Keep in mind also that products with high levels of luminous efficiency may also have relatively low output power (see Sec. 5.1.4).

Luminous efficiency is a special case of spectral (or color) efficiency in which the spectrum of interest is the visible band (400–700 nm). Example 5.2 shows that tungsten-halogens have a visible emission efficiency of only 10 percent when the spectral response of the human eye is included in the calculation. Of course, other wavelength bands are useful as well, and for any wavelength the spectral response of the detector affects overall system performance. The detector used in Example 5.2 is the human eye, whose response is a specific $V(\lambda)$. Chapter 6 describes many other types of detectors, with different $V(\lambda)$ values that are used in optical systems.

5.3.7 Power Consumption

Electrical power consumption can often be derived from the other lamp specs, but it is useful to state it explicitly on the spec sheet. This reinforces the fact that, whereas normalized numbers such as luminous efficiency are useful for comparisons, absolute numbers make or break the viability of the system. It also allows for a check on the calculations: if you find that 2 watts of optical power are output from only 1 watt of electrical power, then there must (by the laws of thermodynamics) have been computational errors.

5.3.8 Stability

Even though a source is rated as having a certain output power, this power will change over time. There are many reasons for this: warm-up time, wear and tear due to lamp design and construction, power supply fluctuations, and long-term aging. Low-level output-power fluctuations can also be due to quantum-mechanical effects such as photon noise (see Chap. 6).

Warm-up time is determined by how quickly a thermal source reaches its operating temperature. A large lamp with a big thermal mass takes longer to heat up, but is also more stable with respect to voltage or current fluctuations in the power supply. As a result, blackbodies and tungsten-halogens are generally the quietest sources available. Unfortunately, all the parameters that contribute to output power—optical transmission of the bulb, emissivity of the filament, and so forth—also contribute to a slow reduction in output power over the lamp's lifetime.

Arc-lamp discharge is an inherently noisy process, and output power can vary by as much as ±10 percent for unregulated lamps (Fig. 5.21). The designer can use a feedback loop to control stability by monitoring the output power and feeding the signal back to the power supply; in this way, power values that deviate less than 1 percent are possible for regulated arc lamps. Output-power stability is also a function of lamp temperature, so convective (forced-air) cooling is often used to prevent arc lamps from losing output power because of overheating.

Quantum sources such as LEDs and lasers do not rely on temperature to generate photons; however, their center wavelength

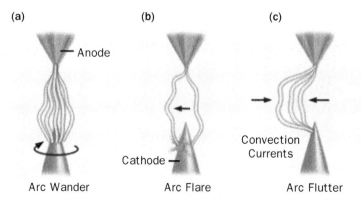

Figure 5.21 Sources of instability in an arc lamp's output power. (*Credit: Carl Zeiss, Inc.*)

does shift as the operating temperature changes. Over time, output power drops slowly because of an unavoidable decrease in wall-plug efficiency (a result of various semiconductor degradation mechanisms).[13]

5.3.9 Uniformity

The field illuminated by a collimator may contain some areas that are brighter than others. This variation may be due to vignetting in the optical design, but it may also be due to a nonuniform source (Fig. 5.22). That is, the irradiance emitted by a source may vary as a function of *where* on the source it is measured.

The "hot spots" where the irradiance is higher should be avoided. By paying careful attention to temperature distribution, designers can create blackbodies that emit nearly the same power over any small area; typical uniformities are on the order of 4°C out of 800°C. However, the arc in an arc lamp cannot be controlled in this way, and the lamp's irradiance may vary by as much as ±10 percent.

If the radiance varies with angle, then the nonuniformity may indicate that the source is not Lambertian. Initial systems engineering calculations often start with the Lambertian assumption, which is reasonable enough for thermal sources but is not valid for lensed LEDs (whose emission is highly directional; see Fig. 5.15). In actual optical systems, very few sources are Lambertian and so the designer must use spatial distributions for actual sources. These distributions can be calculated from measurements and are also available from the

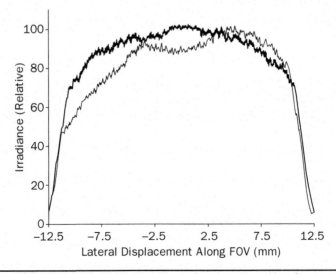

FIGURE 5.22 Nonuniformities in the output of arc lamps used in a fluorescent microscope. (*Credit: C. F. Dimas et al.*, Proceedings of the SPIE, *vol. 5620.*)

libraries of optical design software packages (e.g., Breault Research Organization's ASAP).

Factors that affect the uniformity of a source-plus-optics system include aperture size, source type (thermal versus quantum, blackbody versus tungsten-halogen, etc.), source design (e.g., the temperature distribution of a thermal source), source size, and source *spatial* stability (since the emitting area in an arc lamp drifts slowly from one electrode toward the other as the cathode wears during burn-in, changing the position of the source with respect to the optics). The best possible uniformity can be obtained by combining a source with an optical component known as an *integrating sphere*; see Labsphere's Web site (www.labsphere.com) for more details. Other components such as light pipes and diffusers are also commonly used to improve source uniformity.[1]

One final note: the term *collimated* sometimes carries the popular connotation of "uniform illumination." The origins of this optical "urban myth" are not clear, but the concepts are not the same. "Collimated" denotes planar wavefronts across an aperture, whereas uniformity denotes constant irradiance. It's possible for a source to be collimated but not uniform (laser beams are a good example) or to be uniform over a small area but not collimated. Both properties are often required in practice, but there is no inherent connection between the two. In fact, creating a system that meets both requirements often requires a significant design and engineering effort.

5.3.10 Size

Example 5.4 shows that a halogen source with a larger emitting area does not necessarily result in a brighter collimator, though it will output more power. Reimaging applications may also fail to benefit from a larger source area, since the bigger source just places more power into a bigger spot (with the same irradiance) on the target.

Although the size of the emitting area may not affect the brightness of a particular type of source, the overall mechanical size of the source is an important specification. The primary concern here concerns the mechanical "packaging" aspect of fitting the source into the space available; more time than you can imagine will be spent on this particular task. It's important to include not only the lamp in this design effort but also the size of the optics and power-supply electronics and thermal management components (e.g., heat fins, cooling fans, water pumps, and so on). Such "extras" can easily increase the size of the source-plus-optics system to 10 times the size of the lamp itself (Fig. 5.23).

In addition, the size of the source-plus-optics aperture from which the power is emitted affects other source specifications. The uniformity of the output beam, for example, varies with aperture

226 Chapter Five

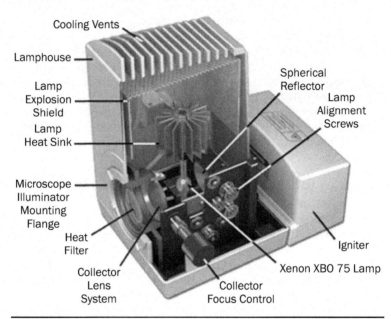

FIGURE 5.23 An optical system that uses an arc lamp as its source to create collimated light; the size and weight of the system are significantly greater than the lamp itself. (*Credit: Carl Zeiss, Inc.*)

size; for apertures larger than about 25 mm, it may be difficult to find an off-the-shelf product that suits the application.

5.3.11 Lifetime

A source that lasts only a week may be acceptable for a lab experiment, but is completely unusable for a commercial product. What's meant by "lasts" needs to be defined; it's usually taken to mean that output power has not yet dropped to half of its beginning-of-life (BOL) value. Such end-of-life (EOL) definitions vary, so it's important know which one the vendor is using. It's also important to put an additional term (with some margin) in the power budget (Sec. 4.2) to accommodate the expected BOL-to-EOL change in performance specification—be it power, uniformity, efficiency, or some other spec.

Typical lifetimes range from 750 hours (for a common tungsten-filament incandescent bulb) to about 2000 hours (for tungsten-halogens and arc lamps) to 50,000 hours (for LEDs). Everything that contributes to output power—transmission of the bulb, filament emissivity, filament diameter (which is reduced by tungsten evaporation), and so on—changes over time to reduce the output power at the EOL. As illustrated in Fig. 5.24, the lifetime for tungsten halogens also depends on operating voltage and thus can be significantly increased by derating the voltage to a value below nominal.

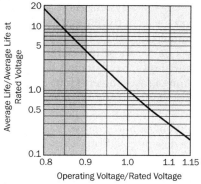

FIGURE 5.24 Degradation of a tungsten filament after several hundred hours of use *(left panel)*. The degradation of a tungsten-halogen lamp decreases significantly as the operating voltage is reduced *(right panel)*. (Photo credit: Wikipedia author Rolf Süssbrich. Graph credit: Newport Corp.; used by permission.)

There may also be special conditions attached to the lifetime such as continuous operation or limits on the number of on–off cycles, which can lower lifetime owing to thermal shocks associated with turn-on. Lifetime is also shorter than advertised because of such "small" things as contaminating the glass bulb with finger grease. Source vendors often provide application notes with guidance on these topics.

5.3.12 Cost

A source which meets all of the technical and lifetime specifications won't make it past management if it's too expensive. A number of metrics are used to specify cost, including cost per lumen, cost per watt, and total cost. Total cost includes the price of the source(s), the cost of the housing and optics, operating costs (e.g., electricity and cooling water), maintenance costs (e.g., replacement of failed lamps and corroded plumbing), and instrument downtime over the life of the instrument (not just the source). It's common *not* to include maintenance costs as part of the overall cost budget; this omission is misleading, and being up-front with customers about such costs at the outset will prevent headaches later on.

5.4 Source Selection

In this section we bring together information from Secs. 5.1 to 5.3 to help with the decision of choosing an optical source. The goal is not to define system requirements or what the source must do (i.e., its output power, radiance, size, cost, etc.) but rather to provide a methodology for selecting sources—one that applies equally well to other components, such as optics and detectors—given the requirements.

Once the system and source requirements have been settled on, source selection often starts with this question: What is available as a commercial off-the-shelf (COTS) product? Source manufacturers ("vendors") must therefore be identified through online searches, trade shows, word of mouth, and so on. In turn, a vendor will ask the following questions: How much power is required, and at what wavelengths? How big can it be? What's the budget? And so forth and so on. Thus, it's a good idea for these questions to be addressed beforehand and summarized in a list of important properties known as a design specification (or "spec"). In practice, a spec is often a long, detailed document containing lots of legalese, but it's always best to start out with a simple, one-page summary.

One purpose of a specification is thus to summarize the best available numerical estimates for these properties. Table 5.5, for example, shows a list of source properties along with the unit of measurement that might be used when quantifying each property. These numbers will typically result from the process of source-plus-optics requirements flowdown that ends in component specifications. For example, in Sec. 5.2.2 we saw how etendue calculations affect source selection. The requirement on etendue performance then "flows down" to the source-plus-optics spec as an acceptable aperture area and solid angle.

Property	Units
Output power	W
Output spectrum	µm or nm
Spectral irradiance	$W/m^2 \cdot \mu m$
Spectral radiance	$W/m^2 \cdot sr \cdot \mu m$
Emitting area	m^2
Irradiance stability	%
Irradiance uniformity	%
Electrical power	W
Wall-plug efficiency	%
Luminous output	lm
Luminous efficiency	lm/W
Total size	m^3
Weight	kg
Lifetime	hours
Cooling requirements	N/A
Total cost	US$, €, ¥, £, etc.

TABLE 5.5 Typical properties and units used when specifying optical sources.

Stepping back a bit from the list of properties, there are three larger issues that must be considered in source specification: (1) normalized numbers (e.g., lumens per watt); (2) spec conditions; and (3) missing information. Perhaps the most important of these is going *beyond* the normalized numbers when choosing a source. The ultimate criterion for evaluation is system performance, and this almost always comes down to an actual number, not a normalized one. For example, with a fixed budget with a limited amount of money to spend on a source, the cost per lumen is useful for comparing different sources but not for determining the final answer. It's easy enough to obtain that final answer, but that step is sometimes left out of the design process.

The second issue concerns the operating conditions under which the specs are quoted. Is the vendor using a limited range of conditions that "cherry pick" the best specs under the best conditions? What's needed is a set of clearly stated conditions that can be extrapolated and interpolated to other conditions; only then can an intelligent choice be made. The pertinent information is not always easy to find, and vendors cannot be expected to have it. In such cases, measurements must be taken or a decision must be made based on experience—despite the fact that all the required information is not available.

Finally, it's important to ask what information is *not* being talked about. For example, a recent advertisement for a high-power LED states that the luminous efficiency is 70 lm/W at 350 mA of input current, "with more than 160 lumens at higher drive currents." In a single sentence the topic has changed from luminous efficiency to output lumens, leaving the efficacy at these higher drive currents unstated. Are we to assume the efficiency is the same as that at the lower drive current? Perhaps some readers will, but it's an uninformed assumption because efficiency will decrease as the LED grows hotter at higher drive currents (Fig. 5.20).

With these precautions in mind, the rest of this section summarizes the important properties needed to specify an optical source. It's not possible to cover everything needed for specific designs, so some topics (e.g., modulation rates and linearity, operating temperature, total subsystem weight) are not reviewed in detail. Source selection is usually an involved process requiring comparison of source specs from different vendors, none of whom has exactly what's needed. It's useful to put everything in a *comparison matrix* table. It's also useful to focus in on the most important three or four requirements first, which forces consideration of the design in terms of what is critical. The table can be refined with more detail and specifications in the subsequent iteration.

Therefore, the first step is to compare and contrast the different types of sources—working from system requirements to source specifications. As shown in Table 5.6, the comparison matrix lists COTS specifications for the various types of sources covered in this

Property	Blackbody	Tungsten-halogen	Arc lamp	LED
Source type	Thermal	Thermal	Thermal/Quantum	Quantum
Visible color	Black	Yellow-orange	White (xenon) Green (mercury) Orange (sodium)	Blue → Red
Output spectrum	MWIR-LWIR	VIS-SWIR	UV-NIR	Blue → SWIR (~ 20–50 nm FWHM)
Output power	180 W	800 W	~ 150 W	0.01–3 W
Wall-plug efficiency	90%	80%	15%	2–10%
Luminous output	~ 0 lm	~ 28,000 lm [11]	10^3–10^6 lm	~ 100 lm (white-light HB-LEDs)
Luminous efficiency	~ 0 lm/W	15–25 lm/W	30–60 lm/W	50–100 lm/W
Emitting area	50–350 mm dia.	5 mm × 18 mm	1 mm × 3 mm	1 mm × 1 mm
Photometric radiance (luminance)	~ 0 lm/mm^2·sr	10–20 lm/mm^2·sr [10]	400 lm/mm^2·sr (xenon)	~ 10 lm/mm^2·sr
Uniformity	±0.01–0.05°C	Fair	±10%	Directional with lens
Stability	±0.1–0.25°C	±1%	±10% [19]	$\Delta\lambda \sim \Delta T$
Power consumption	200 W	1000 W	1000 W	20–5000 mW
Current	1–2 A	8–10 A	40–50 A	10–1000 mA
Cooling requirements	Fan or H$_2$O cooling	Natural cooling	Fan or H$_2$O cooling	Good heat sink
Lifetime	Long	50–2000 hours	1500 hours	50,000 hours
Total cost	Moderate	Low	Moderate	Low

TABLE 5.6 Comparison matrix for typical optical sources operating in a room-temperature environment. Photometric radiance (i.e., the radiance emitted in the visible spectrum) is more commonly known as luminance. All entries are beginning-of-life values.

chapter. It is intended to summarize available performance in one page and to clarify the advantages and disadvantages of each type of source for a particular application and set of requirements.

Looking at the most critical three or four requirements for each source, the comparison matrix results can be distilled into the following list:

- *Blackbodies*—relatively large, uniform, stable sources of MWIR-to-LWIR photons for laboratory calibration of optical performance
- *Tungsten-halogens*—high-power, low-cost, visible-to-NIR sources with good wall-plug efficiency but poor visible (luminous) efficiency
- *Arc lamps*—extremely bright, high-power sources with UV output; they usually require cooling hardware
- *LEDs*—low-power, low-cost with high spectral brightness but a limited output spectrum

The list descriptors should be expanded as the options narrow down from the four types listed here to perhaps one or two. If high brightness is critical, then an arc lamp (or a laser) is the likely choice; if low cost is the most important requirement, then tungsten-halogens or LEDs are good options.

Summary tables such as these can often be misleading and therefore must be interpreted carefully. For example, it's important to use the specifications for COTS parts and not for "vaporware" or research-ware that won't be available for years (unless that's what is being evaluated). There is also the issue, mentioned previously, of the listed specs applying to a single product and not being "cherry picked" from various sources under different operating conditions. LEDs, for instance, are often specified at a temperature that is lower than their operating temperature, artificially enhancing their efficiency. To the extent that they are also specified at lower-than-operating drive currents, LEDs will seem to be cooler (and thus more efficient) than they are in practice. If you fail to ask the vendor the right questions, the result may be a source that's not as efficient, as bright, or as powerful as expected. Naturally, this will affect other parts of the instrument power budget, cost, and so on.

It's fine if the initial entries in the comparison matrix include such qualitative descriptors as "low-medium-high" or "red-yellow-green." Quantitative data are often not available, and extracting them from known data can take some time; to start off quickly, qualitative comparisons may be required.

When using such tables for source selection, many designers get themselves into trouble by focusing exclusively on one requirement at the expense of others. For example, wall-plug efficiency is sometimes selected as the most important requirement but without consideration given to total power consumption or the size of the entire system, including cooling and optics. Choosing a source is an intuitive "juggle" of many counterbalancing parameters in a large trade-off space. Unfortunately, many engineers are taught to think in terms of overly mathematical, "theory of everything" solutions and so are poorly equipped for the task of efficiently making design trade-offs such as these.

Finally, the source must be specified to be used in exactly the same configuration as the design. Again using the LED as an example, its efficiency, output, and/or brightness may not match what is obtained from the LED when mounted in its housing or fixture. In this case, heat is again the culprit: an LED in its fixture will become hotter than one in open air, and LED efficiency decreases with temperature.

After completing the table to compare different source technologies, let's suppose that LEDs are the obvious answer for a particular system. At that point, it makes sense to make a new table, with different LED manufacturers as the columns, in order to compare available vendors. This is an important part of what systems engineers do: knowing who sells what, knowing what a product can do for them, and knowing the limitations of each vendor's products as well as their cost, delivery time, et cetera. It pays to be thorough when researching each vendor's product specs and limitations, and it is this information that should be summarized in a vendor matrix table.

With newer technologies such as LEDs, it's often possible to find some specification that makes one vendor's product better than another's. With older technologies such as blackbody sources, it's often not clear which product is better and so price, delivery time, and vendor stability become important discriminators. Many designers don't have a strong preference for the least expensive product or the fastest delivery but do prefer a vendor that is stable and can be trusted. For many applications, source selection will not come down to technical specifications; instead, it will likely be based on something like cost or delivery time. Nevertheless, in such cases it is still necessary to go through the "juggle" of technical specifications in order to convince yourself (and others you work with) that the specs for two competing products are, in fact, more or less the same.

Problems

5.1 The total exitance emitted by the Sun is approximately 1270 W/m² at the top of the Earth's atmosphere. Obtain this number from Eq. (5.1); use 0.5 degrees for the angular size of the Sun and a temperature of 5900 K at its surface.

5.2 Using Eq. (5.2), plot the $1/\lambda^5$ dependence of the spectral exitance, including the constants in the numerator, at a temperature of 1000 K and over a spectral range of 0.1–10 μm (in steps of 0.1 μm). Next plot the

$1/(\exp\{E_p/kT\} - 1)$ dependence, being careful with the units of wavelength. Finally, plot the product of the two as it appears in Eq. (5.2). Does the peak in the plot match Wien's displacement law? Is it clear *why* there is a peak in the curve?

5.3 How much greater is the photon energy than the thermal energy at the peak of the spectral exitance distribution? Is the result the same at 300 K as it is at 6000 K?

5.4 Problem 5.2 showed that the solar irradiance at the top of the Earth's atmosphere is 1270 W/m². How much of this irradiance is found in the visible wavelengths (400–700 nm)?

5.5 How much of the power emitted by a blackbody at 300 K is in the LWIR (8–12-µm) band? In the MWIR (3–5-µm) band? Are the numbers the same for a graybody with an emissivity of 0.5?

5.6 There are stars that appear blue. What temperature are they? Does your answer depend on whether they are blackbodies, graybodies, or spectral emitters?

5.7 In Example 5.2, an integration step size of 10 nm was selected without explaining why. The idea is to choose a step size small enough to obtain the right answer—that is, one that does not change much when we use a smaller step size. How do the results of Example 5.2 look if a 100-nm step size is used? A 1000-nm (1-µm) step size?

5.8 Can a lens attached to an LED increase the source radiance? After all, isn't radiance conserved? Compared with a Lambertian source, is an LED brighter when a lens is attached that emits into a total angle (circular) of 20 degrees? Does the concept of basic radiance introduced in Chap. 4 affect your answer?

5.9 Summarize the similarities of imaging and collimation for infinite conjugates in terms of wavefront transformations, FOI versus FOV, and spot size (or image size) versus divergence angle.

5.10 In Example 5.4 we found the radiance from two sources using f/2 collimating optics. What is the radiance for the same two sources with f/1 optics? Assume the same 100-mm focal length.

5.11 In Example 5.4, we calculated as equal the radiance from two tungsten-halogen sources using f/2 collimating optics. Using the Newport Oriel catalog (or its equivalent), create a table that compares the radiance of each tungsten-halogen lamp listed in the catalog. Are any of the lamps significantly brighter than the others?

5.12 In collimating a narrowband point source, spherical aberration is one of the lens designer's many concerns. Is the focal length of a spherical lens bigger or smaller at the edges of the lens than it is at the center? Is the divergence angle from a collimator then bigger or smaller at the edges than at the center?

5.13 Is the wall-plug efficiency of a tungsten-halogen lamp used in a cold refrigerator higher or lower than its efficiency at room temperature?

Notes and References

1. Illumination engineering is a large field, with much more content than can be covered in this book. For more details on the topic, see: Robert E. Fischer, B. Tadic-Galeb, and Paul Yoder, *Optical System Design*, 2nd ed., McGraw-Hill (www.mcgraw-hill.com), 2008; or A. V. Arecchi, T. Messadi, and R. J. Koshel, *Field Guide to Illumination*, SPIE Press (www.spie.org), 2007.
2. M. Bass, E. W. Van Stryland, D. R. Williams, and W. L. Wolfe (Eds.), *Handbook of Optics*, vol. 1, McGraw-Hill (www.mcgraw-hill.com), 1995, chaps. 10–13.
3. Ross McCluney, *Introduction to Radiometry and Photometry*, Artech House (www.artechhouse.com), 1994.
4. This equation uses the same symbol for temperature as is used for optical transmission in Chap. 4. In systems engineering it often happens that symbols used in one discipline are the same as those used for a different concept in another discipline. Context is important in reading equations; once you have used Eq. (5.1) a few times, this instance of the variable T will clearly signify something other than transmission.
5. Ed Friedman and J. L. Miller, *Photonics Rules of Thumb*, 2nd ed., McGraw-Hill (www.mcgraw-hill.com), 2004, chaps. 5, 14.
6. M. Bass, E. W. Van Stryland, D. R. Williams, and W. L. Wolfe (Eds.), *Handbook of Optics*, 2nd ed., vol. 2, McGraw-Hill (www.mcgraw-hill.com), 1995, chaps. 25, 37.
7. Arnold Daniels, *Field Guide to Infrared Systems*, SPIE Press (www.spie.org), 2007.
8. Richard D. Hudson, Jr., *Infrared Systems Engineering*, Wiley (www.wiley.com), 1969.
9. The lifetime is based on temperature via the Arhenius relationship $\exp\{E/kT\}$, an expression that shows up in many areas of engineering.
10. Warren J. Smith, *Modern Optical Engineering*, 4th ed., McGraw-Hill (www.mcgraw-hill.com), 2008.
11. Newport Corp., *The Newport Resource* (www.newport.com), "Light Sources" section.
12. 3M Optical Systems Division, "Light Source Performance Data," in *The Photonics Handbook 2005*, Laurin (www.photonics.com), 2005, p. H-256.
13. Mitsuo Fukuda, *Optical Semiconductor Devices*, Wiley (www.wiley.com), 1999.
14. William J. Casserly, "High-Brightness LEDs," *Optics and Photonic News*, 19(1), 2008, pp. 18–23.
15. E. F. Schubert, *Light-Emitting Diodes*, 2nd ed., Cambridge University Press (www.cup.org), 2006.
16. F. A. Jenkins and H. E. White, *Fundamentals of Optics*, 4th ed., McGraw-Hill (www.mcgraw-hill.com), 1976, chap. 2.
17. Ian Ashdown, "Solid-State Lighting: A Systems Engineering Approach," *Optics and Photonic News*, 18(1), 2007, pp. 24–30.
18. Jeff Hecht, *The Laser Guidebook*, 2nd ed., McGraw-Hill (www.mcgraw-hill.com), 1992.
19. Philip C. D. Hobbs, *Building Electro-Optical Systems*, Wiley (www.wiley.com), 2000.

20. The f/5 image cone for the fiber implies a finite-conjugate blur spot $B = 2.44\lambda s_i/D = 2.44 \times 1.5$ μm $\times 5 = 18.3$ μm for a Lambertian source. This is inconsistent with the fiber's aperture of 10 μm because fibers are designed to be used with Gaussian beams, for which the blur $B = 4\lambda s_i/\pi D = 4 \times 1.5$ μm $\times 5/\pi = 9.5$ μm. For more details, see Hobbs.[19]
21. Dennis Derickson, *Fiber Optic Test and Measurement*, Prentice Hall (www.phptr.com), 1998.
22. The term "wall-plug efficiency" is used even when the source is powered by batteries or some other energy source.

CHAPTER 6
Detectors and Focal Plane Arrays

Not too long ago, photographic film was the most common detector for visible light. Masters of the medium, such as Ed Land of Polaroid and George Eastman of Eastman Kodak, became wealthy developing this "last stop for photons" into successful businesses. From the black-and-white prints of Ansel Adams to the kaleidoscopic colors of album covers in the sixties, film brought cameras to the consumer market. Despite its popularity, there were also problems—mainly the slow, messy process of developing and printing needed to render the photographed scene available for another look.

Around the same time, Chester Carlson used a different type of detector to transfer the images from a sheet of paper onto another sheet, an invention that was later bought by Xerox. These detectors are known as *photoconductors*, which change their conductivity when exposed to light and thus allow a photocopy to be made. The semiconductor selenium was used as the material in the original photocopiers, although more recent versions use organic photoconductors. Even though they were developed for the mass market in parallel with film, photoconductors marked the beginning of the modern era in detectors and have since found use in other applications such as laser printers.

Semiconductors can also be used to make a third type of detector, known as a *photodiode*, and it is here that recent developments in the detection of photons are most obvious. One recent application for photodiodes—close relatives of the LEDs reviewed in Chap. 5—is the efficient conversion of solar energy into electricity (Fig. 6.1). Another is megapixel digital cameras that use integrated arrays of detectors to capture images without the problems of film. An additional advantage is the ability to place telescopes in remote locations such as space, and have the images sent back to Earth via radio—a process not even possible with photographic film.

Because it responds to visible light, the most common semiconductor for detectors is silicon. This is presumably a lucky

238 Chapter Six

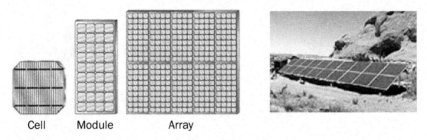

Cell   Module   Array

FIGURE 6.1 The low-cost conversion of sunlight into electricity uses arrays of silicon solar cells. Although it is relatively inefficient, the array shown can generate up to 6.5 kW. (*Credit: U.S. Department of Energy.*)

coincidence, since silicon has also proven itself to be an excellent material for electronics. This allows detectors and electronics to be integrated on the same semiconductor wafer. The manufacturing processes developed for silicon have also resulted in new developments of an older type of device known as a *thermal detector*. Such development is enabled by another recent technology: microelectromechanical systems (MEMS), which is a silicon-based process for the micromachining of micron-scale features used for thermal isolation.

Section 6.1 reviews the technical background needed to use photoconductors, photodiodes, and thermal detectors; some details on the use of film can be found in the *Handbook of Optics*.[1] Silicon is the most common detector material, but several other materials are used for wavelengths outside the visible band; these are listed in Table 6.1. As with the other tables in this book, the list is not exhaustive and shows only the materials used most often. Note that the materials in **bold** are the first options usually considered for each band.

The materials listed in the table are used because of their high absorption coefficient in the given band, enabling the efficient conversion of photons to either electrons or heat while generating the large response needed for a sensitive detector. The absorption may not cover the entire band listed in Table 6.1 or it may extend outside the band, but the wavelengths given are where the detector is most efficient.

Detectors are available as either individual elements or integrated side-by-side in two-dimensional devices known as focal plane arrays (FPAs), so named because they are usually located at the focal plane of an imaging system. Figure 6.2 shows such an array; it contains more than 4 million detectors known as *picture elements* (or *pixels*), each 18 microns square. The HgCdTe FPA shown is among the largest currently available and was designed for the James Webb Space Telescope; it operates at a temperature of 40 K to collect digital images

Detectors and Focal Plane Arrays 239

Wavelengths	Band	Detector Materials
0.2–0.4 μm	UV	GaN, **Si**, SiC
0.4–0.7 μm	VIS	CdS, Ge, HgCdTe, **Si**
0.7–1 μm	NIR	CdSe, **Ge**, HgCdTe, InGaAs, **Si**
1–3 μm	SWIR	**Ge**, HgCdTe, **InGaAs**, InAs, PbS
3–5 μm	MWIR	HgCdTe, **InSb**, PbSe
8–12 μm	LWIR	α-**Si**, **HgCdTe**, QWIP, Si:As, **VOx**
12–30 μm	VLWIR	**HgCdTe**, Si:As, Si:Sb

TABLE **6.1** Materials commonly used for detectors in given wavelength bands. In alphabetical order, these materials are: amorphous silicon (α-Si), cadmium sulfide (CdS), gallium nitride (GaN), germanium (Ge), indium antimonide (InSb), indium gallium arsenide (InGaAs), lead selenide (PbSe), lead sulfide (PbS), mercury cadmium telluride (HgCdTe or MCT), quantum-well infrared photodetector (QWIP), silicon (Si), silicon carbide (SiC), silicon doped with antimony (Si:Sb), silicon doped with arsenic (Si:As), and vanadium oxide (VOx).

FIGURE **6.2** The FPA used on the James Webb Space Telescope; it consists of an array of 2048 × 2048 pixels, each 18-μm square. (*Credit: Teledyne Scientific & Imaging, LLC.*)

of the beginning of the universe over wavelengths ranging from 0.6 to 5 microns.

Digital imaging using FPAs presents many opportunities in addition to unique engineering challenges. The most important challenge is that the image is not a point-to-point reproduction of the scene; instead, the pixels "sample" the scene in small, pixel-size blocks across the FOV. Depending on how small the pixels are in comparison with the spatial resolution of the lens, this sampling can result in poor image quality and contrast as well as image artifacts such as jagged lines and moiré patterns.

As with any complex technology, there are many different FPA "flavors" from which to choose: charge-coupled devices (CCDs), which use a structured method to read the signal from the array; complementary metal-oxide semiconductor (CMOS) devices, which is also a silicon-based technology but is more flexible in terms of how the signal is read out; and a hybrid technology that combines the best aspects of each. Section 6.2 reviews the properties of these different types of FPAs.

Regardless of detector material and the choice of single-pixel or FPA geometries, the most important property of a detector is its ability to measure the "signal." The signal is any photon that's useful, ranging from solar energy for electricity generation to pulse returns for a laser radar system. The per-pixel signal of an FPA also includes the effects of sampling in that spreading the blur over too many pixels reduces the amount of power on each pixel, making it more difficult to distinguish the signal from the noise.

Noise is anything that prevents the signal from being read accurately and thereby reduces the probability of a correct read and increases the probability of a false positive. Noise sources can include the signal itself, out-of-field scattering and background light, detector noise, and electrical noise; these all reduce the value of a key metric in detector design: the signal-to-noise ratio, or SNR. As reviewed in Sec. 6.3, the fundamental radiometric question in all optical systems ("Is there enough light?") is answered in part by this ratio.

Noise also affects a system's *sensitivity*, which is the smallest signal that the optics-plus-detector-plus-electronics system can detect. Lower sensitivity is not necessarily better, because achieving low noise levels to improve sensitivity can be expensive. The designer of an optical system must consider both these costs and the benefits, a task for which the metrics of noise-equivalent power (NEP) and detectivity are useful. Section 6.3 also includes details on these metrics.

Given the different types of detectors (photoconductor, photo-diode, thermal, and others not covered in this book), the several options for materials (silicon, InGaAs, InSb, HgCdTe, etc.), the possible geometries (single pixel versus FPA), and the available FPA technologies (CCD, CMOS, and hybrid), detector selection is one of the most challenging aspects of optical systems engineering. Yet before a detector can be selected it's necessary to understand the limitations on their use; these include saturation (how much power can the detector absorb?), uniformity (is the output from all the pixels in an FPA the same?), operability (are all the pixels in an FPA "operating"?), and so on. These limitations and others are reviewed in Sec. 6.4, whose goal is an understanding of the detector properties required for evaluating and writing detector specs.

Section 6.5 describes the detector selection process, first with a few guidelines and then by reviewing the system trades that are an important part of the process. These trade-offs illustrate that selection involves more than just the detector; in fact, detector performance can be traded against the performance of other subsystems. For example, if a photon-starved design seems to need an extremely sensitive (and expensive) detector, then perhaps another look at the size of the entrance pupil, the IFOV of the pixels, the integration time, or the source brightness will result in a different conclusion. Increasing any one of these factors will reduce the requirements on detector sensitivity. Thus, detector selection is not an isolated process; rather, it is an integrated part of designing the source-plus-optics-plus-detector system.

This chapter is not intended to be a comprehensive review of all detector technologies. There are many types of specialized detectors—including photomultiplier tubes (PMTs), image intensifiers, electron-multiplying CCDs (EM-CCDs), and others—that are not covered in this book; however, the *Handbook of Optics*[1] and Donati[3] include excellent descriptions of many such detectors. There are also specialized geometries that are not covered here such as quad cells, lateral-effects cells, and position-sensitive detectors; for more details on these geometries, see Donati[3] and Hobbs.[5]

At some point, the measurement of detected electrons also becomes less of an optical system topic and more of an electrical one. Some of the electrical issues will be reviewed as they relate to noise, but this is not an electrical engineering text. See Hobbs[5] as well as Horowicz and Hill[6] for more information on this aspect of detectors.

Research directions for the next generation of detectors include improvements in the performance specs of available products in addition to developing a number of advanced detector technologies. High-efficiency solar cells, two-color FPAs (e.g., MWIR and LWIR), uncooled and high-operating-temperature (HOT) materials,[7,8] materials for the UV, quantum-dot infrared photodetectors (QDIPs), and Type-II superlattice detectors[7] are currently being developed. To understand the advantages of using these products in an optical system, we first review some of the fundamental building blocks on which these advanced technologies are based.

6.1 Detector Types

Just as there are a number of ways to classify the different types of sources, there are at least as many ways to classify detectors. The most useful distinction from an engineering perspective is again in terms of available products—although in the case of detectors, this naturally divides into the categories of photon (i.e., quantum) and thermal detectors.

As its name implies, a *thermal detector* measures optical power through an increase in detector temperature. There are many different mechanisms for detecting this temperature increase, resulting in a wide variety of thermal detectors.[2,3] A thermal detector known as a *bolometer* measures the increase through a change in the detector's electrical resistance. These detectors are proving to be most useful for a number of applications; their use as uncooled FPAs for LWIR wavelengths enables thermal imaging that's not possible with visible-wavelength FPAs.[9] Bolometers will be reviewed in Sec. 6.1.3.

In contrast, a *photon detector* measures the number of incident photons. This can range anywhere from single-photon sensitivity to more than 10^{21} photons per second, corresponding to a kilowatt of optical power in the visible. What allows the measurement of photons is the use of a semiconductor such as silicon. For these materials, absorbed photons can transfer their energy to electrons that are normally bound to an atom, giving those electrons enough energy to move around the detector and ultimately be measured as moving charge (i.e., current).

Each incident photon ideally produces one electron, converting photons per second into current (electron charge per second). In practice, the number of electrons created from an incident photon is less than one, with losses occurring because of Fresnel reflection at the detector surface (the reflected photons never make it into the absorption layer), incomplete absorption (due to a thin absorption layer), and poor material quality (which causes electrons to be lost before they can be measured as current).

A detector specification which takes all of these effects into account is known as *responsivity* (symbol $\mathcal{R}$). It is a direct measure of how efficiently a watt of optical power at a specific wavelength produces detector current, and it is often measured in units of amps per watt (A/W) for photon detectors. Figure 6.3 shows how responsivity varies with wavelength for indium gallium arsenide (InGaAs) devices, a typical material for detecting SWIR wavelengths.

There are a couple of features in Fig. 6.3 that must be understood when designing photon detectors into optical systems. The first is the initial increase in responsivity with wavelength. Chapter 4 showed that material absorption depends on wavelength; since the absorption of semiconductors is smaller at longer wavelengths, one would expect their responsivity to be smaller as well. Instead, responsivity *increases* with wavelength (up to a certain cutoff point), so responsivity is influenced by something we haven't reviewed yet.

That "something" is photon energy. Specifically, photon energy is smaller at longer wavelengths, which means that more photons per second are needed to create a watt of optical power. Quantitatively, the number of photons per second at a given wavelength is given by

$$N_p = \frac{\Phi}{E_p} \quad \text{[photons/sec]} \tag{6.1}$$

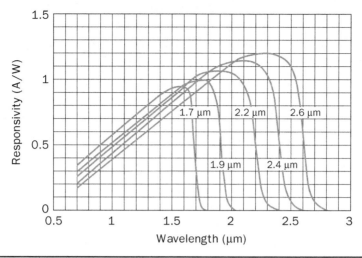

Figure 6.3 The current produced per watt of incident photons depends on wavelength for photon detectors. The numbers shown are the cutoff wavelengths for different varieties of InGaAs. (*Credit: Teledyne Judson Technologies.*)

Here the optical power Φ is for a specific wavelength, and the photon energy associated with this wavelength is $E_p = hc/\lambda$ joules per photon. For example: 1 kW of optical power at λ = 0.5 µm ($E_p = 3.98 \times 10^{-19}$ joules/photon) in the visible corresponds to approximately 2.5×10^{21} photons per second; whereas 1 kW of power in the MWIR (λ ≈ 5 µm) requires 10 times as many photons per second, with each photon having only one-tenth the energy of those in the visible.

Since the number of electrons created is directly proportional to the number of incident photons, it follows that 1 watt of longer-wavelength photons creates more photocurrent (i.e., a higher responsivity) than 1 watt of shorter-wavelength photons. This behavior is completely different from that found in thermal detectors, where the temperature rise depends only on optical power and where responsivity is (in principle) independent of wavelength.

The second feature of interest in Fig. 6.3 is that the increase in responsivity with wavelength for photon detectors doesn't go on forever. The peak occurs when the photon energy, as it becomes less and less at longer wavelengths, is no longer sufficient to give electrons enough energy to conduct. At that point, the drop in material absorption dominates responsivity and hence ℛ quickly decreases to zero.

The point at which responsivity is essentially zero is controlled by a semiconductor material property known as *bandgap energy*. This concept was introduced in Chap. 5, where the peak emission wavelength of LEDs was near the bandgap energy. For detectors, the photon energy must be greater than the bandgap energy in order for the semiconductor's electrons to have enough energy to conduct. The photon wavelength associated with this bandgap energy is the *cutoff* wavelength, beyond which detector responsivity is reduced to zero.

An equally important point on the curves plotted in the figure is the peak responsivity, which occurs at about 90 percent of the cutoff wavelength.[10] For example, the cutoff wavelength for InGaAs with a bandgap energy of 0.75 eV (1.2 × 10^{-19} J) is (1.24 eV·μm)/(0.75 eV) = 1.65 μm. The peak responsivity then occurs at a wavelength of approximately 0.9 × 1.65 μm ≈ 1.5 μm, as seen in the curve labeled "1.7 μm" in Fig. 6.3.

An efficient optical system matches, as best it can, the wavelength of the peak responsivity with the source spectrum. Silicon and InSb are photon-detector materials with a fixed bandgap and peak wavelength, but the bandgap for other material systems (e.g., HgCdTe and InGaAs) can be controlled. Using HgCdTe as an example, there is no such thing as a unique "HgCdTe" detector. The complete description is actually $Hg_{1-x}Cd_xTe$; here x denotes the fraction of cadmium atoms in the semiconductor which determines the bandgap of the material. For example, 200 atoms of the semiconductor $Hg_{0.7}Cd_{0.3}Te$ would consist of 70 atoms of mercury, 30 atoms of cadmium, and 100 atoms of tellurium; its net bandgap energy is approximately 0.25 eV at 77 K.[2] Similarly, "InGaAs" is most often $In_{0.53}Ga_{0.46}As$—although, as seen in Fig. 6.3, other compositions are also used to shift the peak responsivity to different wavelengths. Commercial devices generally don't have the option of selecting a bandgap energy, but paying a vendor to fabricate a detector with a specific peak responsivity is a common practice in specialized areas such as astronomy and remote sensing.

The wide range of materials for photon detectors—GaN, silicon, germanium, InGaAs, InSb, HgCdTe—make it possible to manufacture a variety of detector types. The simplest is a thin layer of semiconductor, which creates a photon detector that we call a photoconductor (Sec. 6.1.1). Section 6.1.2 describes a more sensitive type of detector, the *photodiode*, which is formed by combining two layers to create a p-n junction. There are three types: those that are not biased (photovoltaics such as solar cells), those that are biased for faster speed, and those that amplify the detected signal (avalanche photodiodes).

6.1.1 Photoconductors

Photoconductors (PCs) are the simplest and least expensive of photon detectors. They consist of a thin layer of semiconductor (e.g., HgCdTe or germanium) whose conductance varies with photon flux (Fig. 6.4).

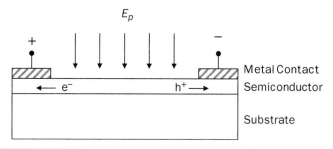

Figure 6.4 A photoconductor consists of a thin (~ 10-μm) layer of semiconductor deposited on a substrate. Electrons given enough energy by incident photons migrate toward the positive terminal.

When the semiconductor absorbs light, its conductivity increases (resistance decreases) because of the increased number of mobile electrons. Conductivity is measured as a change in the voltage drop across the photoconductor or a load resistor as more current flows in response to the higher photon flux.

The reason for the higher conductivity is that absorbed photons transfer their energy to the semiconductor's electrons. When this happens, electrons that are normally bound to an atom now have enough energy to move around the photoconductive layer, with the electric field created by the applied voltage giving these conduction electrons a preferred direction of motion. When the applied voltage is zero, one absorbed photon of the correct wavelength produces one conduction electron. In practice, not all photons are absorbed: one incident photon usually produces an average of 0.1 to 0.9 electrons—depending on the semiconductor's absorption coefficient, the layer thickness, and the photon energy in comparison with the operating wavelength.

However, the current produced by a photoconductor may be *greater* than would be expected from each photon producing fewer than one conduction electrons. If voltage is applied across these detectors, each photon can produce on the order of 10 to 10^3 electrons! The extent of this photoconductive gain depends on the photoconductor material; semiconductors that have slowly moving, opposite-polarity charge carriers known as "holes" exhibit more gain than semiconductors with faster-moving holes.

To understand why, observe that any electron that moves across the photoconductor film and makes its way to the anode is measured as current. Holes typically move more slowly toward the cathode than electrons toward the anode, so the electron's earlier arrival at the anode results in the material having a charge imbalance. Conservation of charge implies that an additional electron must then be injected into the semiconductor from the cathode to restore charge balance.[11] If this additional electron also makes its way to the anode before the hole makes it to the cathode (or otherwise dissipates), then two electrons will be measured for every photon absorbed and yield a 2× photoconductor gain. In many devices, the hole lifetime is longer than the electron travel time by factors that range from 10 to 1000.

Because holes have such long lifetimes, photoconductors are not used for measuring signals that change quickly. With long-lifetime holes registering as current *after* a signal has passed, the speed at which signals can be measured depends on the gain; high-gain photoconductors such as silicon have a speed of about 10^3–10^6 Hz, and low-gain materials such as InGaAs have speeds exceeding 10^9 Hz.

Depending on the detector's temperature and electrical resistance, photoconductors also generate noise due to the random thermal motion of electrons and also exhibit statistical fluctuations in the total number of electrons in the conductive layer. Thermal noise dominates at frequencies higher than that determined by the carrier lifetime ($f = 1/\tau_c$), at which point the gain mechanism has been lost. The noise due to the number of electrons—known as generation–recombination (G-R) noise—is dominant during normal photoconductor operation and is nearly double the inherent noise found in photodiodes.[13] As a result, photoconductors are typically low-performance detectors that are most often used when low cost is an important factor in detector selection.

The wavelengths to which a photoconductor is sensitive depend on the semiconductor used. The most common are cadmium sulfide (CdS), cadmium selenide (CdSe), and CdSSe for visible and NIR applications such as nightlights, camera exposure meters, and dimmers for automobile headlights. For SWIR, MWIR and LWIR applications, PbS is a low-cost option for 1–3-µm wavelengths; PbSe and InSb are used for 3–5-µm wavelengths and HgCdTe for 1–12 µm.

As with the LEDs reviewed in Sec. 5.1.4, the photon energy associated with each wavelength in Fig. 6.3 is $E_p = hc/\lambda$. To obtain significant detector output current, the energy transferred to an electron must be large enough to allow it to conduct; this energy is known as the semiconductor's *inherent* (or *intrinsic*) bandgap energy. For conventional InGaAs, it is equal to 0.75 eV at room temperature, which corresponds to λ = (1.24 eV·µm)/(0.75 eV) = 1.65 µm—the approximate wavelength at which the detector output peaks in Fig. 6.3 for the curve labeled "1.7 µm."

Semiconductors can also be "doped" with additional materials that don't require as much energy to convert their bound electrons into ones that conduct (Fig. 6.5). As a result, doped photoconductors are often used because they are sensitive to longer wavelengths than photodiodes and thermal detectors. Despite their relatively poor performance, they are popular with astronomers who can sometimes

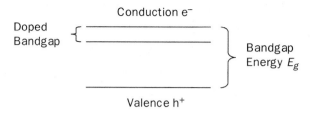

FIGURE 6.5 Relative energy levels for undoped and impurity-doped photoconductors. Detection of longer wavelengths requires a smaller energy than the intrinsic bandgap, which can be obtained with doped materials.

afford the high cost of the complex engineering needed to measure wavelengths not normally detectable by other materials.

Doped silicon has proven to be the most useful long-wavelength photoconductor detector. Typical dopants (also known as impurities) are arsenic (the combination is written as Si:As) with an IR-wavelength coverage of 5–28 μm, and antimony (Si:Sb) with a wavelength range of 14–38 μm. If the concentration of impurities is high then the electrons have a range or "band" of available energies, which leads to what are known as *impurity-band conductors* (IBCs). These are state-of-the-art photoconductive detectors—the Si:As IBCs are more sensitive than conventional photoconductors—that can be used for wavelengths of up to 28 μm. More details can be found in Dereniak and Boreman[2] as well as Richards and McCreight.[12]

Whereas photon energy can create mobile electrons, thermal energy can also create conduction electrons known as *dark current*. As shown in Sec. 6.3, these electrons—whose thermal generation contributes to generation noise in photoconductors—make it more difficult to sense the photon signal. This is particularly problematic for long-wavelength detectors, where the room-temperature thermal energy [$kT = 8.617 \times 10^{-5}$ eV/K $\times$ 300 K = 0.026 eV (26 meV)] is the same order of magnitude as the impurity bandgap energy [$E_i = hc/\lambda = (1.24$ eV·μm$)/(25$ μm$) = 0.05$ eV when measuring a 25-μm wavelength, for example]. As a result, IBCs are cooled to cryogenic temperatures, often as low as 4 K. This significantly increases the size and cost of any system that uses these photoconductors.

In sum, photoconductors are usually low-cost devices with low-end performance, although more expensive detectors offer a better selection of available IR wavelengths. They are also slower and noisier than photodiodes, a comparison that is examined in more detail in the next section.

6.1.2 Photodiodes

Photodiodes (PDs) are more difficult to fabricate than photoconductors and thus are generally more expensive, but they also offer certain benefits. This detector type is the most common for applications requiring higher performance—for instance, higher speed, lower noise, and/or better sensitivity. Examples include the detection by laser radar instruments of very weak, single-photon return signals, receivers for fiber-optic communications that measure data rates faster than 10 GHz, and space-based telescopes for collecting Earth science environmental data.

The operating principle behind a photodiode is similar to that of photoconductors: photons are absorbed, which gives electrons enough energy to conduct. However, as shown in Fig. 6.6, the simplest PD design has two layers—one deposited on top of the other. These layers are often made of the same semiconductor (e.g., silicon), but each layer is doped with additional materials that have either extra electrons (the "n" layer in Fig. 6.6) or extra holes (the "p" layer). Without going into the details, this produces a large electric field across the interface where the p-layer and n-layer meet (the "p-n junction"); this field is in response to a charge redistribution and resulting voltage drop across a thin layer surrounding the junction from which charge is depleted (the "depletion region")—shown as the dashed lines in Fig. 6.6.[13]

The large electric field pulls electrons (and holes) away from the junction, where they are collected as current at electrical contacts. This phenomenon is referred to as the *photovoltaic* (PV) effect and is useful for converting photons into electrical power. Unfortunately, the efficiency with which photons are converted to conduction electrons is limited in part by the low absorption of the thin junction. Photons are also absorbed outside the junction, but those photoelectrons can be lost ("recombine") in the time it takes for them to move ("diffuse") toward the contacts. Reducing this inefficiency at

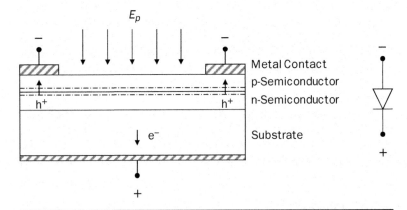

Figure 6.6 Photodiodes rely on a semiconductor p-n junction to deliver faster speed and less noise than a photoconductor; the electrode geometry also allows efficient integration into focal plane arrays.

low cost is a major area of research, with billion-dollar ramifications for the utilization of solar energy to generate electricity.

For better performance, the electric field is made larger by applying a negative voltage (a "reverse bias") across the junction. This results in a slightly thicker depletion region, which yields a modest improvement in conversion efficiency; the larger field near where photons are absorbed also increases the speed at which the electrons are moved away from the junction and measured at the contacts as current. In this case, we could say that the photodiode is being used in "photoconductive" mode. But there is obviously some potential for confusion with photoconductive detectors (Sec. 6.1.1), so a better term to use is *reverse-bias* mode.

Somewhat better efficiency and much higher speed can be obtained by adding a third layer between the p and n layers (Fig. 6.7). This extra layer is not doped and thus creates an *intrinsic* layer where absorption occurs; the result is a three-layer p-i-n photodiode (often written as "PIN"). This additional layer is quite a bit thicker than the p-n junction depletion region; it thus provides more absorption in the vicinity of the electric field, resulting in a shorter diffusion time, lower capacitance, and higher speed.

Less common is a third type of PD known as an *avalanche* photodiode (APD). These detectors add yet another layer that provides gain to the basic PIN structure, with one photon producing more than one electron (Fig. 6.8). After high voltage is applied, the additional layer has a high electric field strength across it; this forces conduction electrons to move quickly through the layer, where they may impact nonconduction electrons with enough energy to make them conductive. Each impact can double the number of conduction electrons, a process that can "avalanche" to produce extremely high

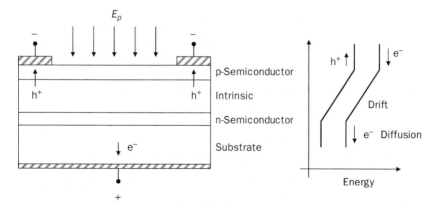

FIGURE 6.7 By reducing the junction capacitance, the thick intrinsic layer in a PIN photodiode exhibits higher speed and greater responsivity than do simple p-n junction detectors.

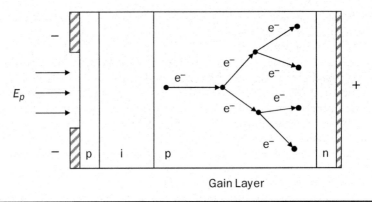

Figure 6.8 Avalanche photodiodes (APDs) rely on momentum transfer via mechanical impact to give electrons enough energy to conduct. Each impact doubles the number of electrons, thereby creating an "avalanche."

gain—a function of the applied voltage. For most semiconductors, the disadvantage is the excess noise associated with electrons and holes avalanching independently of each other.

Photovoltaics

Photovoltaic (PV) detectors are photodiodes operating without any battery or wall plug to apply a reverse-bias voltage. The most common application of these detectors is in solar cells designed to generate electricity directly from sunlight—most typically for homes and industrial buildings, but also for consumer electronics such as calculators. These detectors clearly offer enormous opportunities for commercial development, but meeting this potential requires that their cost be kept low. As a result, consumer-market solar cells consist of a simple silicon p-n junction.

In practice, the current generated from PV solar cells is most often used to charge batteries. To illustrate their use in converting electrical power to heat (e.g., an electrical heater for a room), it's simpler to consider the solar cell as directly driving an electrical resistor. How much heat is generated depends both on the resistance and on the current–voltage (I-V) characteristics of the diode. Figure 6.9 shows a typical I-V curve with a resistor load line to illustrate the efficiency of converting photons to electrical energy.

Depending on the electrical resistance of the heater, the power conversion may or may not be efficient. A very large resistance, for example, will prevent too much current from flowing, whereas a very small resistance will allow current to flow but will only create a small potential difference (voltage). The optimum answer is an intermediate resistance, where the power generated—which is given by $I \times V$, the product of current and voltage—is a maximum.

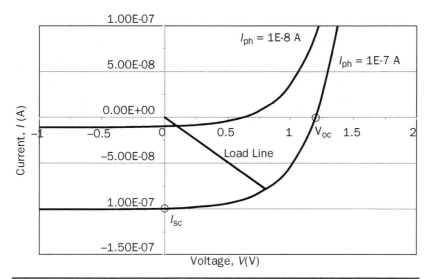

FIGURE 6.9 Voltage (V) and current (I) characteristics of a photodiode exposed to two levels of illumination-generated current. For the larger illumination (I_{ph} = 1E-7 A), also illustrated are the photovoltaic mode of a solar cell (which generates an infinite-resistance open-circuit potential V_{oc}), the photocurrent mode (which generates a zero-resistance short-circuit current i_{sc}), and the reverse-bias mode (applied V < 0).

Mass-market solar cells are not efficient at converting photons to electrical energy. Their efficiency is limited in part by the thinness of the absorption region near the p-n junction. Whereas photons can also be absorbed outside the thin depletion region, electrons created at any appreciable distance from the electric field must diffuse toward the field before they can be carried away, during which time electrons and holes are lost ("recombine") before they can reach the contacts as current.

More importantly, the solar spectrum is not well matched to the spectral responsivity of the solar-cell material (silicon) used for consumer electronics, so much of the solar power is not actually converted to electricity. In particular: the bandgap energy of silicon is 1.12 eV, which corresponds to a peak wavelength of (1.24 eV·µm)/(1.12 eV) = 1.1 µm. Unlike "direct" bandgap materials, silicon is an "indirect" material: its responsivity curve drops off slowly on the long-wavelength side of the bandgap (Fig. 6.10). The net effect for silicon is an overall optical-to-electrical conversion efficiency of only about 25 percent over the entire solar spectrum. Taking additional losses into account and given a maximum solar irradiance of about 1 kW/m² on the ground, each square meter of silicon cells can produce only about 100 to 150 watts of electrical power on a sunny day—barely enough to run an incandescent lightbulb or two.

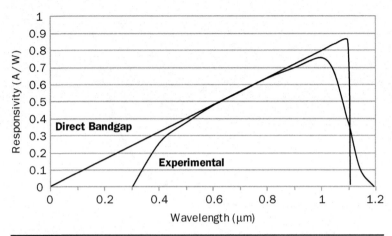

FIGURE 6.10 Comparison of the responsivity of a direct bandgap material with experimental data for typical silicon solar cells. *(Experimental data courtesy of pveducation.org.)*

Example 6.1. In this example we walk through a back-of-the-envelope calculation to estimate the conversion efficiency of a silicon solar cell. This efficiency depends in part on how well the Sun's output spectrum matches the responsivity of a silicon PV detector. Numerically, the photocurrent of a cell is given by the sum over the solar spectrum: $i_p = \Sigma\, E_\lambda(\lambda) A \times \mathcal{R}(\lambda) \times \Delta\lambda$. Chapter 5 showed how to calculate the solar spectral irradiance E_λ for the Sun modeled as a 5900-K blackbody; for more accurate calculations based on the actual solar spectrum and the absorption of the atmosphere, you should use tabulated data for E_λ on the surface of the Earth.[30]

Figure 6.10 shows that the responsivity of the ideal silicon cell increases linearly from 0.16 A/W at $\lambda = 0.2$ μm to 0.88 A/W at the bandgap wavelength $\lambda = 1.1$ μm, after which it drops quickly back to zero. Table 6.2 shows that approximately 480 A/m² are generated by the silicon cell; comparing this value with an incident solar irradiance of 1 kW/m², we find that the average responsivity over the solar spectrum is 0.48 A/W. This is a highly simplified model, and responsivity actually begins its drop to zero at a shorter wavelength; this is due to silicon's unique property of dropping off slowly after reaching peak responsivity, a result of its being an "indirect" bandgap material.[13]

The photocurrent i_p calculated from the overlap summation is equal to that generated by the solar cell if there is negligible resistance in the circuit and hence a small voltage developed across it. This is known as the *short-circuit current*, shown in Fig. 6.9 as i_{sc}. In contrast, if the resistance is extremely large, then the current generated would be small and the potential developed across the resistance is called the *open-circuit voltage*, shown as V_{oc} in Fig. 6.9. Using the diode equation plotted as the curves in Fig. 6.9, these variables can be quantified as follows:[13]

$$i = i_s \left[\exp\left\{\frac{qV}{kT}\right\} - 1 \right] - i_p \qquad (6.2)$$

λ [µm]	E_λ [W/m²·µm]	R [A/W]	$\sum E_\lambda R(\lambda) \times \Delta\lambda$ [A/m²]
0.2	10.7	0.0	0.0
0.3	514.0	0.0	0.0
0.4	1429.0	0.32	0.0
0.5	1942.0	0.40	60.7
0.6	1666.0	0.48	140.1
0.7	1369.0	0.56	219.0
0.8	1109.0	0.64	293.3
0.9	891.0	0.72	361.3
1.0	748.0	0.80	423.6
1.1	593.0	0.88	480.0

TABLE 6.2 Parameters and data needed to calculate the short-circuit current generated by a silicon solar cell; a smaller step between wavelengths gives a more exact calculation. The average spectral irradiance E_λ at any two wavelengths λ_1 and λ_2 is used in the summation for each $\Delta\lambda = \lambda_2 - \lambda_1$, as is the average responsivity $\mathcal{R}$.

Here i_s is the saturation current created by the diffusion of thermally generated electrons, V is the voltage generated across the photodiode, k is Boltzmann's constant, and T is the diode temperature.

When solar energy is incident on the cell, Eq. (6.2) shows that an open-circuit ($i = 0$) voltage $V_{oc} = (kT/q) \ln[1 + (i_p/i_s)]$ is generated, illustrating that a reduction in the saturation current—a reverse current that opposes the photocurrent i_p—has a large effect on the open-circuit voltage. The need to minimize this current—along with a junction leakage current that also opposes the photocurrent—is why solar cells require high-quality materials for high efficiency.

The short-circuit current and open-circuit voltage place bounds on the electrical power that can be generated by the solar cell. In practice, the circuit resistance should be neither very large nor very small and will have a linear load line as shown in Fig. 6.9. The intersection of the load line with the photodiode curve leads to an operating current i_{op} that is somewhat less than i_{sc} and to an operating voltage $V_{op} < V_{oc}$. Hence the power generated is given by the product $i_{op}V_{op}$, and the resulting optical-to-electrical conversion efficiency is $i_{op}V_{op}$ per kilowatt for each square meter of silicon. Note that this efficiency depends on the resistive load (the slope of the load line in Fig. 6.9), where the optimum is found by varying the resistance.

It is reasonable to suppose that using a semiconductor (e.g., CdS) whose spectral responsivity closely matched the peak of the solar spectrum would yield the best conversion efficiency. However, this reasoning fails to take into account that the voltage component (V_{oc}) of the electrical power generated depends on the dark current through the semiconductor bandgap voltage [see Eq. (6.3)]. The net result is that a bandgap energy smaller than CdS (but larger than silicon) is optimal for solar power conversion[3]—a level closely matched by the responsivity of gallium arsenide (GaAs) and cadmium telluride (CdTe), each with a bandgap of approximately 1.4 eV (bandgap wavelength = 1.24 eV divided by 1.4 eV or approximately 0.9 µm). The GaAs cells are expensive and are typically reserved for high-performance applications such as solar arrays on satellites. Silicon has a much lower cost and is used for the flat-panel arrays found on residential rooftops; CdTe is the clear winner for industrial-scale flat-panel arrays, since its use results in more efficient and more cost-effective solar cells than are possible with silicon, GaAs, or CdS.

Efficiency also improves when multijunction designs replace those that rely on a single p-n junction. As many as three junctions can be stacked on top of one another to convert photons to electrons over a broader range of the solar spectrum. Efficiencies of greater than 40 percent have been obtained with triple-junction cells that use III-V semiconductors (InGaP, InGaAs, and Ge) to create different p-n junctions in the same cell; the associated bandgap wavelengths are 0.69 µm, 0.89 µm, and 1.77 µm, respectively. These cells are even more expensive than GaAs cells and are too costly at this point for consumer use.

One approach to lowering cost is to use nonimaging optics to concentrate the solar power down to a smaller area so that the size of the solar cell can be reduced (Fig. 6.11). Damage to the cell is avoided by placing it before or after the focused image of the Sun. Such concentrators reduce the material cost of the detector by the *concentration ratio*: the ratio of the optical area to the detector area. Systems currently under development feature concentration ratios as high as 1000 : 1 ("1000 Suns").[14] There is a cost associated with the optics and (possibly) a solar tracking mechanism, but the savings on the cost of the PV detectors can more than compensate and

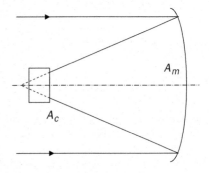

Figure 6.11 A simplified solar concentrator illustrating the reduction in the size of the solar cell by using a nonimaging optic (mirror). The irradiance from the on-axis point on the Sun is magnified by the ratio of mirror area to solar cell area (A_m/A_c); the off-axis irradiance (not shown) is also magnified.

potentially allow the use of very efficient III-V multijunction cells. This combination of optics-plus-detector systems engineering will likely prove to be a major contributor to future utilization of solar energy.

Reverse-Bias Photodiodes

A photodiode with a greater range of applications than that of the photovoltaic is the p-n junction or PIN structure operated in reverse bias.[15] These PDs are designed not to generate large amounts of electricity but instead to measure signals that provide some sort of information about a source or the path between it and the detector. The information may be, for example, an image, a radiometric measurement, a spectroscopic analysis, an interferogram, a fiber-optic data stream, or a laser radar signal. It's common to use a simple p-n junction photodiode in reverse bias for some of these measurements, but the reverse-biased PIN photodiode has the advantages of slightly higher responsivity and much faster speed.

Both of these improvements are due to the relatively thick intrinsic layer between the p and n layers (Fig. 6.7). Equation (4.3) shows that a thicker layer absorbs more photons than a thin one, thus giving more electrons enough energy to become electrically conductive. As a result, peak responsivities for PIN photodiodes are on the order of 0.5–1 A/W, depending on material and operating wavelength.

The thick layer also reduces the detector capacitance, which allows the detector to measure changes in the signal at speeds on the order of 10 GHz and higher for materials such as InGaAs. However, as the intrinsic layer is made thicker, the detector speed becomes limited not by its capacitance but rather by the time it takes (the "drift" time) for an electron to be carried across the thick layer to the contacts by the electric field. With capacitance and drift time competing against each other, detector speed reaches a peak that depends on the absorption coefficient and the thickness of the intrinsic layer.

Responsivity thus initially improves with speed but ultimately trades against it. In a comparison of different detector materials, for example, silicon does not absorb nearly as well as InGaAs at a wavelength of 1 μm, which means that silicon's intrinsic layer must be much thicker to yield a responsivity that's competitive with InGaAs. As a result, silicon's PIN speed is generally 3–10 times slower than that of InGaAs.

Increasing the reverse bias also increases the speed somewhat as the larger electric field forces conduction electrons to move faster toward the electrodes. Unfortunately, the detector noise also increases. As described in Sec. 6.3, there are many different types of noise that reduce the detection system's sensitivity. For the detector to *not* be the limiting factor, its dark current must be lower than other sources of noise (e.g., the background and the electrical subsystem).

Dark current—a result of both reverse saturation current and leakage current—is the current that flows even when there aren't any photons incident on the detector; it is the consequence of the thermal excitation of electrons to higher-energy, conductive states. Because this excitation depends on an Arhenius relationship, it is quite sensitive to temperature:

$$i_d = i_o \exp\left\{\frac{-E_g}{kT}\right\} \quad [A] \qquad (6.3)$$

Here i_d is the dark current at room temperature (RT) and depends on the detector size, bias, material, quality, and design parameters.[13] Ignoring the dependence of bandgap energy E_g on temperature, the important physics in Eq. (6.3) is the exponential increase of dark current with temperature, often pushing the system design to a cooled detector. This was seen in Sec. 6.1.1 for photoconductors, where temperatures as low as 4 K are used to reduce the dark current for VLWIR astronomy.

Cryogenically cooled MWIR and LWIR p-n junction detectors are also common in situations requiring high sensitivity. Neither p-n junction nor PIN photodiodes are as noisy as photoconductors, and operating temperatures of 80 K to 120 K are usually sufficient for high-performance devices. Compared with visible-wavelength detectors that do not require cooling, the increase in system size, weight, and complexity for cooled photodiodes is significant; cooling techniques range from liquid nitrogen and Stirling-cycle coolers (80 K) to thermoelectric coolers (250 K). Dark current and detector cooling requirements are reviewed in more detail in Sec. 6.3.2 and Sec. 6.4.

Once dark current is reduced to acceptable levels, it's often the electrical circuit *following* the detector that limits performance. Reading out the current from a PIN detector, for example, usually requires an amplifier to convert detector current to a voltage. These amplifiers have noise of their own, and amplifier noise may limit the sensitivity of the detection system. In this case, cooling the detector to reduce dark current will not improve sensitivity. One way around this is to use a detector with gain, such as the avalanche photodiode discussed next.

Avalanche Photodiodes

If detection performance is limited by amplifier noise, then an APD can give better sensitivity than a PIN photodiode. These detectors add a fourth layer, which provides gain to the basic PIN structure (Fig. 6.8) because one photon produces *more* than one electron. The resulting ratio of electrons to photons is known as the gain (symbol M), with optimum values of $M \approx 100$ for a quiet material such as silicon and $M \approx 10$ for a noisier material such as InGaAs. Hence the gain-multiplied responsivity can be large, with a peak value of $RM \approx 0.7$ A/W × 100 = 70 A/W for AR-coated silicon APDs.

There are two prices to pay for this gain. The first is that obtaining gain with a large electric field requires a higher voltage than is needed for a PIN, with 350–400 volts for $M = 100$ being common for silicon APDs (cf. the 2 to 50 volts needed for silicon PINs). The second price is that APDs add their own type of noise to the system, which becomes much bigger at higher gains.[16] This means that the gain can't be increased indefinitely; instead, there's an optimum gain whose value is determined by the material, amplifier noise, and the electrical bandwidth. Below that optimum, the APD noise is still lower than the bandwidth-dependent amplifier noise (see Sec. 6.3.2); above the optimum, the APD noise is higher. Both noises are equal at the optimum, which results in a detector with a much higher responsivity than a PIN photodiode. For high-bandwidth systems with large amplifier noise, the sensitivity of an optimum-gain APD can thus be better than that of a PIN.

6.1.3 Thermal Detectors

In contrast with photoconductors and photodiodes, the absorption of optical power by a thermal detector creates a measurable increase in temperature, not a change in electron mobility. There are a number of different mechanisms for detecting this temperature increase and hence a wide variety of thermal detectors.[2,3] For example, the temperature increase may change the detector's electrical resistance (bolometers), voltage (thermoelectrics), or capacitance (pyroelectrics), all of which can be measured with appropriate circuitry.

Of these, bolometers are perhaps the most useful; we noted earlier that they serve as FPAs for LWIR wavelengths. Either photon or thermal detectors can be used to measure temperature differences;[17] bolometers can measure thermal differences that are impossible to sense with visible-wavelength FPAs. The design rules for determining how well a slightly overheated target stands out from its background will not be addressed here; they can be found in Dereniak and Boreman[2] and Holst.[19]

The reason for emphasizing infrared applications is that such arrays can be made reasonably sensitive without cooling—a huge advantage over the typical IR system, which requires cryogenics or Stirling-cycle coolers. The LWIR band in particular has a large change in spectral exitance for small temperature changes (dM_λ/dT) near 300 K, which means that this band has the most potential for high-sensitivity thermal detection of room-temperature objects.[9]

The structure of a typical bolometer pixel is shown in Fig. 6.12. The pixel consists of a supporting membrane on which is deposited an absorbing layer—of vanadium oxide (VOx) or amorphous silicon (α-Si)—whose resistance varies with temperature. The absorbing

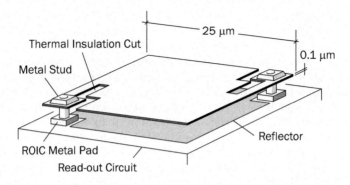

FIGURE 6.12 Mechanical structure of a microbolometer pixel. Shown are a thermally sensitive α-Si membrane whose temperature increases with irradiance; a connection to a readout circuit below; and thin, narrow thermal isolation legs to prevent heat transfer away from the membrane. (*Credit: E. Mottin et al.*, Proceedings of the SPIE, *vol.* 4650.)

material faces toward the thermal source and heats up when exposed to an optical signal, a result of its good absorption over the LWIR wavelengths. The VOx layer is a commonly used bolometer material because it has a large *temperature coefficient of resistance* (TCR)—that is, the percentage change in resistance for a given temperature increase—of about 2–3 percent per degree K]. Because each pixel in the FPA is a bolometer on the order of a 25 μm × 25 μm square, the detectors are also known as *microbolometers.*

The temperature increase is relatively large because the VOx layer has been thermally insulated from the environment by placing the FPA in a vacuum (reducing convection) and isolating the pixels via long, thin support legs (reducing conduction). The ideal design absorbs all of the incident optical power, gives a large change in resistance as it heats up, and transfers no heat to the surroundings. In practice, actual designs can't reach this ideal and involve a compromise among design parameters.

Microbolometer arrays are fabricated using microelectromechanical systems (MEMS) technology, the process used to micromachine such micron-size features as the supports used to thermally isolate the VOx or α-Si layer. To collect more optical power, other microbolometer designs use an "umbrella" structure that also requires the use of MEMS technology, which has become a critical part of fabricating thermal detectors.

The responsivity of bolometers is typically measured in units of volts generated per watt of incident power (V/W). Even though they are thermal detectors, their responsivity is not necessarily flat with wavelength because the absorption layer may not absorb uniformly over all wavelengths.

Because they are thermal detectors, the response time of bolometers is much slower than photon detectors such as photoconductors or photodiodes. As a result, a typical pixel size cannot respond to changes in the signal any faster than about 10 milliseconds; hence, they measure an average power instead. This is not a problem for video rates (30 Hz), which is another reason—in addition to its uncooled convenience and low cost—why imaging FPAs based on this technology have become increasingly common.

The loss of heat during the time it takes to measure a signal lowers the temperature of the absorbing layer below what would be measured if it were perfectly insulated. Along with $1/f$ noise, such "temperature fluctuation" noise is one of the biggest sources of measurement uncertainty in a bolometer. As the thermal isolation is improved, noise due to radiation (both to and from the background) also lowers the sensitivity of these detectors, ultimately limiting their performance. Nonetheless, microbolometers are sensitive enough to detect temperature differences in a 300-K blackbody scene on the order of 0.05 K (50 mK) with 25-μm square pixels.

In short, microbolometer arrays are relatively inexpensive, low-performance FPAs for uncooled LWIR imaging. Even though they are slower and less sensitive than cooled photon detectors, their availability in FPA format with integrated silicon electronics make them a good choice for applications such as day-and-night video surveillance. As shown in the next section, creating a focal plane array using single-pixel detectors is not straightforward and involves consideration of several engineering trade-offs.

6.2 Focal Plane Arrays

Microbolometer arrays are useful for LWIR thermal imaging, yet visible-wavelength FPAs using photon detectors made from silicon are even more common. These are found in a wide array of consumer "electronics," including camcorders, camera phones, digital cameras, video-production cameras used to record high-definition TV, and so on. Moreover, high-performance FPAs for scientific, industrial, and medical applications such as astronomy, pollution monitoring, and cancer research are allowing discoveries that wouldn't be possible without them. The images shown in Chap. 1 from the Hubble Space Telescope, for example, would have been much more difficult to obtain if FPAs were not available.

What's common to all FPAs is the integration of a large number of individual detectors onto the same chip, thereby allowing digital imaging of a scene. It is possible to image a scene onto a single detector, but what's measured in that case is the scene's total irradiance and not an image. Before the development of detector arrays, the choices for images were either photographic film or some sort of scanning mechanism to expose a single-pixel detector to the entire FOV over time. Both options had drawbacks, which became more evident with the advent of digital computers and high-speed signal processing.

The use of an FPA thus allows optical systems to "stare", precluding the need for large, heavy, mechanically complex scanning systems. This trend has continued over the last 20 years with the development of large-format arrays that feature more pixels (to cover bigger FOVs) and smaller pixels (to obtain better resolution), yielding smaller, lighter optical systems. Typical pixel counts are currently on the order of 10 million pixels (megapixels) for consumer cameras, with large-format arrays common for IR imagers (640 horizontal pixels by 512 vertical pixels, or "640 × 512," and larger), and array formats of 4096 × 4096 and higher available for high-end scientific instruments.

Digital imaging also presents unique engineering challenges. The most important is that the image is not a point-to-point reproduction of the scene; instead, the pixels "sample" the scene in small IFOV units. This sampling can result in lower image quality, especially if the scene contains periodic patterns that cannot be resolved by the pixel spacing. At the same time, if the diffraction-limited blur size is spread over too many pixels, then again the image quality is reduced; the amount of power on each pixel is also smaller, which makes it more difficult to distinguish the signal from various sources of noise.

In addition to sampling, the concepts to be covered in this section that are unique to FPAs include: (1) the gaps between pixels that cannot detect photons, which are quantified using the idea of *fill factor*; (2) the slight difference in responsivity from pixel to pixel which results in a *nonuniformity* that can limit FPA performance; (3) the delay in reading out data from all pixels in a "frame," a process that is slower than the response time of an individual pixel; and (4) the manufacture of FPAs in three different types of technologies—CCD, CMOS, and hybrid—each with their own advantages and disadvantages. These items are the most important for a first pass at understanding FPAs; concepts such as noise, operability, and dynamic range, are examined in more detail in Sec. 6.3 and Sec. 6.4.

6.2.1 Sampling

Perhaps the most unique feature of an FPA is that the image is broken up into small detector areas known as *pixels*. Chapter 2 shows how this leads to the concept of an instantaneous field-of-view (IFOV), where the FOV of the entire scene is divided into smaller units determined by the number of pixels and the effective focal length (EFL) of the optical system. A 512 × 512 FPA with an FOV of 4 degrees, for example, has an IFOV of approximately 4 degrees ÷ 512 pixels = 0.0078 degrees per pixel (136 μrad per pixel).

More importantly, this *sampling* of the FOV by discrete detectors has two important consequences for optical system design. The first is that the IFOV defined by the pixels can be different from the optical resolution determined by the wavelength and entrance-pupil diameter. The second is that, depending on the size of the diffraction-limited blur in comparison with the pixel size, the resulting image quality may be reduced owing to an effect known as *aliasing*.

Chapter 3 showed that the smallest possible spot size for a diffraction-limited optical system is given by Eq. (3.11). For example, the angular size of the Airy blur for an imager with a 100-mm entrance pupil is $\beta = 2.44\lambda/D = (2.44 \times 0.5\ \mu m)/(0.1\ m) = 12.2\ \mu rad$ at a wavelength of 0.5 μm. The blur is often designed to be somewhat larger than the IFOV and depends on how much ensquared energy is required, which in turn depends on the image quality and SNR requirements (Sec. 6.3). It's also common for the blur to be smaller than ("underfill") a pixel in designs for which a large, fast aperture is required to collect enough photons.

Although it is sometimes necessary to match the IFOV of an FPA to the optical spot size, there are a number of reasons for making the blur larger than a pixel. The first is that a smaller telescope can be used, given the $1/D$ dependence of the blur size. Another is that spreading the blur over multiple pixels makes it much easier to track the motion of a point object. For example, if the blur size covers a 2 × 2 array of pixels on the FPA—where the IFOV for each pixel in our example is 12.2/2 μrad = 6.1 μrad in the x and y directions—then relative motion can be detected by looking at the changes in irradiance on each of the four pixels. Even if the object seems not to be moving—as in the case of a distant star—the satellite on which a space-based telescope is mounted will be moving.

One disadvantage of spreading the blur into multiple pixels is that the intensity from the point source is spread out over more than one detector, yielding less irradiance per pixel. This makes it more difficult to detect the source intensity against any noise that may be present. It also reduces the image contrast for area sources, an effect that can be understood as follows.

Section 3.3 illustrated how a lens affects the image contrast of the various spatial frequencies in the object. We can summarize that section by saying that lenses transmit low-frequency spatial patterns fairly well whereas high-frequency patterns are blurred and lose their "crispness"—until they reach a cutoff frequency where they are not reproduced at all. This behavior is capture by a modulation transfer function (MTF) curve, as in Fig. 3.46 with a typical cutoff frequency for an f/2 lens of $f_c = 1/(\lambda \times f/\#) = 1/[(5 \times 10^{-3}\ mm) \times 2] = 100$ lp/mm at a wavelength $\lambda = 5$ μm.

When an FPA is included in the optics-plus-detector system, the detector array often limits the spatial frequencies that can be measured because an FPA with closely spaced pixels can measure a higher-frequency spatial pattern than one with a coarser spacing. The detector's sampling capability thus depends on the pixel spacing (or *pitch*, denoted x_p), with a 10-µm pitch (for example) giving a detector cutoff of $f_d = 1/x_p = 1/10 \times 10^{-3}$ mm = 100 detector pairs per millimeter (or lp/mm). This corresponds to the example shown in Fig. 6.13(a), where the small pixels sample only the peaks of a periodic image, leading to the erroneous conclusion that the image has no contrast and consists only of a gray blur over the entire focal plane.

By reducing the distance between pixels, higher spatial frequencies can be detected. This is shown in Fig. 6.13(b), where the same periodic image is again sampled by an FPA but now over the peaks and valleys both. The figure shows that reducing the pixel spacing to 5 µm, yielding a detector cutoff of $f_d = 1/x_p = 1/5 \times 10^{-3}$ mm = 200 lp/mm, allows the peaks and valleys to be resolved.

To reproduce the spatial frequency of a sinusoid, it's necessary for the FPA to sample it at a rate of twice per cycle (e.g., at the peak and the valley). The quantitative metric that allows the resolution of spatial frequencies is known as the FPA Nyquist frequency f_N, which is determined by twice the pixel spacing so that $f_N = 1/2x_p$ for one pixel measuring a peak and one pixel measuring a valley. In order to avoid incorrectly detecting (i.e., "aliasing") higher frequencies as lower [Fig. 6.13(a)], the detector Nyquist frequency must be equal to (or greater than) the spatial frequencies resolved by the lens. If this is not the case, then higher spatial frequencies imaged by the optics will be aliased as lower frequencies by the FPA; therefore, to remove aliasing requires that we use a smaller pixel spacing [Fig. 6.13(b)] such that $f_N \geq f_c$.

Even when not measured by the FPA, higher spatial frequencies can thus create imaging problems by showing up as zero-frequency blurs with no modulation; against this constant background,

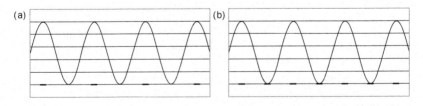

FIGURE 6.13 The spacing (or "pitch") of a periodic array of detectors determines the spatial frequencies that can be measured. By measuring only the peaks, the image in (a) is undersampled, while that in (b) has a sufficient number of detectors to measure the image modulation when the peaks and valleys are as shown.

detecting an image is much more difficult. Usually the sampling frequency is not exactly half the Nyquist frequency, so the energy in the image's high-frequency components produce some low-frequency components with artificially heightened contrast. These latter components cause effects such as low-resolution ("pixelated") elements and moiré patterns in the image (Fig. 6.14) that do not actually exist in the object.[18]

Aliasing is generally not a problem for consumer cameras; photos of periodic tweed patterns on a jacket may look strange, but with most cameras the patterns are not noticeable unless photographing picket-fenced homes from far away. However, aliasing is a serious issue for applications that rely on humans or computer image processing to interpret scenes that include periodic patterns— typically scientific and aerospace applications requiring high-precision imagery and radiometry.

This problem can be alleviated by using detectors that are spaced more closely to sample the image. A smaller detector pitch results in a larger Nyquist frequency, increasing the spatial frequency that can be sampled without aliasing. Stated differently, for a given spatial frequency in the image, more detectors per millimeter allows measurement of the peaks, valleys, and points in between. Because more samples of a sinusoid are taken than are required by Nyquist (i.e., more than twice per cycle to reproduce a given frequency), the image is now said to be "oversampled."

Figure 6.15 shows oversampling of the Rayleigh-resolved images of two side-by-side point sources. The cost associated with oversampling is that, with more detectors to measure the scene, each detector ends up with fewer photons, and this makes it more difficult to extract the image from the background noise (see Sec. 6.3). This disadvantage must be traded off against the aliasing that occurs when the image is undersampled.

Just as a telescope has an aperture that determines the MTF of the optics, pixels can be thought of as "apertures" that determine the

FIGURE **6.14** Comparison of low-resolution and higher-resolution images; pixelation is evident *(left panel)* when using a large-pitch FPA. *(Credit: Wiltse and Miller,* Optical Engineering, *vol. 45.)*

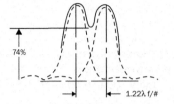 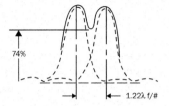

FIGURE 6.15 Sampling two Rayleigh-resolved point source images with detectors spaced at about 1.22λ × f/# is insufficient to resolve the high-frequency dip between the points. More detectors (right image) produce better spatial resolution but fewer photons per pixel, illustrating the trade-off between resolution and sensitivity when f/# is held constant.

MTF of the detector. The effects of pixel size on detector MTF are summarized by the following equation:[2]

$$\mathrm{MTF}_{\mathrm{det}} = \left|\frac{\sin(\pi f d)}{\pi f d}\right| = |\mathrm{sinc}(\pi f d)| \qquad (6.4)$$

Here f is any given spatial frequency (lp/mm) present in the image up to the optical cutoff, and d is the detector width or height (rectangular pixels have different MTFs in the x and y directions). This equation shows the effects of the finite size of each individual pixel collecting photons from a region of the image. When placed in an array with a pixel pitch x_p equal to the detector size d, the detector MTF goes to zero at the detector cutoff frequency $f_{\mathrm{det}} = 1/d = 1/x_p$; observe that $\mathrm{MTF}_{\mathrm{det}}$ approaches zero at higher spatial frequencies for smaller pitches (because then there are more detectors per millimeter).

Equation (6.4) does not take into account that—when detectors are arrayed into an FPA—the point spread function (PSF) will not be sampled, on average, exactly at the center of an $m \times n$ subarray consisting of pixels separated by a pitch x_p. If $x_p \approx d$ (which results in the least amount of energy being aliased into lower frequencies), then the approximation

$$\mathrm{MTF}_{\mathrm{det}} = \left|\frac{\sin(\pi f d)}{\pi f d}\right|\left|\frac{\sin(\pi f x_p)}{\pi f x_p}\right| = |\mathrm{sinc}(\pi f d)||\mathrm{sinc}(\pi f x_p)| \approx \mathrm{sinc}^2(\pi f d)$$

is sometimes used to take this additional sampling into account.[2] This effect is often ignored during FPA comparisons, when only the detector MTF is used as a metric.

Figure 6.16 shows plots of Eq. (6.4) for one FPA with a detector pitch $x_p = 10$ μm and for a different FPA with $x_p = 20$ μm. Assuming an optical cutoff frequency of 100 lp/mm, the figure shows that the FPA with the 10-μm pitch has the same cutoff frequency as the lens

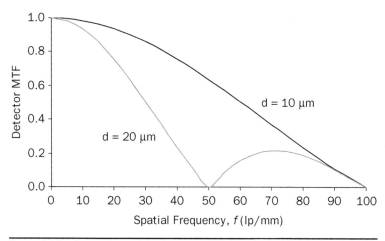

Figure 6.16 Comparison of the detector MTF given by Eq. (6.4) for pixels with a 10-μm pitch and pixels with a 20-μm pitch. The smaller pitch is able to measure higher spatial frequencies with better contrast (MTF).

whereas the FPA with the 20-μm pitch can detect spatial frequencies only half as big. As expected for a sinc function, there is some residual contrast for the 20-μm pitch at spatial frequencies higher than the detector cutoff, but these should be avoided because they represent reversals in MTF phase (white-on-black patterns at these spatial frequencies are seen as black-on-white, and vice versa).

In summary: with a lens, higher spatial frequencies are not *transmitted* very well; with an FPA, higher spatial frequencies are not *measured* very well. Therefore, the MTF for the optics-plus-detector system depends on both the transmission and the measurement:

$$\text{MTF} = \text{MTF}_{opt} \times \text{MTF}_{det} \quad (6.5)$$

where MTF_{opt} is given by Eq. (3.17) for a diffraction- or aberration-limited optical system. Just as there are many variables that can make the optical MTF_{opt} worse than diffraction limited, there are variables that can make the detector MTF_{det} worse than the ideal given by Eq. (6.4). These variables include pixel-to-pixel electron (or hole) diffusion, amplifier crosstalk, charge-transfer inefficiencies, and others.[19] Note also that some optical systems (e.g., TV) involve human viewing; in such cases Eq. (6.5) must also reflect the MTFs of the display and the eye, which further reduces the overall MTF of the optics-plus-detector-plus-display-plus-human system.[19]

Equation 6.5 is plotted in Fig. 6.17 for an ideal FPA and an f/20 lens with WFE = 0 and a cutoff frequency of 100 lp/mm at a wavelength of 0.5 μm. The detector MTF_{det} is for the case of 20-μm

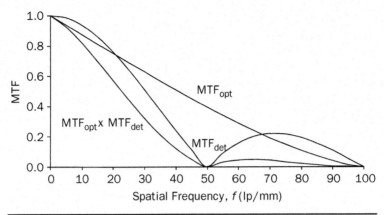

Figure 6.17 System MTF given by Eq. (6.5) for an optics-plus-detector system. The cutoff frequency of this system is that of the detector, which is given by $f_d = 1/x_p$ for an FPA.

pixels, which yields a detector cutoff frequency of $f_d = 1/x_p = 50$ lp/mm. The detector Nyquist frequency is then $f_N = 1/2x_p = 25$ lp/mm; any spatial frequency lower than this will be oversampled by the 20-μm pitch, while any frequency that is larger will be undersampled. It is the undersampled frequencies that will be aliased to look like lower-frequency patterns in the detected image, increasing their contrast at the expense of image quality and resolution. An exact analysis for identifying specific objects proceeds by quantifying the effects of FPA sampling through a construct known as the "MTF squeeze."[10,19]

It is common to specify an optical system's MTF at the FPA Nyquist frequency. For the detector component alone, the FPA Nyquist MTF ≈ $2/\pi$ and occurs at a spatial frequency of $f_N = 25$ lp/mm in Fig. 6.17 (half the detector cutoff frequency of 50 lp/mm). The system MTF given by Eq. (6.5) is then approximately 0.44 at the Nyquist frequency, a value that is specific to the ideal f/20 optic assumed here. System MTF must also incorporate the effects of aberrations, detector diffusion, and line-of-sight motion, which further reduces MTF at the Nyquist frequency; see Chap. 7.

Example 6.2. Designing an optics-plus-detector system is an unconventional dance of leader and follower, with the roles of the optics and FPA changing as the design process proceeds. To illustrate the difficulties, this example looks at designing a consumer-market digital telescope to image the night sky with a 512 × 512 silicon FPA.

We start by looking at the requirements: crisp images of galaxies are required over the visible band, which implies a design with the best possible resolution over the 400–700 nm range of wavelengths. We must also collect as much light as possible, but given that we will be imaging through the resolution-limiting atmosphere, an aperture of 100-mm diameter is chosen as the starting point.

Finally, the marketing group for this commercial product believes that a 0.1 degree (1.75 mrad) field of view is appropriate and insists that there not be any aliasing of the image.

Given the aperture and field, some initial properties of the optical system can be determined. First, the angular size of the diffraction-limited blur is $\beta = 2.44\lambda/D = (2.44 \times 0.5~\mu m)/(0.1~m) = 12.2~\mu rad$. Second, the FOV of 1.75 mrad is divided by the FPA's pixels into 512 instantaneous FOVs, giving an imaging IFOV = 1.75 mrad ÷ 512 = 3.4 µrad per pixel, or approximately 3.6 × 3.6 pixels per blur (from 12.2 µrad ÷ 3.4 µrad). As shown below, the diffraction-limited blur needs to be sampled with at least 4.88 pixels in order to avoid aliasing with a circular aperture. Approximating this to n_s = 5 pixels, each square pixel requires a blur sampling of β/n_s = 12.2 µrad ÷ 5 pixels = 2.44 µrad per pixel in the x and y directions.

However, the blur sampling must be consistent with the requirement on imaging IFOV. Given the requirements for the FOV and no aliasing, it's clear that there's an inconsistency between the two, a surprisingly common situation. To get around this problem, the design must be iterated—either by decreasing the imaging IFOV or by increasing the blur sampling to something less than 5 pixels.

For the first option, the imaging IFOV needs to be decreased to 2.44 µrad in order to meet the aliasing requirement. The total FOV is thus a factor of 2.44/3.4 = 0.72 smaller, which the marketing department may not like, but at least the FOV is only 30 percent smaller and not 50 percent or 90 percent smaller. Alternatively, the FOV can be maintained by increasing the number of pixels in the FPA to 712 × 712, thus reducing the imaging IFOV to 2.44 µrad per pixel. This requires either an FPA with smaller pixels (to match the smaller IFOV) or a telescope with a longer focal length, neither of which may be feasible.

Using fewer pixels than Nyquist to sample the optical blur does lead to some spatial frequencies being aliased, but it also increases the irradiance per pixel. This is a common operating point for imagers: a trade-off between the unaliased image quality obtained with 4.88-pixel sampling and the high radiometric accuracy of using 1-pixel sampling to measure the number of incident photons. In practice, a 2 × 2 subarray is often used for the design of the optics-plus-detector sampling; then, once the instrument is assembled, optomechanical tolerances increase the blur size to (it is hoped) no more than a 3 × 3 subarray. Chapter 7 has more details on this tolerancing.

There are many other aspects of this design in need of further development. For example, the EFL for a 100-mm aperture and an FPA with 20-µm pixels is 5.88 meters (EFL = d/IFOV = 20 µm ÷ 3.4 µrad). This is a difficult size to package into a consumer-market telescope, but it can be reduced by using smaller pixels. At the same time, radiometric calculations must also be performed to ensure that the pixel size is large enough to collect enough photons in a "reasonable" amount of time and with enough dynamic range.

Thus, when including an FPA in the optical-system design, the concepts of aperture, field, and etendue are not enough to specify an instrument. To fully design the optics-plus-detector system, it is also necessary to include detector sampling.

Detector sampling also requires a balance between the requirements for imaging and those for radiometric accuracy. A common metric for measuring over- and undersampling is the Q-parameter. Physically, this metric compares the blur size with the pixel pitch and leads to the definition illustrated in Fig. 6.18. Mathematically, it is written as

$$Q = \frac{B}{2.44 x_p} = \frac{\beta}{2.44 \times \text{IFOV}} = \frac{\lambda \times f/\#}{x_p} \quad (6.6)$$

The right-hand side of this expression assumes a detector pitch x_p equal to the detector size d, as is the case for FPAs with a 100 percent fill factor (see Sec. 6.2.2).

Typical values of Q are different for imagers than for radiometers. For example, a Nyquist-limited imager whose detector Nyquist frequency equals the optical cutoff frequency [or $f_N = f_c$, implying that $1/2x_p = 1/(\lambda \times f/\#)$] has $Q = 2$, where Eq. (6.6) gives the blur $B = 4.88 x_p$ as used in Example 6.2. It's also possible to define Q in the language of MTF, since Eq. (6.6) implies the equivalent definition $Q = 2f_N/f_c$; in this case, $Q = 2$ requires that $f_N = f_c$ (see Fig. 6.19). Since all spatial frequencies that are passed by the lens in this case are smaller than

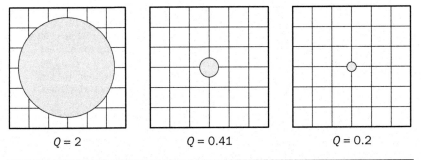

Figure 6.18 Blur size in comparison with pixel pitch. Here, Q [given by Eq. (6.6)] is changed by keeping the pixel pitch constant while varying the aperture and blur size.

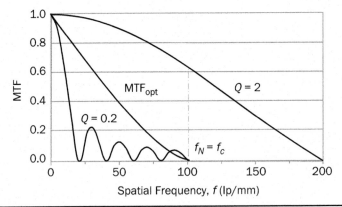

Figure 6.19 Detector MTF curves associated with $Q = 2$ and $Q = 0.2$ sampling obtained by varying the pixel pitch. The optical MTF is for an f/20 lens at $\lambda = 0.5$ μm; it has zero MTF for spatial frequencies $f \geq f_c = 100$ lp/mm. Therefore, the $Q = 2$ system cannot alias frequencies $f > f_N = f_c/2 = f_c$.

the detector Nyquist frequency, it follows that aliasing is not physically possible for $Q \geq 2$. The cost of avoiding aliasing is the reduction in optical MTF at all frequencies as compared with the case of using a faster lens to extend the optical cutoff f_c beyond the detector Nyquist. This cost must be traded off against the possible loss in image quality that may occur due to aliasing. Because most natural scenes do not contain periodic patterns that can create aliasing, this trade is usually resolved in favor of higher optical MTF, with the resulting optimal $Q \approx 1$.[42] This concept is explored in more detail in Sec. 6.5.2.

Radiometers, on the other hand, must accurately measure the optical power collected from the object. To do this, they must measure more than the Airy disk and do so on a single pixel. Table 3.9 shows that measuring only the Airy disk throws away 16 percent of the light from a point source; as more of the spot is measured, the radiometric measurement of a point source improves. At the same time, some light will be lost between pixels for a sampled blur, which reduces radiometric accuracy. Because the goal of a radiometer is not to reconstruct an accurate image of all the spatial frequencies present in the object, undersampling of the blur is not a problem. Instead, as much power as possible from a point source must be imaged onto a single pixel; this implies $B < x_p$, for which $Q < 0.41$ (see Fig. 6.18).

For both imagers and radiometers, a smaller Q puts more optical power on a single pixel, affecting how well the signal can be distinguished from the detector noise. A more direct influence on detected signal is the fill factor, which is discussed in the next section.

6.2.2 Fill Factor

We have just shown that the pixel size can be smaller than the pitch. This results in undetected light falling on the FPA, reducing the signal in an imager and the measurement accuracy in a radiometer. This is generally not a problem when the pixels can be packed close to each other with only a small gap between them. However, microbolometers—because of their intricate MEMS architectures—have traditionally required pixel sizes smaller than their spacing, although recent designs have a fill factor of nearly 95 percent.

As illustrated in Fig. 6.20, *fill factor* is the ratio of sensitive pixel area to pitch area. This ratio approaches unity (i.e., 100 percent) for closely packed photodiode arrays but can be much less for high-speed pixels requiring a small size. A fill factor of 80 percent, for example, signifies a ratio of pixel area to pitch area of 0.8; this, in turn, implies that the pixel size is $(0.8)^{1/2} \approx 0.89$ times the pitch, leaving gaps between pixels that cannot detect 20 percent of the incident photons. This is sometimes specified as an "effective responsivity" that is 80 percent of the "inherent material responsivity" of a single pixel.

For visible FPAs with small pixels, microlens arrays are sometimes used to improve the fill factor. This is shown in Fig. 6.21,

FIGURE 6.20 The FPA fill factor (FF), defined as the ratio of the sensitive pixel area (shown in white) to the pitch area, determines the amount of light than can be measured.

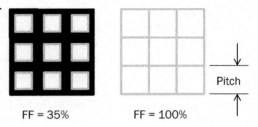

FF = 35% FF = 100% Pitch

FIGURE 6.21 The performance of FPAs with low fill factors can be improved with a microlens array for moderate-to-large f/# optics. (*Photo credit: JENOPTIK Optical Systems, Inc.; formerly MEMS Optical Inc.*)

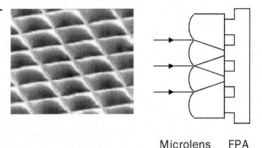

Microlens   FPA

where light that would normally fall between pixels is redirected by submillimeter-scale lenses back onto the pixel. Like all lenses, these reduce throughput (owing to Fresnel and absorption losses). Because of how the microlenses are manufactured, they are not of diffraction-limited quality and so should not be relied upon for use with pixels that are too small, optics with small f/# (< f/8 or so), or large FOVs.[20] Despite these limitations, microlens arrays are commonly used in consumer CMOS FPAs with small pixels, where a large fraction of the pixel area is taken up by the support electronics integrated on the FPA.

6.2.3 Uniformity

Even with over-sampling and a fill factor of 100 percent, the image obtained with an FPA can still be horrendous. A perfectly uniform scene, for example, is often detected as highly nonuniform, with some pixels producing more output current than others (Fig. 6.22). If not corrected, this nonuniformity in pixel output can prove to be the limiting factor in image quality and signal-to-noise ratio (Sec. 6.3).

The reason for this *photo-response nonuniformity* (PRNU) is the manufacturing process used to make FPAs. The thickness of the layers, the concentration of the dopants, and even the relative

FIGURE 6.22 Inherent differences in the current generated by each pixel in an FPA create nonuniformities, as seen here, that must be corrected. (*Credit: J. L. Tissot et al.*, Proceedings of the SPIE, *vol. 6395.*)

proportions of the semiconductors (such as indium, gallium, and arsenic for an InGaAs device) can vary significantly across the dimensions of an FPA (~ 50 mm × 50 mm). Electronic gain following the detector can also be different from pixel to pixel, and this nonuniformity is often greater than the detector's. Since these fundamental parameters differ from one pixel to the next, it follows that the attendant FPA properties—such as photo-responsivity, dark current, amplifier offsets, and APD gain—also vary. This variation is what we call *nonuniformity*.

It is possible to correct for nonuniformities by exposing the FPA to a highly uniform scene and then measuring the responsivity (or some other property) for each pixel individually. Such nonuniformity correction (NUC)—also known as "flat fielding" in the astronomy community[21]—requires careful engineering. In general, FPAs cannot be fully corrected, and it's common for FPA manufacturers to specify an array's nonuniformity both with and without NUC. For example, InSb FPAs from a commercial manufacturer have a PRNU of 5 percent before correction and a residual nonuniformity (RNU) of no more than 0.1 percent after correction. Section 6.4.7 reviews the correction process in more detail.

6.2.4 Frame Rate

Because FPAs are used to capture an image of an entire scene without scanning, it's necessary to know how quickly that capture can occur to ensure that the scene does not change "much" during imaging. An individual pixel may be able to respond within microseconds to changes in the scene, but stitching together the signal from a megapixel's worth of data can take a lot longer. This collection of data from all the pixels on an FPA is known as a *frame*; the time it takes to collect and read out this data is the frame time, which occurs at the frame rate.

In principle, smaller arrays can have larger frame rates, but rates are also determined by the application.[22] A common frame rate for visible FPAs is 30 or 60 frames per second (fps), chosen for compatibility with consumer video equipment. Array formats for computer displays have a rectangular aspect ratio of four horizontal pixels for every three vertical, with 640 × 480 pixels a common format (VGA) that is steadily being replaced by 1024 × 768 for high-definition TV. Scientific and machine-vision FPAs such as infrared imagers typically use square and 5 : 4 rectangular (320 × 256, 640 × 512, etc.) formats. These arrays are available in sizes up to 4096 × 4096 for IR arrays and 8200 × 6150 for high-performance CCDs, with available frame rates currently around 120 Hz for a 1-megapixel (1024 × 1024) array.

Certain types of FPAs use a technique known as *windowing* to read out the data from a small subarray. Because there are fewer pixels in the smaller array, windowed frames can be read out at a faster rate than that of the full FPA. For example, a commercial InGaAs FPA (Goodrich-SUI #SU640SDWH-1.7RT) has a frame rate of 109 Hz for a 640 × 512 FPA, which increases to 15.2 kHz for a 16 × 16 window. Windowing is a significant advantage for situations where the scene of interest can be narrowed down to smaller and smaller regions to capture objects that are moving at faster and faster rates. This capability is available in CMOS and hybrid FPAs, as described in the next section.

6.2.5 CCD, CMOS, and Hybrid Arrays

The available pixel pitch, fill factor, and frame rate all depend on the FPA technology used. For photon detectors, there are currently three types: charge-coupled devices (CCDs), complementary metal-oxide semiconductor (CMOS), and hybrid arrays—each with advantages and disadvantages that must be traded off against each other in order to accommodate multiple optical system requirements.

CCD FPAs are used for high-performance applications in the visible (consumer, scientific, medical, etc.) that require low noise and high responsivity.[23] As shown in Fig. 6.23, each pixel in a CCD has an absorbing semiconductor (a silicon p-n junction) to convert photons to electrons (charge), a metal-oxide semiconductor (MOS) capacitor to store this charge,[24] and a clever electrode scheme to couple the charge between pixels for transfer to the end of a row, where it is read out and processed by electrical circuitry. This transfer is not perfect, and charge-transfer efficiencies (CTEs) of 0.999995 are required (and available) to reduce charge loss for large-format arrays with many pixels along a row.

The CCD was a key step in developing digital cameras to replace photographic film. In addition to the obvious benefits of not requiring chemical development, the primary design advantage over consumer film is responsivity: the CCD is up to 40 times more sensitive.[21] Unfortunately, CCD fabrication is not compatible with high-cost, low-volume production.

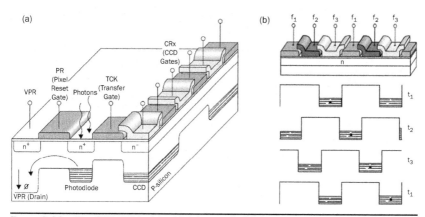

FIGURE 6.23 The CCD FPA in (a), which is usually fabricated in silicon, relies on a unique method of moving ("coupling") charge from one pixel to another (b) until it is transferred to readout circuitry at the edge of the FPA. (*Credit: Images courtesy of DALSA Corp.*)

To address this issue, CMOS FPAs were developed and are preferred for miniaturized ("camera on a chip"), low-power-consumption, low-price products—such as cell-phone cameras—where high performance is not critical. To reduce the size of the FPA's circuit board, readout and processing circuitry are incorporated into each pixel (creating an "active-pixel sensor" or APS), including transistors fabricated using a "complementary" MOS (i.e., CMOS) process. Not surprisingly, reducing this circuitry onto the small size of a pixel results in design compromises, most notably that CMOS noise is higher than that of a CCD at the frame rates found in consumer cameras. The APS circuitry also takes up some of the pixel area, significantly reducing the fill factor and effective responsivity for CMOS FPAs with small pixels (< 10 µm).

Because each pixel has readout circuitry associated with it, the collected charge does not have to be transferred down a row. Hence individual pixels can be read out randomly with CMOS FPAs, which means that no electrons are lost to charge-transfer inefficiencies in the process. Windowing of subarrays—known as *digital zoom* in consumer cameras—is also possible, as it is in a third FPA technology: *hybrid* FPAs.

Hybrid FPAs address a major weakness of CCD and CMOS arrays—namely, their reliance on silicon as the same material for both detection and electronics. The use of silicon for detection, processing, and readout does allow integration of CCD and CMOS FPAs onto the same semiconductor chip (or "die"). Both CCD and CMOS FPAs thus take advantage of the fabrication infrastructure that supports the integrated-circuit industry. The problem is that a different technology is required for detecting wavelengths outside the visible and NIR bands (~ 400–1000 nm) to which silicon is sensitive.

The approach taken by hybrid arrays is shown in Fig. 6.24, where one chip is stacked on top of another. The top chip is the detector array, which is made of the material (HgCdTe, InGaAs, etc.) appropriate for the waveband of interest. The bottom chip is a readout integrated circuit (ROIC), an electronic array typically made out of silicon to leverage the CMOS technology base and capabilities of silicon foundry processes. The two chips are bonded together ("hybridized"), usually with indium balls (or "bumps") that are small enough to connect each detector pixel electrically to each unit cell in the ROIC (also known as a *multiplexer*).

The detector in a hybrid FPA can be a PIN, an avalanche photodiode, or the simple p-n junction used in CCDs and CMOS arrays. As with CCDs, hybrid FPAs are used for high-performance applications in which the advantages (e.g., windowing) of CMOS readouts are useful. In addition, design compromises are reduced because the detector array can be designed more or less independently of the silicon ROIC. The hybrid concept can also be used to create silicon FPAs by combining a CMOS ROIC with a silicon PIN or APD.

Using CCDs, CMOS, or hybrids in an optical system involves many intricacies (antiblooming, correlated double sampling, image fringing, etc.) that are covered in detail by Howell[21] and by Holst and Lohmheim.[25] Properties such as responsivity, dynamic range, and read noise all play an important role. As illustrated in the next section,

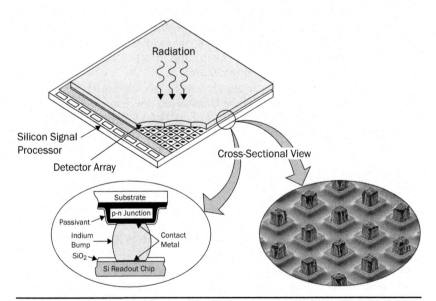

Figure 6.24 A hybrid array consists of a detector array bonded to a silicon readout integrated circuit (ROIC). (*Credit: A. Rogalski*, Proceedings of the SPIE, *vol. 4999.*)

the most important property common to all FPA technologies—and to individual detectors as well—is the signal-to-noise ratio.

6.3 Signals, Noise, and Sensitivity

We all have an intuitive understanding of signals and noise. For example, fans of Philip Glass's dissonant compositions may be listening to a "signal" in the notes while the rest of the world hears noise. Similarly, fireworks exploding at 10 in the evening on Canada Day may be inspiring to a photographer, but the same display at 3 in the morning will keep everyone awake and is nothing more than a bustle in your hedgerow.

Beyond these subjective interpretations, both signals and noise can be measured by the detector subsystem. In themselves these measurements are meaningless numbers; they become useful only when compared with each other by dividing the signal by the noise. This defines the idea of the signal-to-noise ratio (SNR), a key metric for evaluating optical systems.

Figure 6.25 illustrates the utility of the concept, showing images with an SNR of 0.2, 0.5, 1, and 5. For SNR = 0.2, the plot of irradiance on the left of Fig. 6.25 does not reveal the object's existence; nonetheless, it's possible that an excellent pattern-recognition system

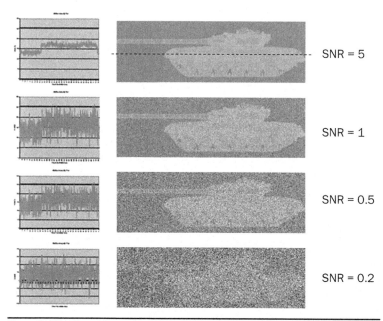

FIGURE **6.25** The signal-to-noise ratio (SNR) determines the quality and hence the utility of a signal. (*Credit: Georgia Tech course in Infrared Technology & Applications.*)

(such as the brain) can discern the tank shape on the right. In contrast, with an SNR of 5 the tank image clearly stands out against the background noise. Further increases in SNR make the image even more visible until the effect eventually saturates; at this point, any increase in SNR is not worth the additional effort and cost to achieve it.

A typical detector subsystem for measuring SNR is shown in Fig. 6.26. It consists of a two-pixel detector array and associated readout electronics, including a low-noise pre-amplifier ("pre-amp"), a multiplexer (MUX) where the signals from pixels are combined, an analog-to-digital converter (ADC), and possibly additional amplifiers. Instead of the pre-amplifier, a simple resistor may be sufficient for large-signal detection, where enough current is created to produce relatively large voltage drops. For lower-level signals, a pre-amplifier such as a high-impedance or transimpedance amplifier (TIA) is often required; low noise is a key requirement in such cases because too much noise will limit the smallest signal that can be measured.

As a first step in quantifying the signal, we observe that the detector output i_d in Fig. 6.26 has three components: $i_d = i_s + i_b + i_{dc}$.

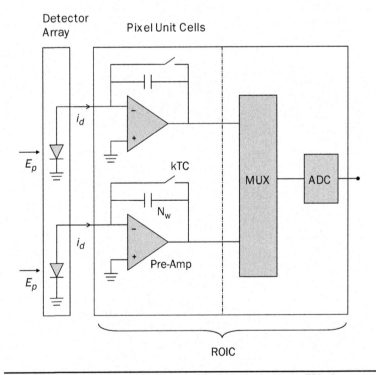

Figure 6.26 Detector circuitry associated with each pixel in an FPA; the schematic shows the detector array and readout integrated circuit (ROIC) with electrical unit cells for two pixels.

These currents correspond to optical power from the signal (i_s) and background (i_b) and to the inherent noise of the detector known as *dark current* (i_{dc}). Large background or dark currents make it difficult to extract the signal and can limit the largest possible SNR that can be measured. Many other types of noise reviewed in this section will also be present in the post-detector electronics, with likewise deleterious effects on the SNR.

At some point in the process of designing an optical system, the following question will come up: Is the SNR good enough? System requirements generally push the SNR to some optimum range—a range below which performance drops dramatically and above which the associated technical and cost difficulties will affect the development schedule. In order to avoid these difficulties, it is useful to understand the noise limitations for special situations where one type of noise is much larger than the others; a review of such cases is included in Sec. 6.3.3.

Finally, the idea of SNR can be looked at from another perspective that results in the concept of *sensitivity* (i.e., the smallest signal that the detector subsystem can measure). There are a number of variations on this theme, which are looked at in Sec. 6.3.4. To keep the length of each section manageable, we focus in all cases on photodiodes—either as used in FPAs or as single-pixel detectors.

6.3.1 Signals

The electrical signal measured by an optical system depends on the optical power collected and on the detector's responsivity. For a point source uniformly emitting into a 4π solid angle with intensity I [W/sr], the signal consists of the current generated by the detector and is given by $i_s = R\Phi_s = RTI\Omega$, where $\Omega = A_{EP}/R^2$ is the solid angle intercepted by the entrance pupil at a distance R [see Eq. (4.10)]. Images of point sources are often distributed over multiple pixels. Hence, the signal current generated by each pixel is lower by what is known as the *point visibility factor* (PVF) or ensquared energy (see Sec. 3.3.4), both of which are less than unity.

For an area source with radiance L [W/m²·sr] emitting into an optical system with solid angle Ω and entrance pupil area A_{EP}, the signal is again the current output of the detector: $i_s = R\Phi_s = RTLA_{EP}\Omega$ [see Eq. (4.8)].

Responsivity is normally a function of the wavelength being measured, in which case the simple product $R\Phi_s$ gives incorrect results for a broadband source with wavelength-dependent output. As shown in Example 6.1, the integrated responsivity multiplied by the spectral power at each wavelength must be used to calculate the signal current:

$$i_s = \mathcal{R}\Phi_s \equiv \int \mathcal{R}(\lambda)\Phi_{sp}(\lambda)\,d\lambda \quad [A] \tag{6.7}$$

Here the wavelength-dependent spectral flux $\Phi_{sp}(\lambda)$ is the optical signal *absorbed by* the detector so that all optical throughput losses of the system are taken into account, including the detector's Fresnel losses.

Signals can also change over time and, as photons are converted to electrons, both the detector and the electrical subsystem must be fast enough to respond. The response time determines an electrical bandwidth Δf, which is usually approximated by $\Delta f \approx 1/2\Delta t$, where Δt is the integration time over which photons are collected. As shown in the next section, this bandwidth plays a large role in determining how much noise is measured by the detector and electrical subsystems.

6.3.2 Noise

All sources have noise. The noise source may be the signal photons of interest or background photons that make it difficult to extract useful information from the signal. Analogously, searching for a small boat in a large ocean on a sunny day is difficult, in part, because of the sunlight ("glint") reflected off the waves. The detection of LWIR signal photons from a distant galaxy can be equally difficult because of the 10-μm peak wavelength of blackbody emission from optical system components at 300 K.

Figure 6.27 shows the output over time of a typical detector, a response to an optical source whose output power was designed to be constant over time. The mean current for such a source is shown in the figure, as is the standard deviation σ. Chapter 4 and Sec. 6.3.1 showed how to calculate the mean signal; in this section, the numerous sources of noise that contribute to the standard deviation are reviewed and quantified.

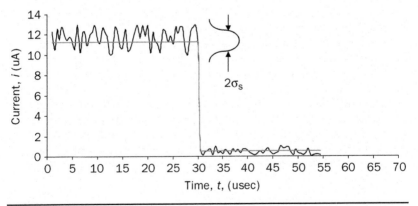

Figure 6.27 Even for a source with a perfectly stabilized average power, low-level random fluctuations in output power are a fundamental limit on source noise.

The most basic source of noise is the source itself. That is, despite the best design intentions and execution, output power from *all* sources changes over time in response to the quantum-mechanical processes that produce photons.[27] When detecting incoherent sources that produce a relatively large number of photons—tungsten-halogens and arc lamps, for example—the photodiode current follows Gaussian statistics as a large-photon approximation to Poisson statistics,[26,27] and the attendant noise is called *shot noise* because the energy can be associated with individual electrons or "shots". An important consequence is that the output fluctuates randomly with a variance σ^2 that grows as the source power increases.

For the same reason, photons originating in the background also have noise associated with them. The detector does not "know" whether the photons it is measuring come from the signal or the background, and the distinction is arbitrary from the perspective of measured noise. In some situations—for instance, trying to identify lightning flashes from above solar-illuminated clouds—the entire background may be considered noise. Yet if such backgrounds can be averaged and subtracted from the measurement, then any remaining fluctuation is due to noise in the background.

There are inherent detector currents that add noise to the signal; typically known as "dark" currents, these are present even in the absence of a signal or background. There are also noises associated with the electrical circuitry; these noises, too, affect the precision in the number of electrons measured. For example, Johnson noise is inherent in the circuit resistance, a result of the random thermal motion of electrons.

In addition, there are noises associated with the electrical readout—specifically, amplifier noise and quantization noise. Amplifier noise is that due to amplifiers (and other circuitry immediately following the detector) that convert the current flow to a voltage. There are many types of such "front end" amplifiers available; located just after the detector, each type is optimized for noise, charge capacity, physical size, speed, and so forth (Fig. 6.28).

Figure 6.26 also shows an analog-to-digital converter following the amplifier. This component converts a range of analog signals into binary numbers with an inherent noise that depends on the charge capacity of the amplifier. Because it depends also on how many bits are used to digitize this capacity, ADC noise is sometimes referred to as *quantization* noise.

Another group of noise sources are specific to FPAs. The primary two are spatial noise and read noise, of which the first was seen previously in Sec. 6.2.3 as array nonuniformity due to pixel-to-pixel variations in responsivity and ROIC gain. In addition, thermally induced variations in the number of electrons in the ROIC leads to identical signals being measured as something slightly different each

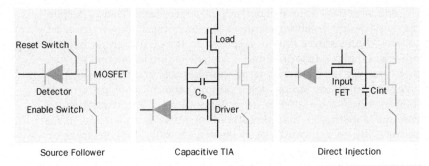

Figure 6.28 Amplifier noise depends on the type of amplifier, and the choice of amplifier type involves system trade-offs with respect to noise, speed, pixel size, and other factors. (*Credit: Teledyne Scientific & Imaging, LLC.*)

time they are read. This random variation is known as read noise, and it occurs along with the other types of noise in the detection process.

Noise is typically expressed as a standard deviation σ. Because the noises from different sources are (usually) independent of each other (i.e., they are uncorrelated or "orthogonal"), they can be added as a root sum of squares [or RSS, shorthand for the "(square) root of the sum of the squares"] of the standard deviations to produce a net noise current σ_n

$$\sigma_n = \sqrt{\sigma_s^2 + \sigma_b^2 + \sigma_{dc}^2 + \sigma_J^2 + \cdots} \quad [A] \tag{6.8}$$

By geometric analogy we can say that this technique is applicable to independent noise terms in the same way that the legs of a perpendicular triangle can be changed without affecting the length of the other, and thus "RSS together" to yield the hypotenuse length. The individual variances σ^2 for each type of noise source are quantified for photodiodes in the material that follows. The procedure involves looking at what the noise depends on, establishing the best ways to reduce it, and discussing the system-level trade-offs involved in doing so. The focus will be on p-n junction and PIN photodiodes; you should consult the references for information on how photodiodes differ from photoconductors and thermal detectors.[2–5]

Signal Shot Noise
The output power from any source will change over time. This change may consist of a slow, long-term drift of the type described in Sec. 5.3 for tungsten-halogens and arc lamps, but it will always include a higher-frequency jitter about a mean value—a phenomenon known as photon noise or *quantum noise*. We saw in Chap. 4 that the mean signal $\Phi_s = TL_\lambda A\Omega\Delta\lambda$; the noise variance about this mean is σ_s^2

for the mean signal current $i_s = \mathcal{R}\Phi_s$ with average responsivity $\mathcal{R}$ over the optical bandwidth $\Delta\lambda$.

At the same time, the very detection of photons by a photodiode also adds noise to the measured signal. This noise is associated with a p-n junction or PIN energy barrier, which electrons must overcome before they have enough energy to conduct. This process is a statistical one, and when the effects of both quantum and junction noise are included, the standard deviation σ_s [root-mean-square (RMS) value] of the signal current produced by a photodiode is given by[13]

$$\sigma_s = \sqrt{2qi_s\Delta f} \quad [\text{A}] \tag{6.9}$$

Here q is the fundamental charge on an electron ($q = 1.602 \times 10^{-19}$ coulombs per electron) and i_s is the mean detector current generated by the signal flux, as calculated in Eq. (6.7). An important feature of this so-called signal shot noise is that it increases with the source power, as illustrated in Fig. 6.27.

The electrical bandwidth Δf in Eq. (6.9) assumes that shot noise occurs as white noise—in other words, the noise is equally distributed over a broad range of frequencies. As a result, the more frequencies that are included in the measurement, the more noise that's collected by the detection process. This pushes the system design to as low a bandwidth as possible while maintaining the requirement that the detector be able to measure changes in the signal. A signal that is flashing at 30 Hz, for example, must be measured with a detector (and following electronics) with an electrical bandwidth of at least 60 Hz, for otherwise the detector subsystem will be unable to follow the changes. Our white-noise assumption breaks down when the detector's responsivity is frequency-dependent (typically at very high or very low frequencies, where $1/f$ noise is dominant), but it is commonly used nonetheless in system-level calculations.

Comparing devices and designs is easier when using a metric that does not depend on Δf. The concept of *noise spectral density* (NSD) addresses this issue by dividing both sides of Eq. (6.9) by $\Delta f^{1/2}$ to yield a signal NSD of $\sigma_s/\Delta f^{1/2} = (2qi_s)^{1/2}$; it is measured in units of A/Hz$^{1/2}$ (pronounced "amps per root Hertz") and is often found on spec sheets for detectors and amplifiers. The square of the NSD is also used, in which case it is known as the *power spectral density* (PSD) and is measured in units of A^2/Hz (proportional to the electrical power per unit bandwidth). The shot noise still depends on the bandwidth in either case, but the NSD and PSD concepts allow us to compare component performance and specs with a metric that's independent of the particular bandwidth required for a specific optical system.

There are almost always not *enough* photons in optical systems, so signal shot noise is usually insignificant for visible systems unless a low-noise, high-gain detector (e.g., a photomultiplier tube) is used.

If the aperture size is so large that the signal noise is excessive for a conventional detector, then the source may be too powerful—or the system may be too large, heavy, and expensive and hence must be scaled down to the point where optical power is again a luxury. Certainly some sources put out more than enough power to ensure a strong signal with reasonably sized optics, but such cases are the exception.

It's also possible to design "quiet" sources with shot-noise-limited (or better) performance yet still not be able to measure down to the noise level indicated by Eq. (6.9) with the detector subsystem. Signal shot noise dominates receiver noise when the signal is relatively large or when all other sources of noise are small. These other sources of noise are described in what follows.

Background Shot Noise

It is more typical for background (rather than signal) noise to be a limiting factor in an optical system. For example, solar background can overpower signals from the UV to the MWIR, and blackbody emission from sources such as the optics themselves can dominate the signal in LWIR systems. Techniques such as spatial, spectral, and temporal filtering can be used to reduce the background level, but despite these techniques the remaining background noise may still be large.

Like signal noise, background noise from an incoherent source is due to the large-mean Gaussian distribution of detected photons. The resulting background shot noise is thus given by

$$\sigma_b = \sqrt{2qi_b\Delta f} \quad [A] \tag{6.10}$$

where i_b is the mean detector current generated by the background. In Chap. 4 we saw how to calculate the optical power that creates this current; for example, out-of-field scattering from rough optical surfaces results in power scattered onto the detector, which is given by Eq. (4.12) as $\Phi_b = E_\lambda \times \text{BSDF} \times TA\Omega \Delta\lambda$ (recall that BSDF denotes the bidirectional scatter distribution function). Even more difficult is background light from an in-field noise source (e.g., the Sun), an interesting example of which occurs when trying to measure lightning storms from a satellite or airplane flying *above* solar-illuminated clouds. In this case, the background power is given by Eq. (4.9) as $\Phi_b = TL_\lambda A\Omega\Delta\lambda$, which results in a background current $i_b = \mathcal{R}_{avg}\Phi_b$ in the detector for an average responsivity $\mathcal{R}_{avg}$ over the optical bandwidth $\Delta\lambda$; given this current, the background shot noise is then calculated using Eq. (6.10).

Besides those for reducing the bandwidth Δf, there are a number of techniques for reducing the background shot noise. For out-of-field background, Eq. (4.12) shows that reducing the BSDF by using smooth, clean optics and/or adding baffles and sunshades, reduces the background current. For in-field background, Eqs. (4.9) and (5.2) indicate that reducing the temperature and emissivity of the optics should be considered for an IR system. Equation (4.9) also shows the importance of reducing the solid angle Ω of the detector (*spatial filtering*) or using an optical filter with a narrower spectral bandwidth $\Delta\lambda$ (*spectral* filtering). This latter technique is useful only up to a point, since a filter bandwidth that's too narrow may cut out some of the signal as well, reducing the SNR.

Dark-Current Shot Noise

When the signal and background are reduced to relatively low levels, there is a third type of shot noise that can limit performance in systems using PIN and p-n junction detectors. This detector "dark current" and its associated shot noise is not due to photons creating conduction electrons; rather, it is the temperature of the detector that give bound electrons enough energy to conduct even in the absence of photons.

Like the photoelectrons and photon noise generated by signal or background light, these dark-current electrons are also randomly generated with a Gaussian probability distribution and an assumed white-noise spectrum. Therefore, dark-current shot noise has the same form as signal shot noise and background shot noise:

$$\sigma_{dc} = \sqrt{2qi_{dc}\Delta f} \quad [A] \tag{6.11}$$

with typical room-temperature values of i_{dc} per unit detector area given in Table 6.3. Silicon is a relatively "quiet" material; one reason that its dark current is small is its larger bandgap, which gives fewer electrons enough energy to conduct. Dark current also depends strongly on the quality of the material growth and etching in the fabrication process. This means that vendor specs should give numbers similar to those reported in Table 6.3. When developing a specialized design, however, the dark current may be significantly higher if fabrication bugs have not all been worked out.

Material	Temperature	Dark Current
Silicon PIN	300 K	2 nA/cm²
InSb	80 K	40 nA/cm²
HgCdTe (at 10 µm)	80 K	7 µA/cm²
InGaAs (at 1 µm)	300 K	10–20 nA/cm²
Germanium	300 K	1–50 µA/cm²

TABLE 6.3 Approximate dark current densities for common photodiode materials at the appropriate bias for the given operating temperature.

Dark current is affected not only by materials selection and bandwidth but also by detector temperature and reverse bias. The benefits of lower temperature on detector performance are seen in Eq. (6.3). Dark current also increases with reverse bias; this is seen in the I-V curve for a typical photodiode, where a larger bias ($-V_b$ in Fig. 6.9) produces more current in the presence or absence of incident photons.

Dark current also depends on surface area, so another method for reducing dark current is to use smaller detectors. A smaller detector has fewer atoms to emit dark current, approximately in proportion to the detector area. Smaller detectors and FPA pixels also have the advantage of enabling a shorter focal length for the same IFOV, which results in smaller, lighter systems. Because of the smaller etendue, the disadvantage of smaller detectors is that fewer photons are collected—unless the f/# of the optics is also decreased to compensate—and the trade between measured signal and detector noise depends on whether the source is extended or a point and on whether the sensitivity is limited by the background or the detector.

Johnson Noise

A type of electrically generated noise that depends on the resistance of the detector and subsequent circuitry is known as Johnson noise. It is a consequence of the random thermal motion of electrons whereby higher temperatures impart more velocity to other electrons during collisions. If we equate the thermal power ($4kT\Delta f$) with the noise-dissipated electrical power in a resistor ($\sigma_J^2 R$), then the standard deviation of the Johnson current is given by[2]

$$\sigma_J = \sqrt{\frac{4kT\Delta f}{R}} \quad [A] \tag{6.12}$$

This expression shows that Johnson noise current depends on the electron's thermal energy kT as compared with its resistance R to current flow in the circuit; hence a lower temperature T or a higher resistance reduces this type of noise. In terms of its noise spectral density, a typical Johnson noise level at room temperature ($T = 300$ K) for a 5-kΩ resistance is 1.8 pA/Hz$^{1/2}$.

The current generated by a photodiode is not always sufficient to generate a measurable voltage drop across a resistor at a fast enough speed.[43] There are many methods for increasing speed; the simplest is to lower the resistance, which unfortunately also increases the Johnson noise current. In this case, a common solution is to increase speed by instead using an amplifier (see Fig. 6.29). The Johnson noise is then determined by the detector resistance R_d and resistance in the following circuitry (e.g., a load resistor R_l, amplifier feedback resistor R_f)[5] and is found by adding these individual resistances in parallel; thus, $T/R = T_d/R_d + T_f/R_f$. For a given temperature, the smallest

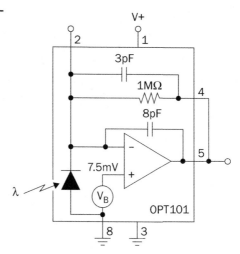

FIGURE 6.29 An integrated PIN detector and transimpedance amplifier; its 1-MΩ feedback resistor is a source of Johnson noise. (*Credit: Courtesy of Texas Instruments.*)

resistance of the two dominates and allows more Johnson current to flow through that resistor.

Whether or not the detector dominates depends on the detector material, detector size, and reverse bias; as seen in Fig. 6.9, a reverse bias decreases the slope of the *I-V* curve, increasing the detector resistance ($1/R_d = di/dV$). The detector resistance (0.1–1 GΩ) for typical silicon, InGaAs, and germanium detectors is much larger than the resistance of typical load or amplifier feedback resistors. When the detector resistance is smaller, there is a trade-off between dark current and Johnson noise: a higher reverse bias decreases Johnson noise but increases dark current.

Intentionally increasing load or amplifier feedback resistance is sometimes used to decrease Johnson noise. Cutting off the Johnson in this way has an additional cost, however, in that a larger resistance reduces the receiver response time in cases where the response time is limited by the resistance–capacitance (RC) time constant and not the detector diffusion time. Since noise controls sensitivity, it follows that there is a trade-off between speed and sensitivity (amplifier gain) when Johnson noise is dominant.

When the detector's resistance controls the Johnson current, cooling the detector is another common way to reduce the effects of this noise. If the controlling resistance is in the amplifier, then it's the amplifier circuit that is cooled. It sometimes happens that placing the detector and electronics close to each other prevents radio-frequency (RF) noise from propagating over the cables connecting them, in which case both the detector and electronics are cooled.

Replacing the feedback resistor in the amplifier circuit with a capacitor is yet another technique for reducing Johnson noise. Such capacitor-based amplifiers (e.g., capacitive transimpedance amplifiers, CTIAs) lower the resistive Johnson noise, thus leaving only the Johnson noise of the detector and the inherent noise of the CTIA itself. The noise-reduction benefits of CTIAs have made them standard components of many ROICs employed for high-sensitivity detection.

Amplifier Noise

Internal amplifier circuitry—whether integrated or discrete—also has an inherent noise—independent of shot noise and the receiver's Johnson noise—that can hide the signal being measured. Unlike shot and Johnson noise, however, amplifier noise is not something that can be easily calculated. Fortunately, it is often listed in manufacturer's spec sheets; typical noise values for high-speed (bipolar) TIAs are $\sigma_{TIA}/\Delta f^{1/2} \approx 1–2$ pA/Hz$^{1/2}$ at room temperature. Low-bandwidth FET-based TIAs are much quieter, with values of $\sigma_{TIA}/\Delta f^{1/2}$ as low as 0.001–0.01 pA/Hz$^{1/2}$.[5]

Amplifier noise generally isn't proportional to $\Delta f^{1/2}$ because high-speed amplifiers designed for larger bandwidths are inherently noisier. As a result, an amplifier with a bandwidth of only 1 MHz may have an NSD of 2 pA/Hz$^{1/2}$ [with total noise current σ_{TIA} = 2 pA/Hz$^{1/2}$ × (10^6 Hz)$^{1/2}$ = 2 nA] whereas a 1-GHz-bandwidth amplifier may have an NSD of 4 pA/Hz$^{1/2}$ [with total noise current σ_{TIA} = 4 pA/Hz$^{1/2}$ × (10^9 Hz)$^{1/2}$ = 126 nA]. The wider-bandwidth amplifier has significantly more noise current, which is due not only to the wider bandwidth but also to the doubling of the intrinsic noise of the higher-speed device. Assuming that amplifier NSD is independent of frequency is thus an approximation that in practice must be checked (Fig. 6.30).

In comparing a low-noise bipolar TIA ($\sigma_{TIA} \approx 2$ pA/Hz$^{1/2}$) with the Johnson noise for a 5-kΩ resistor at room temperature ($\sigma_J \approx 1.8$ pA/Hz$^{1/2}$), you can see that they are more or less the same. This means that replacing resistor in a TIA with a capacitor reduces the total noise by a factor of $2^{1/2}$. Furthermore, if the resistive TIA is cooled down to 240 K (a common temperature for thermoelectric cooling), then its Johnson noise current is only about 1.6 pA/Hz$^{1/2}$ and will have less of an impact on total noise. Even so, using low-noise amplifiers to shift some of the sensitivity burden from the optical subsystem to the detector will likely yield an instrument that is smaller, lighter, and less costly.

Quantization Noise

After the TIA (and perhaps some additional electrical amplifiers), the next component in the receiver electronics chain is often an analog-to-digital converter. The ADC converts the TIA's analog signal into a digital number consisting of a series of bits of 1s and 0s that can be read by a digital computer.

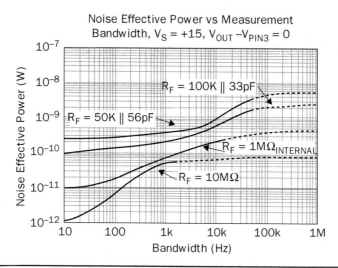

FIGURE 6.30 The noise-equivalent power (NEP) of an amplifier circuit—that is, the power sufficient to produce a signal equal to the detector noise—depends on the feedback resistance. (*Credit: Figure courtesy of Texas Instruments.*)

It should not be surprising that ADCs also add noise to the detection process; the amount of this noise depends in part on the number of bits into which the signal is digitized. A signal that is digitized into 12 bits, for example, divides the number of collected electrons into $2^{12} = 4096$ "bins" and has more truncation and rounding error than a signal digitized into smaller units using 16 bits (or $2^{16} = 65,536$ bins).

The ADC noise also depends on the maximum number of electrons that can be collected (aka the *well capacity* or *well depth*) during a measurement time. For example, if the well depth is large then the noise is high because more electrons are split up into the bin size determined by the number of bits. For an FPA, well depth may be limited by capacitor size or allowable voltage change.

Because a given well depth is divided into a discrete number of electrons that depends on the number of bits, ADC noise is also known as quantization noise. It is typically measured in units of electrons (e⁻) and is given by[10]

$$\sigma_q = \frac{N_w}{2^N \sqrt{12}} \quad [e^-] \tag{6.13}$$

Here N is the number of bits and N_w is the well depth, so $N_w/2^N$ is the mean number of electrons in each measurement bin (Fig. 6.31); the standard deviation of this mean is smaller by a factor of $12^{1/2}$. Although not exactly linear with pixel area, typical numbers are $N_w \approx 1000$ e⁻/μm² for a silicon CCD and $N_w \approx 2000$ e⁻/μm² for hybrid and

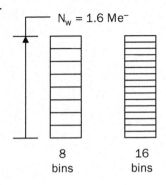

Figure 6.31 Quantization noise is a result of dividing the total charge capacity of the well into discrete units determined by the number of bits used in the analog-to-digital conversion.

CMOS arrays using CTIAs. For 25-µm × 25-µm square pixels, Equation (6.13) then shows that quantization noise with a 12-bit ADC is $\sigma_q = 625 \text{ ke}^-/(4096 \times 12^{1/2}) = 44 \text{ e}^-$ for the CCD, and 88 e⁻ for a high-sensitivity hybrid FPA.

Although CTIAs are used to improve sensitivity, they have a relatively small well depth and cannot collect high-flux signals over long periods of time. As seen in Fig. 6.28, alternative amplifier circuitry is commonly used for designs requiring larger well capacity; this circuitry includes direct injection (DI) and buffered direct injection (BDI). For hybrid and CMOS ROICs, the resulting well capacity is greater by factor of 10 than a CTIA with a well capacity of approximately 2 ke⁻/µm²—although at the price of poorer sensitivity due in part to the increase in quantization noise for the DI and BDI ROICs.[40]

The number of bits available in off-the-shelf ADCs depends on their speed. Digitizing video signals at 30 Hz with 8-bit ADCs is a typical combination for consumer products. Higher speeds and more bits are also available: 16-bit ADCs can be purchased with digitizing speeds up to 80 million samples per second (MSPS) per channel. More common are 12- to 14-bit ADCs with speeds up to 250 MSPS; this illustrates the general trend of faster ADCs not being able to digitize as many bits.

Spatial Noise

Besides dark current, two major contributors to nonuniformities are pixel-to-pixel differences in responsivity and amplifier gain, differences that result from fabrication variations in (respectively) the detector array and the ROIC. These photoresponse nonuniformities cannot be completely removed, and the remaining pattern of spatial noise is specified as a percentage of the responsivity.

With a corrected PRNU of 0.1 percent, for example, the responsivity of one pixel can have an output that's 0.05 percent higher than the mean value of the array (within 1σ RMS) while that of another pixel may be 0.05 percent lower. The difference in responsivity produces a pixel-to-pixel difference in output current, even though the optical source producing the signal or background photons may be perfectly uniform. This phenomenon can limit the receiver's sensitivity because the *fixed-pattern* noise (FPN) depends on the incident power,

$$\sigma_{\text{FPN}} = \text{PRNU} \times \mathcal{R}\Phi = \text{PRNU} \times \mathcal{R}L_\lambda A \Phi \Delta\lambda \quad [\text{A}] \quad (6.14)$$

for a wavelength band $\Delta\lambda$ emitted by an area source. Observe that a greater flux Φ from the source produces a proportional increase in FPN (aka *spatial noise*). This is distinct from signal and background shot noise, which increase with the square root of the photon flux [see Eqs. (6.9) and (6.10)]. The greater sensitivity to optical flux burdens the system designer with reducing the PRNU to low values (~ 0.1 percent or lower). The design may thus be pushed to a specific detector material, some of which (e.g., silicon) have a more mature fabrication process and inherently better uniformity than others (e.g., HgCdTe). We will see in Example 6.3 that the requirement on PRNU will depend on the magnitudes of the other types of FPA noise present, but nonuniform spatial noise can easily be a limiting factor in FPA selection.

Read Noise

A second type of noise found in FPAs is *read* noise; this noise occurs every time a signal from the detector array is read out by the ROIC.[28] Because of thermally induced changes in the number of electrons in the ROIC, the same signal will be measured as something slightly different each time it is read. This random variation is in addition to signal, background, and dark current shot noise, and it has a typical standard deviation of $\sigma_r \approx$ 1–1000 electrons per pixel per read depending on pixel size, frame rate, temperature, and ROIC quality.

Unlike many of the other noises described in this section, read noise is to first order independent of the FPA frame rate. It occurs once per frame and is added as a constant background to the shot-type noises that depend on integration time. For instance, PRNU spatial noise increases with signal and therefore decreases at higher frame rates (where fewer signal electrons are collected during a shorter integration time). Temporal noises (e.g., signal and background shot noise) likewise decrease as integration time is reduced, and so read noise becomes the dominant FPA noise for short integration times (see Fig. 6.32).

Read noise can be removed with a technique known as *correlated double sampling* (CDS), which requires the ability to subtract the "preread" voltage from the signal for every pixel in an array.[10] Given

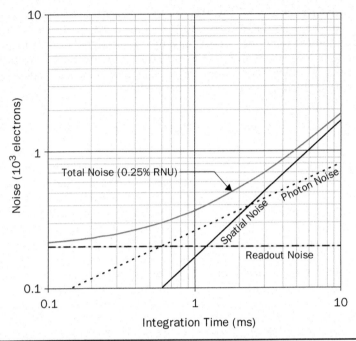

FIGURE 6.32 Depending on the integration time, different noises can dominate. At high frame rates (short integration times), noise that occurs once per readout can be the largest. (*Credit: L. LaCroix and S. Kurzius,* Proceedings of the SPIE, *vol. 5732.*)

the increased cost of these storage and signal-processing requirements, CDS is generally not used with consumer CMOS arrays and is reserved for higher-performance CCD and hybrid arrays.

Example 6.3. By itself, the total noise in an optical system is not a useful number. The more common and useful metric is the noise in comparison with the signal. To make this comparison, the noises must be quantified and added together; this example illustrates the details of these calculations.

An important tool in environmental sensing is the measurement of lightning from above the clouds. Such measurements are needed because lightning storms can prevent airplanes from taking off or landing; hence knowing where these storms are located can keep air traffic moving with minimum delays. The measurements are best made from a geostationary orbit that continuously monitors the same area on Earth using an FPA. However, to measure lightning flashes from above the clouds during the day requires that the optical system distinguish the flash from the large solar background reflected off the tops of the clouds.

Calculating the signal and noises for such an optical instrument quickly indicates this problem's difficulty. The signal current is calculated from Eq. (4.8) as $i_s = \mathcal{R}\Phi_s = \mathcal{R}TL_sA\Omega$. The radiance of lightning that's been scattered by the clouds has been measured as $L_s \approx 0.01$ W/m²·sr in a narrow band at 777 nm and spread out more or less uniformly over a 10-km by 10-km area.[29]

To calculate the signal, we assume a 100-mm aperture for the entrance pupil diameter and an optical throughput of $T = 0.5$; the signal power collected on a pixel from a GEO orbit at approximately 36,000 km above the Earth [with $\Omega = (10/36,000)^2$] is equal to $\Phi_s = TL_s A\Omega = 0.5 \times 0.01$ W/m²·sr $\times$ 0.00785 m² $\times (10/36,000)^2$ sr $\approx 3 \times 10^{-12}$ W. The resulting signal current is $i_s = \mathcal{R}\Phi_s = 1.94 \times 10^{-12}$ A (RMS) with $\mathcal{R} \approx 0.64$ A/W for a silicon photodiode at 777 nm (Fig. 6.10). Because FPAs measure signal via charge collection and not current, it's also necessary to express this as the number of electrons generated by the flash: $N_s = i_s \Delta t_p / q$ for a lightning-flash duration of $\Delta t_p \approx 0.5$ msec, yielding $N_s = (1.94 \times 10^{-12}$ A $\times$ 0.0005 sec)/(1.602×10^{-19} C/e⁻) $\approx$ 6052 electrons; here C denotes coulombs.

The background power depends on the spectral irradiance incident on the clouds from the Sun, which is given in *The Infrared and Electro-Optical Systems Handbook* as $E_\lambda \approx 1170$ W/m²·µm at 777 nm;[30] the spectral radiance scattered off the Lambertian clouds is thus $L_\lambda = \rho E_\lambda / \pi \approx 372$ W/m²·sr·µm for $\rho = 1$. Since the aperture, pixel solid angle, throughput, and pixel responsivity are the same for the background as they are for the signal and since the optical filter bandwidth must transmit the lightning spectrum (with $\Delta\lambda \approx 1$ nm), it follows that Eq. (6.7) gives the background current as

$$i_b = \mathcal{R}\Phi_b = \mathcal{R}TL_\lambda A\Phi\Delta\lambda$$
$$= 0.64 \times 0.5 \times 372 \text{ W/m}^2 \cdot \text{sr} \cdot \mu\text{m} \times 0.00785 \text{ m}^2 \times (10/36,000)^2 \text{ sr} \times 0.001 \text{ µm}$$
$$\approx 7.2 \times 10^{-11} \text{ [A]}$$

or about *37 times greater than the signal current.*

Despite this 37× difference, the noise calculations to follow show that it is still possible to extract the weak lightning signal from this overpowering solar background. Using a silicon detector, these noises include the shot noise associated with both the signal and background currents [Eqs. (6.9) and (6.10)], as well as the other types of noise reviewed prior to this example. These noise sources may be summarized as follows.

- *Signal shot noise*—For a signal current i_s, the signal shot noise is given by Eq. (6.9) as $\sigma_s = (2qi_s \Delta f)^{1/2}$. The electrical-system bandwidth required to measure the flash duration is found from $\Delta f \approx 1/2\Delta t_p = 1/(2 \times 0.0005$ sec) = 1000 Hz. Substituting into Eq. (6.9), we find that $\sigma_s = (2 \times 1.602 \times 10^{-19}$ C $\times 1.94 \times 10^{-12}$ A $\times 1000$ Hz)$^{1/2} \approx 2.5 \times 10^{-14}$ A (RMS)—or, in terms of the number of electrons, $N_s = \sigma_s \Delta t_p / q \approx 78$ electrons RMS for $\Delta t_p = 0.5$ ms.

- *Background shot noise*—The shot noise due to the background is $\sigma_b = (2qi_b \Delta f)^{1/2}$, where the electrical-system bandwidth over which noise is collected must be sufficient to measure the pulse width; hence, Δf again equals 1 kHz. Equation (6.10) shows that the background shot noise $\sigma_b = (2 \times 1.602 \times 10^{-19}$ C $\times 7.2 \times 10^{-11}$ A $\times 1000$ Hz)$^{1/2} \approx 1.5 \times 10^{-13}$ A RMS (or $N_b = \sigma_b \Delta t / q = 475$ electrons RMS for $\Delta t = 0.5$ msec). Note that signal and background noise electrons *both* increase as the integration time is made longer. Although background current i_b does not change with integration time, the total number of noise electrons increases linearly ($N_b = \sigma_b \Delta t / q$). Substituting this equation and the relation between bandwidth and integration time ($\Delta f = 1/2\Delta t$) into Eq. (6.10) shows that $N_b = (i_b \Delta t/q)^{1/2}$, giving the increase in background noise electrons N_b with integration time.

- *Dark-current shot noise*—From Table 6.3, a silicon FPA has a dark current density $\approx$ 2 nA/cm². For 25-µm square pixels with an area of 625 µm², this gives a dark current of 2 × 10⁻⁹ A/cm² × 625 × 10⁻⁸ cm² = 1.25 × 10⁻¹⁴ amps. From Eq. (6.11), the dark-current shot noise is σ_{dc} = (2 × 1.602 × 10⁻¹⁹ C × 6.25 × 10⁻¹⁶ A × 1000 Hz)$^{1/2}$ ≈ 2 × 10⁻¹⁵ A (RMS); the resulting noise charge created over the integration time is $N_{dc} = \sigma_{dc}\Delta t/q \approx 6$ electrons RMS.

- *Johnson noise*—For silicon PDs with a large internal resistance (a "shunt" or parallel junction resistance of about 0.1–1 GΩ), the Johnson noise is governed by the load resistor feeding the amplifier; this leads to the use of a high-impedance amplifier.[13] With a 100-MΩ load resistance added in parallel with the detector's internal resistance of 1 GΩ, Eq. (6.12) shows that $\sigma_J = (4kT\Delta f/R)^{1/2}$ = [(4 × 1.38 × 10⁻²³ J/K × 300 K × 1000 Hz)/(47.6 × 10⁶ Ω)]$^{1/2}$ = 5.9 × 10⁻¹³ A RMS, or $N_J \approx$ 1840 electrons RMS. A larger load resistor would reduce this further; however, depending on the required bandwidth, this may not be a practical solution when one considers the effects of a larger resistance on circuit bandwidth.[5] Since the Johnson noise is significantly greater than that due to the background, the amplifier may need to be cooled in order to prevent the Johnson from dominating.

- *Amplifier noise*—For a high-performance, low-bandwidth (1-kHz) amplifiers, a noise spectral density of $\sigma_{amp}/\Delta f^{1/2} \approx$ 0.001 pA/Hz$^{1/2}$ is typical for amplifiers at room temperature.[5] The resulting noise is then σ_{amp} = 0.001 pA/Hz$^{1/2}$ × (1000 Hz)$^{1/2}$ = 3.16 × 10⁻¹⁴ A RMS. This is equivalent to $N_{amp} \approx$ 99 electrons RMS, so reducing the amplifier temperature to lower its noise may not be necessary.

- *Quantization noise*—This noise depends on the number of ADC bits and on the ROIC well capacity. Typical well capacity for a high-performance silicon CCD is 1.5 ke⁻/µm², which yields N_w = 1500 × 625 µm² = 937.5 ke⁻ for 25-µm square pixels. Then, by Eq. (6.13), a standard 14-bit ADC has quantization noise $\sigma_q = N_w/2^{14}12^{1/2}$ = 937.5 ke⁻/(16,384 × 12$^{1/2}$) = 16 electrons. Given the magnitude of the other noises calculated up to this point, a lower-performance ADC with 12 bits—for which σ_q = 66 electrons—would be acceptable.

- *Spatial noise*—The spatial noise due to detector or ROIC nonuniformities may result either from the signal or the background flux. Because the dominant noise source in this problem is the background, the spatial noise from Eq. (6.14) is σ_{FPN} = PRNU × $\mathcal{R}\Phi$ = PRNU × $\mathcal{R}TL_\lambda A\Omega\Delta\lambda$ = 0.001 × 7.2 × 10⁻¹¹ amps = 7.2 × 10⁻¹⁴ A (or about 450 electrons for a corrected PRNU of 0.1 percent).

- *Read noise*—It's possible to use a CDS or a high-quality FPA with low read noise $\sigma_r \approx$ 1–10 e⁻ per pixel per read; however, with Johnson noise on the order of 3600 electrons, low read noise is not required. As we shall see, a read noise that's about one-tenth the dominant source level won't add significant noise to the final answer.

Stepping back from these calculations, we can see that the Johnson noise of approximately 1840 e⁻ dominates the other noise terms and that the amplifier and other terms contribute only small amounts of noise. Because the noises

Detectors and Focal Plane Arrays

are independent of each other, the root sum square of the individual noise terms is used to quantify the SNR.[31] Rewriting Eq. (6.8) in terms of electrons, the result is that

$$\sigma_n = (\sigma_s^2 + \sigma_b^2 + \sigma_{dc}^2 + \sigma_j^2 + \sigma_{amp}^2 + \sigma_q^2 + \sigma_{FPN}^2 + \sigma_r^2)^{1/2}$$
$$\approx (\sigma_b^2 + \sigma_j^2 + \sigma_{FPN}^2)^{1/2}$$
$$= (475^2 + 1840^2 + 450^2)^{1/2}$$

or approximately 1952 noise electrons RMS. Compared with the number of signal electrons, there is only (6052 e⁻)/(1952 e⁻) = 3.1 times more signal than noise—a result whose meaning is reviewed in Sec. 6.3.3.

Note that obtaining this SNR depends on the ability to subtract the background. Conceptually, the background shot noise—rather than the background itself—is used only if the background can be averaged over "many" frames, and subtracted from the current measurement. What's left is the statistical variance in the background flux, as given by Eq. (6.10). For the task of measuring lightning, the reflected solar background does not change much from frame to frame. If the background is changing "quickly" between frames then it would be difficult to get a good average and so background subtraction wouldn't be accurate. The noise in that case would be due to the background itself, giving a signal-to-background ratio of $i_s/i_b = (1.94 \times 10^{-12}$ A)/$(7.2 \times 10^{-11}$ A$) = 0.027$; this would make the extraction of the signal from the noise an impossible task.

6.3.3 Signal-to-Noise Ratio

The comparison of the signal and noise on a pixel is one of the most important measures of optical system performance. In an imaging system, for example, the SNR affects image contrast and MTF. Tracking systems are similarly affected in that low SNR makes it more difficult to follow an object as its image "dances" around the FPA. We shall see that the lightning detection system described in Example 6.3 relies on a minimum SNR in order not to miss any of the flashes that occur ("probability of detection") or to mistakenly "measure" a flash when there wasn't one ("probability of false alarms"). In all cases, system requirements push the SNR to some optimum range; performance drops markedly below this range and above it, designing an instrument will involve extraordinary cost, schedule, and technical difficulties.

Putting together the results from Secs. 6.3.1 and 6.3.2, the SNR for independent noise sources is given by the ratio

$$\frac{S}{N} = \frac{i_s}{\sigma_n} = \frac{i_s}{\sqrt{\sigma_s^2 + \sigma_b^2 + \sigma_{dc}^2 + \sigma_j^2 + \sigma_{amp}^2 + \sigma_q^2 + \sigma_{FPN}^2 + \sigma_r^2}} \quad (6.15)$$

This ratio could also be expressed (as in Example 6.3) in units of the number of electrons. Because detectors produce current in response to optical power, the SNR formulation of Eq. (6.15) is the convention used in optical engineering.[2] Electrical engineers use electrical power

($\propto i_s^2$) to define SNR; their SNR is the square of that used by optical engineers, so that SNR (electrical) = SNR² (optical).

The probability of a correct detection (P_d) and the probability of false alarms (P_{fa}) for measuring flashes, pulses, and point sources depend strongly on the SNR. Because the Gaussian approximation to Poisson statistics describes sources with large, uncorrelated photon emission, the detection probability P_d is given by the area of the Gaussian probability distribution beyond a given threshold current. This distribution is shown in Fig. 6.33 for both a signal and a background. The signal, with the mean current I_s = 10 mA used in Eq. (6.9) to determine the standard deviation σ_s, may represent either the 1s in a fiber-optic digital pulse stream (where the background is represented by the 0s) or a point source such as a faint star embedded in the cold background of deep space.

Detecting the signal correctly requires measuring as much as possible of the signal distribution and as little as possible of the background. Setting a threshold current I_{th} between the two peaks, the probability of detection depends on how much of the signal is measured past that threshold:[32]

$$P_d = \frac{1}{\sigma_s \sqrt{2\pi}} \int_{I_{th}}^{\infty} \exp\left\{\frac{-(I-I_s)^2}{2\sigma_s^2}\right\} dI \qquad (6.16)$$

Here I_{th} is the threshold current, I_s is the signal determined by Eq. (6.7) or its equivalent, and σ_s is the signal's standard deviation from Eq. (6.9). For background subtraction, the mean signal current I_s is

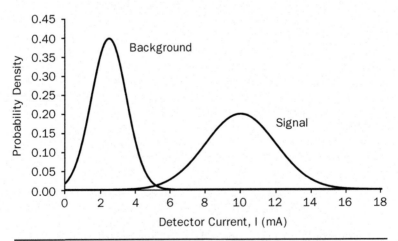

FIGURE 6.33 Gaussian statistics characterize the signal and background variance from Fig. 6.27, where the stronger signal has a larger standard deviation than does the low-level background for shot-noise-limited detectors. The parameters used in the plots are I_s = 10 mA, σ_s = 2 mA, I_b = 2.5 mA, and σ_b = 1 mA. Background subtraction shifts the left curve to zero mean current.

much greater than the mean background current I_b, and the signal-to-noise ratio SNR ≈ I_s/σ_s; the dependence of P_d on SNR is illustrated in Fig. 6.34 for two threshold currents. The figure shows that a larger threshold requires a higher SNR to obtain the same probability of detection. Alternatively, the lower threshold integrates more of the signal probability distribution, yielding a higher P_d for a given SNR.

A low threshold current ensures that the probability of detection will be high ($P_d ≈ 1$), but it also adds background current to the measurement. This current increases the probability of a false alarm—that is, when the detection system indicates that a signal is present when in fact there is none. This probability depends strongly on the SNR; for Gaussian statistics and a shot-noise limited SNR ≈ I_s/σ_s, the resulting probability of false alarms is given by[32]

$$P_{fa} = \mathrm{ERFC}\left(\frac{\mathrm{SNR}}{\sqrt{2}}\right) \approx \frac{\exp\{-0.5(\mathrm{SNR})^2\}}{\mathrm{SNR}\sqrt{2\pi}} \qquad (6.17)$$

Here ERFC = 1 − ERF is the complementary error function,[26] and the approximation on the right-hand side is valid for SNR > 3. Typical design curves for the dependence of P_d and P_{fa} on SNR are shown in Fig. 6.35 for an exact analysis that uses the ERFC. Both curves improve with SNR; if a P_d is specified then the higher-threshold (I_{th} = 6 mA) curve shown in Fig. 6.34 requires a larger SNR and thus has a lower P_{fa}. Mathematically, the higher threshold integrates less of the background in Fig. 6.33 and so leads to fewer false alarms.

The results from Eqs. (6.16) and (6.17) are pure numbers (e.g., 10^{-6}) that apply to individual pixels. For an FPA, the false-alarm

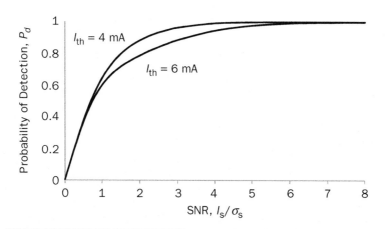

Figure 6.34 The probability of detection P_d as calculated from Eq. (6.16) for the signal and background shown in Fig. 6.33; the graph compares different current thresholds I_{th}.

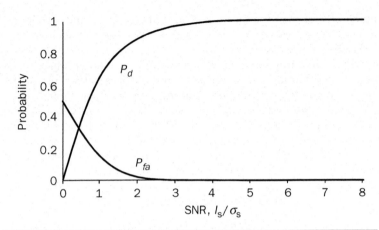

FIGURE 6.35 Comparison of P_{fa} and P_d from Eqs. (6.17) and (6.16) for $I_{th} = 4$ mA in Fig. 6.34. The probability of false alarms seems small for SNR > 3 but the probability of detection is also relatively low; system design often requires $P_{fa} < 10^{-9}$ (SNR ≈ 6) for large-format FPAs to obtain a low false-alarm rate.

rate (FAR) for an entire frame gives a better understanding of performance. This value is found from multiplying the P_{fa} by the number of pixels and the frame rate. With an SNR = 6, for example, the P_{fa} of a pixel is found from Eq. (6.17) to be approximately 10^{-9}; the FAR of a 512 × 512 FPA operating at 30 Hz is then 10^{-9} × 512 × 512 × 30 Hz = 7.86 × 10^{-3} false alarms per second, or a false alarm every 127 seconds.

The detector current i_d has three components ($i_s + i_b + i_{dc}$), but the SNR equation has only i_s in the numerator. A number of approaches have been developed over the years to isolate the signal current from the others and that allow measurement of the SNR. These approaches include the background subtraction technique illustrated in Example 6.3; modulating ("chopping") the source, when it is accessible, to allow the detector to isolate ("lock in") the modulation frequency; and the use of gate (or "boxcar") averaging.

Not to be forgotten are the methods described in Sec. 6.3.2 for reducing the background and dark-current noise terms to ensure that i_s is much greater than i_b or i_d. These methods include spatial, spectral, and temporal filtering as well as low-bandwidth and low-temperature detectors. Employing these techniques may lead to special situations where one type of noise is much greater than the others. As a result, the effects of different design parameters can be isolated, allowing the systems engineer to focus efforts more efficiently on which type of noise requires reduction. These situations are described in the following paragraphs.

Signal-Limited SNR

This is not a common situation, but it occurs when the signal shot noise is much larger than the other noises. It is found mostly in certain classes of visible detection (e.g., reflected sunlight at noon) where the various noises are low and the signal is high. After substituting Eq. (6.9) into Eq. (6.15), we find that the SNR $\approx i_s/(2qi_s \Delta f)^{1/2}$ $\approx (i_s \Delta t/q)^{1/2}$. This expression shows that, even though the noise current increases with the signal, it does so more slowly than the signal itself and therefore yields a net improvement in SNR.

A unique aspect of charge accumulation with FPAs is that a longer measurement time Δt also increases the SNR, and collecting photons for as long as possible has the same advantage as increasing the signal strength with a larger aperture or higher throughput. If the object or instrument are not moving "much," then this approach to increasing SNR results in a smaller, lighter system. It is more typical for the integration time to be set by the motion of the object or optical system, and limits on the size of entrance aperture keep the signal at a relatively low level. Therefore, sensitivity and SNR are limited either by background shot noise or by the detector and its electronics.

Background-Limited SNR

This is a much more common situation—especially in MWIR and LWIR systems, where the background emission from room-temperature objects is high. In the past, an IR detector was known as a *background-limited infrared photodetector* (BLIP) when the noise is determined primarily by the background shot noise. With background noise much greater than the other noises in Eq. (6.15), the SNR $\approx i_s/(2qi_b \Delta f)^{1/2} \approx i_s(\Delta t/qi_b)^{1/2}$.

The key design driver that emerges from this result is *not* that the SNR can be increased with a lower-noise detector. After all, the detector noise has already been decreased as much as possible yet the background is still dominant for this SNR approximation. To increase the SNR, ways must be found to increase the signal current i_s or to decrease the background current i_b. For background sources within the FOV, Eq. (4.9) shows four possible paths: change the throughput, aperture, solid angle, or optical band-pass.

Cutting back on the throughput and aperture size are never good strategies because these options also reduce the signal, resulting in a net decrease in SNR. Reducing the collection solid angle and optical bandwidth may or may not be good options, depending on how much of the signal is cut off by making these changes. It is often possible, for example, to reduce the number of MWIR and LWIR photons with a cold filter that reduces the blackbody background to which the detector is exposed. If the filter does not reduce the signal, then the SNR may improve. Of course, using an optical filter with too narrow a band-pass will simultaneously cut off the signal even faster than the noise and therefore reduce the SNR.

We saw in Chap. 4 that there may also be background sources outside the FOV that nonetheless work their way onto the detector via scattering. In addition to the four variables that affect the in-field photons, scattering has the additional complication of each optical surface's bidirectional scatter distribution function. In this case, a larger BSDF reduces the SNR because more out-of-field light is scattered onto the detector while more in-field light is scattered away.

As with a signal-limited SNR, a longer integration time also improves the SNR when the detector is BLIP. A useful way to think of this is that the smaller electrical bandwidth Δf doesn't pass as much noise [Eq. (6.10)]. Alternatively: a longer integration time collects signals from the object at a faster rate ($\sim \Delta t$) than it collects background photon noise ($\sim \Delta t^{1/2}$), increasing the SNR.

A BLIP situation may arise if the detector subsystem is so quiet that the only remaining noise is a low-level background noise or if the background noise is so high that it overwhelms the high-level noise of the detector and electronics. For instance, the blackbody emission from a 300-K background is much greater than that from a 4-K background, and reducing the detector noise below that from the 4-K background is clearly a much more difficult task. For IR systems, the background temperature thus determines how much detector noise is acceptable; reducing the detector noise further will not increase the SNR.[33]

Backgrounds may not always be visible through the target; a simple example of this is a home-security system imaging a backyard intruder with a LWIR thermal camera. With a "background" of the Earth, whose radiation at 300 K cannot penetrate through the intruder (who radiates at about 305 K), another measure of signal quality is required. In this case, the *difference* in SNR between the target and the background is used as a measure of how well the intruder stands out against the background. This results in the commonly used SNR concept of *noise-equivalent temperature difference* (NETD).[19]

The detector limits on SNR may be due to any of the remaining noises from Sec. 6.3.2—dark-current, Johnson, amplifier, quantization, spatial nonuniformity, or read—as well as others (such as $1/f$) not included in this chapter. Detectors known as quantum-well infrared photodetectors (QWIPs) have a large dark current and must be cooled to cryogenic temperatures before the dark-current noise is less than the 300-K thermal background noise. More generally, all of the detector noises must be evaluated for this comparison in order to identify possible detector-limited (DELI) situations, where the detector noise cannot be reduced below that of the background. For VLWIR systems, such situations are usually determined by Johnson noise, and lead to what is known as Johnson-limited (JOLI) SNR.

Johnson-Limited SNR

This is another of the so-called temporal noises that increase with integration time. If the Johnson noise is much larger than the other noises in Eq. (6.15) then the SNR $\approx i_s (R\Delta t/2kT)^{1/2}$, which increases with $\Delta t^{1/2}$. This expression also indicates that a larger electrical resistance or a lower detector temperature increases the SNR. Increasing resistance comes at the expense of reducing speed, and the cost of lowering the temperature is additional size, weight, power consumption, and cooling expenses. Nonetheless, both of these approaches—as well as increasing integration time—are common in JOLI situations such as deep-space imaging of a very-low-background (3-K) scene, where it is extremely difficult to reduce the detector noise below that of the background.

A Johnson-limited SNR differs from background-limited SNRs in that the JOLI situation is independent of background irradiance. This means that reducing the background current improves detector sensitivity until a JOLI situation is reached, at which point reducing the background is no longer useful. As a result, a lower temperature (or larger resistance) does increase the SNR, as illustrated in Fig. 6.36.

Example 6.4. The design process of reducing noise is an iterative one: reducing one noise source allows another to dominate, opening the door to design changes that result in a different noise being the largest, and so on. This example illustrates the process by looking at the design of a MWIR imager collecting photons at night against a 300-K background.

Using Eq. (5.2) for the spectral irradiance of the 300-K background, we find that $M_\lambda = 2.04$ W/m²·µm for an assumed emissivity $\varepsilon = 0.25$ at a wavelength of 5 µm; the spectral radiance is thus $L_\lambda = M_\lambda / \pi \approx 0.65$ W/m²·sr·µm. If the instrument has a 100-mm aperture, a 10-nm band-pass filter, and no cold stop (with collection solid angle $\Omega = 2\pi$ steradians), then the background power $\Phi_b = TL_\lambda A\Omega\Delta\lambda = 0.65$ W/m²·sr·µm $\times$ 0.00785 m² $\times$ 2π sr $\times 10^{-3}$ µm = 0.32 mW.

If we use an InSb detector at 77 K with an average responsivity R = 3 A/W at λ = 5 µm and an imager with optical throughput of 75 percent, then the detector

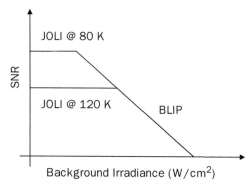

FIGURE 6.36 Log-log plot showing how, as the background is reduced, SNR increases until a point is reached where Johnson noise dominates.

current due to the background can be calculated as $i_b = R\Phi_b T = 3 \text{ A/W} \times 0.32 \times 10^{-3} \text{ W} \times 0.75 = 7.2 \times 10^{-4}$ A. A blackbody background probably changes slowly in comparison with the 30-Hz frame rate; as a result, it can be averaged and subtracted. Then the remaining shot noise is given by Eq. (6.10) as $\sigma_b = (2qi_b\Delta f)^{1/2} = (2 \times 1.602 \times 10^{-19} \text{ C} \times 7.2 \times 10^{-4} \text{ A} \times 30 \text{ Hz})^{1/2} = 8.32 \times 10^{-11}$ A.

The detector's internal resistance, which depends on its diameter, is on the order of 1 MΩ at 0-V bias for 0.5-mm-diameter InSb detectors. Comparing T_d/R_d (= 77 K ÷ 1 MΩ) with T_f/R_f (= 300 K ÷ 5 kΩ) for a detection circuit with an uncooled (T = 300 K) 5-kΩ amplifier resistor shows that the Johnson noise $\sigma_j = (4kT\Delta f/R)^{1/2} \approx [(4 \times 1.381 \times 10^{-23} \text{ J/K} \times 300 \text{ K} \times 30 \text{ Hz})/(5 \times 10^3 \text{ Ω})]^{1/2} = 9.97 \times 10^{-12}$ A. This is approximately 8 times less than the background shot noise and so the detector—*based only on the comparison with Johnson noise*—is background limited. Comparisons with other types of detector noise are also necessary and depending on the amount of noise present, the conclusion may be different in those cases.

Clearly, efforts to reduce the temperature or increase the amplifier's resistance in order to reduce the size of the Johnson will be only marginally productive at reducing the total noise. Alternatively, the background can easily be reduced by using a cold stop to reduce the solid angle; in that case the detector can become JOLI, and reducing the amplifier temperature may then be used to restore the detector's BLIP status.

The distinction between background-limited and detector-limited SNR can be fuzzy, and it's best simply to calculate all the noise terms in Eq. (6.15) to determine which has the most influence on SNR. Design changes can then be made to reduce the largest noise factor, in which case the detector may flip from BLIP to DELI and back again as different noise terms dominate. Ideally, the largest detector noise terms should have about the same magnitude; this will limit the effect on SNR if one term increases over time. There are, of course, other detector parameters that are useful for indicating which direction to take a design. These metrics are described in the next section.

6.3.4 Sensitivity

In addition to SNR, another important measure of performance is the system sensitivity—that is, how small a signal the optics-plus-detector-plus-electronics system can detect. The language to describe sensitivity can be a bit confusing, in that "high" sensitivity translates into a system that can accurately detect a smaller amount of power. Fortunately, "better" and "poor" sensitivity mean what they sound like they should mean, so these terms can be used in cases where confusion could arise.

In any case, the need for better sensitivity must be traded against the use of a somewhat larger telescope to collect more photons; the latter puts more of the burden of this requirement on the optical subsystem and less on the detector. Other factors affecting sensitivity are the pixel size, background flux, and integration time. In this section, two metrics for comparing sensitivity are reviewed: noise-equivalent power and detectivity. Additional issues associated with the practical use of this concept are reviewed in Sec. 6.4.

Noise-Equivalent Power

The idea behind NEP is straightforward: it is the signal power that just equals the noise power and so results in an SNR = 1. The NEP will thus depend on the type of noise that's dominant. There are limiting cases such as BLIP NEP and JOLI NEP to help with the pedagogics, though combinations using many of the noise types described in Sec. 6.3.2 are common. Hence it is not useful to specify a number for "the" NEP for a system unless the limiting noise (typically BLIP) is also specified.

Using the NEP for a BLIP situation as an example, the SNR was given in Sec. 6.3.3 as SNR $\approx i_s/(2qi_b\Delta f)^{1/2}$. Working the signal and noise power into this equation along with the responsivity, we obtain SNR $\approx \mathcal{R}\Phi_s/(2q\mathcal{R}\Phi_b\Delta f)^{1/2}$. Setting this expression to SNR = 1 and then renaming Φ_s as the new variable NEP_{BLIP} yields

$$\text{NEP}_{\text{BLIP}} = \sqrt{\frac{2q\Phi_b\Delta f}{\mathcal{R}}} \quad [\text{W}] \tag{6.18}$$

for cooled photodiodes. Similar equations are available for uncooled photodiodes as well as for cooled and uncooled photoconductors.[34] Smaller numbers indicate a more sensitive detector, with typical values on the order of 25 pW for a 1-mm-diameter InSb detector at the peak of the responsivity curve measured over a 1-kHz bandwidth Δf.

To avoid specifying the bandwidth, it is also common to use NEP/$\Delta f^{1/2}$ as a metric with units of W/Hz$^{1/2}$. At higher bandwidths the NEP/$\Delta f^{1/2}$ starts to increase—usually as a result of more amplifier noise. In this case the detector is not purely BLIP but rather limited by some combination of background and amplifier noise, which is sometimes expressed as a percentage of BLIP.

The NEP also depends on detector area via the background flux Φ_b in Eq. (6.18). The backgrounds used to determine Φ_b are typically a 300-K blackbody for MWIR and LWIR thermal detection or a 2856-K tungsten-halogen lamp for the visible and NIR.[2] These produce an output current over the entire range of wavelengths to which the detector is sensitive; this is known as *blackbody responsivity*. The responsivity $\mathcal{R}$ used in Eq. (6.18) can also be at a specific wavelength, in which case the NEP is known as spectral NEP. Details on converting between blackbody and spectral responsivity can be found in Dereniak and Boreman[2] or Holst.[19]

Not surprisingly, Eq. (6.18) shows that a greater background flux makes it more difficult to distinguish the signal from the noise. Higher bandwidth has the same effect because more electrical noise is then included in the SNR measurement. However, an increase in responsivity results in a smaller NEP and allows manufacturer claims of a more sensitive detector.

There are some related metrics that allow the same claim: noise-equivalent irradiance (NEI), which is the NEP divided by the area of either the entrance pupil or pixel (which one is used must be specified); noise-equivalent electrons (NEE), which is the NEP expressed in units of electrons; noise-equivalent temperature difference (NEDT or NEΔT) for the detection of temperature differences; and so on. In all cases, a common drawback of these metrics is that a smaller number indicates better performance. The resulting language ambiguities (e.g., "high" sensitivity) engender a new metric, known as *detectivity*.

Detectivity

In its simplest form, detectivity $D = 1/\text{NEP}$ [1/W]; hence a large detectivity value indicates a detector that is more sensitive. Detectivity has been used to evaluate IR detectors in BLIP situations where the background noise is large, although NEP is more common for visible and UV detectors where background noise is not the limiting factor in system performance.

Going beyond the simple "bigger is better" philosophy, there are variations on the detectivity concept that enable us to compare different detectors. With "raw" detectivity it is still possible for manufacturers to manipulate their specs by using small-bandwidth signals and small-diameter detectors. The small bandwidth limits the electrical noise, and the small pixel restricts both the dark current and the solid angle from which the detector collects background photons (assuming the same focal length). In order to remove the effects of area and bandwidth—and thereby allow a comparison of detectors on an equal basis—the concept of *specific detectivity* D^* (pronounced "D star") was introduced:

$$D^* = \frac{\sqrt{A_d \Delta f}}{\text{NEP}} \quad [\text{cm} \cdot \text{Hz}^{1/2} / \text{W}] \quad (6.19)$$

Here A_d is the detector (or pixel) area and an additional fill-factor term is included in the numerator for FF < 1.[9] Figure 6.37 shows how D^* varies with wavelength for a number of detector materials. Typical values are given in Table 6.4 for the most common materials at the temperature and wavelength where the peak D^* occurs. Large values of D^* are preferred, as they result in a smaller NEP.

Just as the NEP is defined for the special cases of a particular type of dominant noise, so is D^*. For BLIP situations, the best possible D^* available from a detector is given by Eqs. (6.18) and (6.19) as

$$D^*_{\text{BLIP}} = \sqrt{\frac{A_d R}{2q\Phi_b}} \quad [\text{cm} \cdot \text{Hz}^{1/2} / \text{W}] \quad (6.20)$$

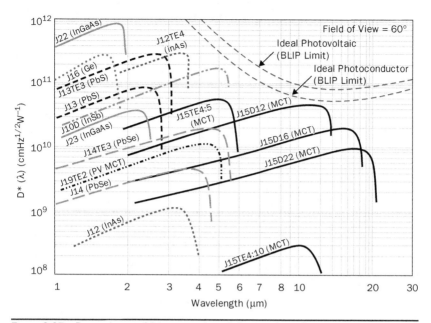

FIGURE 6.37 Dependence of D^* on wavelength for a variety of photodetector materials and types. (*Credit: Teledyne Judson Technologies.*)

Detector Material	Typical D^* [cm·Hz$^{1/2}$/W]
Silicon (VIS)	5×10^{13} (300 K)
HgCdTe (10 μm)	2×10^{10} (80 K)
InGaAs (SWIR)	10^{13} (300 K)
InSb (MWIR)	10^{11} (80 K)
VOx bolometer (LWIR)	10^9 (300 K)

TABLE 6.4 Approximate specific detectivities (D^* values) for photodiodes and microbolometers at their peak spectral responsivity for a hemispherical (2π-steradian) FOV and a 295-K blackbody background temperature.

This expression shows that a background flux must be specified in order for D^* to be calculated; common backgrounds for MWIR and LWIR thermal detection are 300-K blackbodies with emissivity $\varepsilon \approx 1$. Observe that the pixel area in the numerator of Eq. (6.20) cancels that found in the denominator's term (Φ_b) for background flux, which makes D^*_{BLIP} independent of detector area.

Another common specific detectivity is that for Johnson-limited situations, where

$$D^*_{JOLI} = \mathcal{R}\sqrt{\frac{RA_d}{4kT}} \quad [\text{cm} \cdot \text{Hz}^{1/2}/\text{W}] \qquad (6.21)$$

[note that R and T were defined in Eq. (6.12)]. In this case, the detector area does *not* cancel with any other term in Eq. (6.21) *even though* the D^* concept was originally intended to be independent of area.

For a given material, however, D^*_{JOLI} is independent of detector resistance multiplied by cross-sectional area of the photodiode junction (which is typically based on pixel size)—given that an increase in area decreases electrical resistance. This *dynamic resistance* is measured at bias voltage $V_b = 0$, written as $R_0 \, [= 1/(dI/dV)_0$ on the I-V curve at zero bias] and as shown in Fig. 6.9, depends on the photon background. Hence $R_0 A_d$, which is usually written as $R_0 A$, depends on the material and thus allows for an equal comparison of detectors. From Eq. (6.21) it follows that a higher $R_0 A$ indicates a more sensitive material with lower noise, the latter a result of the reduction in Johnson current for a larger resistance. For semiconductors, $R_0 A$ depends also on temperature[10] because resistance increases at lower temperatures, a result of fewer thermally excited electrons available for conduction.

Astronomers designing JOLI optical systems for deep-space imaging often depend on $R_0 A$ rather than D^* as the fundamental metric for comparing detector sensitivity. Typical values for common detector materials are given in Table 6.5; Fig. 6.38 shows the dependence of the dynamic resistance $R_d = (dI/dV)^{-1}$ on temperature and bias, including the zero-bias condition that defines R_0.

Example 6.5. The equations for NEP and SNR do not explicitly show how these performance metrics depend on optical system parameters such as f/# and number of pixels. In the case of BLIP, for example, is the NEP expected to get better or worse for a faster f/#? Does the SNR change in the same way?

Looking first at the NEP for a BLIP situation, the system parameters show up in the flux term in Eq. (6.18), which can be written as $\Phi_b = TL_b A\Omega$ for an area source [see Eq. (4.8)]. Rewriting the etendue as $A\Omega \approx \pi D^2 A_d /4f^2$ for small angles and then substituting into Eq. (6.18), we find that $NEP_{BLIP} = (1/f/\#)(\pi q T L_b A_d \Delta f /2R)^{1/2}$. Therefore, using faster optics (smaller f/#) *increases* the NEP. This unexpected result follows because for a BLIP situation, a faster system collects more background noise, which increases the signal power that's needed to equal it.

The SNR for a BLIP situation behaves differently; recall from Sec. 6.3.3 that $SNR_{BLIP} \approx i_s / (2qi_b \Delta f)^{1/2}$. Converting the optical flux to current via the responsivity ($i = \mathcal{R}\Phi$) and then substituting into the equation for flux from the previous paragraph, we find that $SNR_{BLIP} = [L_s /2(f/\#)][(\pi \mathcal{R} T A_d)/(2qL_b \Delta f)]^{1/2}$. Hence the SNR increases for bigger pixels and faster optics, which is more in line with expectations. That this trend differs from the one for NEP follows because bigger pixels and faster optics—while they collect more background

Detector Material	Typical R_oA [$\Omega \cdot cm^2$]
Silicon (VIS)	10^7 (300 K)
HgCdTe (5.2 µm)	10^7 (80 K)
HgCdTe (10 µm)	10^3 (80 K)
InSb (MWIR)	10^5 (80 K)

TABLE 6.5 Approximate resistance–area products (R_oA) for various detector materials. A higher detector temperature than that shown in parentheses will decrease R_oA and thereby reduce sensitivity.

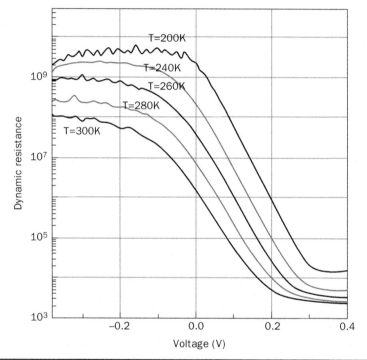

FIGURE 6.38 Dependence of dynamic resistance $R_d = (dI/dV)^{-1}$ on temperature and voltage for a SWIR HgCdTe photodiode with an area of 110 µm × 110 µm. (Credit: Jin-Sang Kim et al., Proceedings of the SPIE, vol. 4795.)

noise—also collect more signal. The increase in signal with pixel size and f/# is faster than the increase in noise, and the SNR increases as a result.

Although metrics are useful for initial comparisons, relying on a single one (e.g., R_oA, D^*, or NEP) for detector selection is inadvisable given the number of important properties that would not be taken into account. The full evaluation of a detector's suitability for an optical system must also include additional specs such as responsivity, response time, power consumption, and so forth. These metrics are discussed further in Secs. 6.4 and 6.5.

6.4 Detector Specifications

In Secs. 6.1–6.3, three different types of detectors were reviewed along with the unique properties of FPAs and the concept of SNR as a common metric for detector performance. Beyond these concepts, however, are a number of engineering details that directly or indirectly affect the SNR over a range of operating conditions. For example, the SNR depends not only on noise limitations but also on an effect known as *saturation*. The dependence of dark current on bias voltage and temperature—neither of which can be held to precise values—also affects the point at which a detector becomes background limited. Even details such as how to fully describe responsivity—for example, in the context of wavelength dependence, optical coatings, fill factor, and temperature shifts—become important when specifying a detector.

Furthermore, there are additional evaluation criteria that don't affect the SNR but are nonetheless an important part of detector selection. An example in this category is electrical power consumption, which affects the size, weight, reliability, and cost of the detector subsystem. Hence, before addressing selection in detail, we first review the properties and typical values found in single-pixel and FPA specs so that informed decisions can be made regarding suitability for a particular optical system. Our goal is to understand the information required for either evaluating or writing detector specs. The ability to read detector specs critically—and understand a vendor's capacity for meeting them—is a crucial skill for optical systems engineers.

The specs covered in this section are responsivity, sensitivity, dark current, integration time, pixel and array size, dynamic range, uniformity, operability, and power consumption. However, reliability, maintenance, pixel crosstalk, and many other factors are not explicitly covered, and neither are the specs associated with the electrical subsystem. Of the three types of detectors reviewed in Sec. 6.1, photodiodes (in one form or another) are the most likely option for many designs and thus are the focus of this section. In Sec. 6.5 we summarize the chapter with a few comments on detector selection.

6.4.1 Responsivity and Quantum Efficiency

The concept of responsivity has been used a number of times in this chapter as a measure of detection efficiency. In Sec. 6.1, for example, it was shown that a watt of longer-wavelength power has more photons, and therefore produces more current in a photon detector, than does a watt of shorter-wavelength photons (Fig. 6.3). It's also possible to measure the output voltage in response to the optical power, but the standard commonly used in spec sheets is that of *current* responsivity (in units of A/W).

In Sec. 6.3.4 we also made a distinction between peak and blackbody responsivities. Peak $\mathcal{R}$ measures output current per watt of absorbed power over a narrow spectrum centered around a specific wavelength, whereas blackbody $\mathcal{R}$ measures the total current produced by a broad range of wavelengths such as that emitted by a blackbody source. Although it is possible to convert between the two approaches, the focus in this chapter is on peak responsivity—the number typically supplied by manufacturers in their data sheets.

An equally common measure of detector output for photon detectors is the quantum efficiency (QE). Rather than measuring current per watt of optical power, QE measures the related quantity of how many electrical quanta (electrons) are produced for each optical quantum (photon) absorbed. Even if the electrons are later amplified, as in an APD detector, the number of electrons produced by photon absorption can be no more than 1 and is typically less than 0.95. As shown in Fig. 6.39, the QE is fairly constant with wavelength across the absorption range of the detector material (1–1.5 µm); this is because, unlike the responsivity plotted in Fig. 6.3, QE does not depend on the absolute number of absorbed photons.

Responsivity is an engineering spec that's easily measured, whereas QE is usually derived from that measurement. Given that QE is defined as the number of electrons generated (N_e) per number

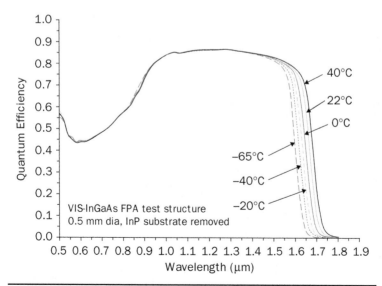

Figure 6.39 In contrast to the responsivity shown in Fig. 6.3, the quantum efficiency (QE) is typically more or less constant with wavelength. (*Credit:* H. Yuan et al., Proceedings of the SPIE, *vol. 6940*.)

of photons absorbed (N_p), it can be converted to responsivity once we realize that current is coulombs per second ($i = N_e q/\Delta t$) and optical power is photon energy per second ($\Phi = N_p h\nu/\Delta t$). Thus, the relation between responsivity $\mathcal{R}$ and quantum efficiency η for photodiodes can be expressed as follows:

$$\mathcal{R} = \frac{i}{\Phi} = \frac{N_e q/\Delta t}{N_p h\nu/\Delta t} = \frac{\eta q}{h\nu} = \frac{\eta q \lambda}{hc} = \frac{\eta \lambda}{1.24} \quad [\text{A/W}] \qquad (6.22)$$

where the wavelength λ must be measured in units of microns to yield the constant $hc/q = 1.24$ J·μm/C. As expected, the responsivity for photodiodes increases linearly with wavelength for a constant QE, since proportionally more photons are required to produce a watt of optical power at longer wavelengths. For photoconductors, Eq. (6.22) includes an additional factor G in the numerator for the photoconductive gain.

Because of this wavelength dependence, comparing responsivity and QE can lead to incorrect conclusions regarding detection efficiency. For example, InGaAs has a QE of 80 percent at a wavelength of 1.55 μm, which Eq. (6.22) shows corresponds to a responsivity $\mathcal{R} = 1.0$ A/W. Yet a comparable visible detector with the same QE at 0.5 μm has a responsivity of only 0.32 A/W. Quantum efficiency is thus the more fundamental measure for detector comparison.

Photon absorption is a key factor in determining both responsivity and QE, since not all the photons that make their way through the optical system will actually be absorbed by the detector. Photons not absorbed cannot produce electrons, and even the electrons that are produced may not survive long enough to be measured as current. At low electrical frequencies, the biggest difference between incident photons and measured electrons is due not to lost electrons but rather to Fresnel reflection losses of photons at the detector surface.[35] For example, a high-index material such as InGaAs ($n \approx 3.2$) has a Fresnel loss $R = [(3.2 - 1)/(3.2 + 1)]^2 = 28$ percent. This is a huge fraction of photons to be losing, and it reduces the responsivity by a factor of 0.72.

Much as with internal and external transmittance (see Chap. 4), a distinction is thus made between internal quantum efficiency η_i and external quantum efficiency η_e. The internal QE is the best possible detector performance in the absence of Fresnel losses ($R = 0$), and the external QE is the actual performance obtained after taking the Fresnel losses into account:

$$\eta_e = \eta_i(1-R) \approx (1-e^{-\alpha L})(1-R) \qquad (6.23)$$

The internal QE η_i depends primarily on the material absorption αL of the region surrounding the p-n junction. The external QE used in

Eq. (6.22) includes Fresnel losses and can be improved dramatically by the application of AR coatings to reduce R. Single-layer AR coatings will increase the external QE over a small band of wavelengths; multilayer coatings are designed to yield smaller reflectance over a larger spectral band. The lower cost of a single-layer coating makes it more common in commercial detectors, and its effect is typically incorporated into the specs listed on manufacturer data sheets.

Not captured in Eq. (6.23) is that, in a CCD, the circuitry (see Fig. 6.23) may block photons from reaching the p-n junction. The QE for front-illuminated CCDs is thus lower in proportion to the fraction of pixel area covered by the circuitry. It is possible to illuminate pixels from the *back* of the FPA, but in that case the wafer is relatively thick; hence it must be thinned in order to prevent too much absorption by the substrate and not enough absorption by the p-n junction's depletion region. The result of this technique is an approximate doubling of the average QE and an associated increase in UV response (Fig. 6.40).

For focal plane arrays, the fill factor also affects the measured responsivity and QE. Although pixels have a responsivity equal to that of individual detectors, the loss of photons between pixels results in an effective pixel responsivity that's smaller by the fill factor. For instance, a material QE of 80 percent and fill factor of 95 percent, for example, results in an effective QE = 0.8 × 0.95 = 0.76 for the SNR estimates. This is sometimes listed in manufacturer spec sheets as QE × FF.

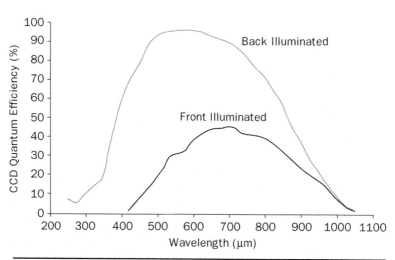

FIGURE **6.40** Back-illuminated silicon CCDs have lower optical losses and therefore higher quantum efficiency. (*Credit: Princeton Instruments.*)

Temperature is an important consideration when selecting a detector, and a shift in cutoff wavelength with temperature also affects the responsivity. Atoms move closer together when detector temperature is lowered, which increases their interaction energy and associated bandgap. As illustrated in Fig. 6.39, this results in a shift of the bandgap to shorter wavelengths (e.g., by approximately 0.2–0.3 nm/°C for InGaAs), reducing responsivity near the high-temperature cutoff wavelength. This effect is not a concern when the detector is operated at the temperature recommended by the manufacturer; however, as shown in the next section, it needs to be considered when operating near the bandgap at a different temperature.

6.4.2 Sensitivity

Detectors that use AR coatings and a large fill factor to increase the effective QE also improve the sensitivity. The D^*_{BLIP} and D^*_{JOLI} that can be obtained with Eqs. (6.20) and (6.21), for example, both increase with QE. The concept of sensitivity was developed in Sec. 6.3.4, but there are additional issues that must be considered when specifying a detector. The most important of these is the dependence of sensitivity on detector temperature. This is illustrated in Fig. 6.41, where the D^* for an InSb detector is 300-K-background limited until about 120 K, at which the point the specific detectivity rolls off.

This behavior is common for cooled photodiodes; as the dark current increases at higher temperatures, it eventually overcomes the D^*_{BLIP} situation in which background shot noise dominates at lower detector temperatures. The important point made by the plot in Fig. 6.41 is that the detector temperature may be increased somewhat beyond that recommended by the manufacturer (here, 77 K) with minimal effect on sensitivity. This is an extremely

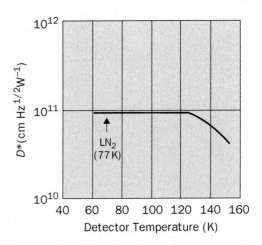

FIGURE 6.41 Dependence of D^* on temperature for an InSb detector, where the drop is a result of an increase in dark current at higher temperatures. (Credit: Teledyne Judson Technologies.)

important system-level trade, since a higher operating temperature can yield significant savings in the size, weight, and cost of the cooling components.

Detector temperature also affects a system's largest achievable D^*_{BLIP} value. A cooled photon detector collects background from at most a 2π-steradian FOV; however, an uncooled LWIR photodetector also absorbs photons emitted by itself and this results in a 4π-steradian FOV.[36] The net result is a factor-of-2 increase in D^*_{BLIP} for a detector cooled to a temperature where its spectral emittance is on the order of one-tenth of the background.

Not only detector temperature but also stops and filters can affect system sensitivity. This is illustrated in Fig. 6.42 for an InSb detector, where the FOV over which the detector collects 300-K background photons is reduced from 180 degrees (2π steradians) down to $\theta = 60$ degrees with a cold stop;[37] the D^* then increases by a factor of $1/\sin(\theta/2)$. The value of D^* also increases when a cold filter is used to reduce the broadband blackbody background emitting into the detector FOV. Figure 6.42 plots the effects of two such filters, one (the SP35) with a transmission of 85 percent and the other (the SP28) with a transmission of 95 percent. In each case the cold filter reduces background spectral radiance and so the D^* is correspondingly higher.

For JOLI situations, detector sensitivity is governed by the temperature dependence of $R_0 A$ in Eq. (6.21). Here the resistance

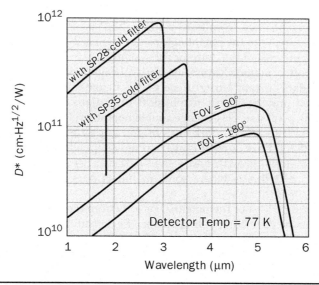

FIGURE 6.42 Maximum D^* increases when a smaller FOV and lower cold-filter emissivity restrict the background flux. (*Credit: Teledyne Judson Technologies.*)

decreases at higher temperatures in response to the increased mobility of electrons. As a result, R_0A decreases at a rapid rate, yielding an increase in NEP. Empirically, the decrease follows an exponential law reasonably well.[10] Note that D^*_{JOLI} is also increased by the detector's self-emittance, as we saw for D^*_{BLIP}.

Given the effects of various system parameters—including wavelength, background temperature, cold-stop f/#, cold-filter transmission, integration time, and electrical bandwidth—it is important to specify the conditions under which sensitivity is measured. The manufacturers of COTS parts may fail to do this. Nonetheless, it is critical to include the operating conditions when writing specs for NEP, D^*, or R_0A; otherwise, the achievable sensitivity may differ from what's expected. We saw in Chap. 5 that normalized specs are helpful metrics for comparison but must be used with caution when quantifying a source's utility in a system. The same caution applies to D^*; in particular, a large D^* may be the result of a small NEP or a large detector. This means that the NEP obtained with a larger detector may not meet the system's sensitivity requirements.

6.4.3 Dark Current

The most important reason for cooling a detector is that dark current is reduced at lower temperatures. As described in the previous section, the dark current determines the temperature at which a MWIR detector is no longer background limited and hence D^* starts to drop. Dark current is equally important in UV-to-SWIR systems, in which dark current is often the dominant source of noise. Reducing dark current, then, results in a lower NEP or a higher operating temperature, with the potential for significant reductions in the size, weight, and cost of the detector subsystem.

Detector manufacturers are thus always looking for ways to reduce dark current. Unfortunately, this current cannot be arbitrarily specified and is subject to physical limitations such as detector material, pixel size, bias voltage, and temperature. One rule of thumb is that LWIR HgCdTe dark current doubles for every 5-K increase above 80 K.[10] Dark current is also strongly dependent on the quality of the material growth and etched surface area in the fabrication process, and this is where some manufacturers are placing their efforts.

To get an idea of the potential payoffs from these efforts, we can estimate the dark-current-limited specific detectivity D^*_{dc}. Using Eq. (6.11) and Eq. (6.19), we find that $D^*_{\text{dc}} = R(A_d/2qi_{\text{dc}})^{1/2} \approx R/(2qk)^{1/2}$, where we have used that the dark current is proportional to the pixel area ($i_{\text{dc}} \propto kA_d$). The ad hoc constant k depends on the material and fabrication quality; marginal reductions in k lead to slight reductions in dark current and so only gradually improve D^*_{dc}.

6.4.4 Bandwidth, Integration Time, and Frame Rate

Although a larger reverse-bias voltage increases the dark current, it also increases the detector speed. This is one of the fundamental trade-offs in detector selection and, for a system with scenes that are changing, plays a large role in determining which detector type is appropriate. To understand why, note that there are different time scales for detectors: detector response time (or speed), integration time (aka shutter time or dwell time), and frame time. Any one of these factors can determine the noise bandwidth Δf.

The shortest time scale is the response to pulses that may occur in the scene, such as lightning flashes or laser radar pulses, and the entire detector subsystem must be fast enough to respond to these signals. Fourier analysis indicates that creating a rapidly changing signal requires a spectrum extending out to higher-frequency sinusoids.[38] Likewise, measuring these signals requires a detector and electronics that can respond to the same high frequencies.

Optical systems to measure such pulse energy therefore require a *detector response time* that is fast enough (and an associated bandwidth that is wide enough) to pass the pulse spectrum and not attenuate it. This means that shorter pulses require more bandwidth Δf. Depending on the pulse shape, the bandwidth is usually approximated by[39]

$$\Delta f \approx \frac{1}{2\Delta t} \quad [\text{Hz}] \qquad (6.24)$$

In this case, Δt is the pulse width ($\Delta t = t_{pulse}$). The use of this expression is illustrated in Example 6.3: lightning flashes of 0.5-ms duration result in an electrical-system measurement bandwidth $\Delta f \approx 1/(2 \times 0.5 \times 10^{-3}$ sec$) = 1000$ Hz. Measuring the pulse shape (and not just its energy) requires a faster response time so that the details of the pulse spectrum can be extracted.

Because it is difficult to time the measurement of a pulse to the exact moment it occurs, a second time scale, known as the *integration time*, is used. This is the time—which is longer than the pulse itself—during which the "shutter" is open to collect pulse photons; note that background photons are also collected before and after the pulse's actual occurrence. If the electrical-system circuitry has been designed to measure the pulse, then the bandwidth Δf must in this case be large enough to measure the pulse width, too, and so $\Delta t = t_{pulse}$.

Measuring pulses within a frame is less common than a scene or object that is relatively static, which allows the signal to be collected over an entire frame. As the details of pulse shapes are thus no longer relevant, then the *frame time* determines the bandwidth that the electrical system must support. For example, video cameras operating at a 30-Hz frame rate have a frame time $\Delta t = 33.3$ ms and a frame bandwidth $\Delta f = 1/2\Delta t = 1/(2 \times 0.0333 \text{ sec}) = 15$ Hz. There are also many situations in which the integration time is less than the frame time. This occurs with IR systems, for example, where integration times are commonly on the order of 10 milliseconds to meet sensitivity and charge well requirements—only one-third or so of the frame time. In this case, then, it is the integration time that determines the bandwidth ($\Delta t = t_{int}$).

Scenes that do not contain pulses may still have objects that are moving, and sometimes at high speeds. To avoid the blur seen in photos of cars flying through Boston—that is, to "freeze" a scene with rapidly moving objects—the shutter must close before the object has time to move and thus "blur" or "smear" to an adjacent pixel (or more) on the FPA. In contrast, far-away galaxies seldom change much across (or within) the FOV, and collecting enough photons may require integration times on the order of days. For either fast or slow objects, the integration time is once again the time scale ($\Delta t = t_{int}$) that determines the electrical-noise bandwidth Δf in Eq. (6.24).

Even though the integration time can be almost as long as the frame time, there may be good reasons for instead integrating over a shorter time. The primary reason is that it is possible to collect too many photons (and therefore electrons) during the integration time, overloading the voltage swings or charge storage in the ROIC. However, too short an integration time results in not enough photons being collected and also increases the noise bandwidth in Eq. (6.24), again reducing the SNR. Integration time is thus a delicate balance of not-too-short and not-too-long—a trade that's usually addressed by allowing it to vary and evaluating the SNR for different time durations.

The integration time applies to individual pixels in an FPA, yet there is also a time associated with reading out the current that's been collected as charge. This time can be relatively short even for large-format FPAs with millions of pixels. If necessary, the resulting integration time can be approximately equal to the frame time.

Commercial detectors are available with a wide range of pixel speeds, integration times, and frame rates. Of the detector types reviewed in Sec. 6.1, PIN photodiodes are the fastest; they have rates of 10 GHz and higher, depending on material type. The InGaAs PINs in particular have been optimized for very high speeds, given their high responsivity at a wavelength of 1.5 µm and resulting applicability for use in high-speed fiber-optic networks.

Photoconductor speed depends on the gain–bandwidth trade-off, with maximum speeds up to a few gigahertz possible using optimized electrodes.[13] These are typically slower than PINs owing to their higher capacitance, and the transit time in a typical photoconductor is longer than the RC time constant of a photodiode. Bolometers, on the other hand, rely on heating and cooling to measure temperature differences; even with low thermal mass, this is a much slower process than electron transport, which gives thermal detectors a response time on the order of 10 milliseconds.

Finally, many FPAs are designed with frame rates of 30 to 60 Hz for compatibility with video equipment. When video output is not a requirement, the maximum frame rate depends on the size of the array: it takes more time to read out more pixels. Typical rates for large-format arrays (512 × 512 and larger) are 30 to 120 Hz, with highly specialized large-format arrays available with rates of about 1000 Hz.

Depending on the type of FPA, the frame rate can also be variable. For example, the frame rate of a CCD may be fixed at 30 Hz to accommodate the "bucket brigade" that transfers charge from the FPA.[23] The ROIC on CMOS and hybrid arrays, however, allows for variable frame rates via a process known as *windowing*. Software-driven selection of small subframes of the FPA—that is, subarrays that are smaller than the maximum size of the array—make it possible to increase the windowed frame rate because of the smaller number of pixels that must be read out. For instance, commercial InGaAs FPAs have a frame rate of 109 Hz for a 640 × 512 FPA, yet this rate increases more than a hundredfold (to 15.2 kHz) for a 16 × 16 window.

6.4.5 Pixel and Array Size

The integration time that establishes the electrical bandwidth is determined in part by the detector size. Those who are familiar with photographic film may recall that some films were said to be "faster" than others. Physically, fast film has bigger silver-halide grains to detect photons; the larger grains collect photons from a wider IFOV (i.e., they have a larger etendue) and therefore allow a shorter exposure (or integration) time than the smaller-grain films. As shown in Chap. 4, the same concept applies to pixel-based FPAs, where a larger pixel collects more photons than a smaller one and thus allows a shorter integration time for a given IFOV.

Pixel size can also affect the focal length of the optics. The trend in mass-market consumer optics (e.g., cell-phone cameras) is toward smaller pixels, currently on the order of 3 microns. For a given angular resolution (pixel IFOV), smaller pixels result in a shorter focal length, yielding small, lightweight products.

The disadvantage of smaller pixels is the shutter lag, which is due to the smaller aperture required to maintain the same f/# as the focal length is reduced. This results in reduced etendue and an increase in the integration time needed to collect enough photons for a reasonable SNR. Observe that it is etendue that controls the absolute integration time, and the use of *relative* aperture (f/#) in this situation leads to incorrect conclusions when comparing designs (see Sec. 4.3.1).

Another design option with smaller pixels is to maintain the IFOV, again reducing the focal length but now keeping the same aperture. The etendue thus stays the same, but the smaller f/# places a limit on how small a pixel and focal length can be used. As a third option, smaller pixels can be used to attain better resolution, which is the trend for high-performance cameras. In this case, the focal length stays the same, reducing the IFOV; as compensation, the aperture can be increased to give the same etendue (with a faster f/#).

Any changes in pixel size and f/# that affect the etendue also have an impact on the SNR. The extent of this impact depends on the noise floor that determines the SNR. It is shown in Example 6.5, for example, that a larger detector size increases the background-limited SNR in proportion to $A_d^{1/2}$; this dependence pushes designs in the direction of a larger pixel size if the integration time cannot be increased to compensate for smaller pixels. In contrast, for a quantization-limited detector—whose noise floor is set by the well size—we have $SNR_q \propto A_d$, which leads to an even greater emphasis on the trade between pixel size and integration time.

Pixel size also determines the maximum number of pixels that can fit on an FPA. With square pixels that are 25 µm × 25 µm in size, for example, a 640 × 512 array is approximately 16 mm × 12.8 mm in size (or $N_p d$ in each dimension). Not surprisingly, there are manufacturing limitations—wafer size and flatness, for instance—on how large an FPA can be; their overall size is currently limited to about 60 mm on a side. This means that, with pixels larger than (say) 25 µm × 25 µm, fewer pixels can be squeezed onto the FPA. Hence larger pixels either restrict the total FOV that can be imaged (FOV ≈ N_p × IFOV) or reduce the available resolution because they increase the IFOV if the focal length of the optics is kept the same.

Smaller pixels can therefore be used for better resolution or smaller array sizes. A smaller array has the additional advantage of lower cost due to the higher yields that result from better uniformity and operability—a consequence of material deposition and wafer processing over a smaller area. Uniformity and operability will be explored in more detail in Secs. 6.4.7 and 6.4.8 (respectively), but first we must look at some additional effects of pixel size: saturation and dynamic range.

6.4.6 Linearity, Saturation, and Dynamic Range

In the previous section we saw that smaller pixels require an increase in integration time in order to compensate for the drop in SNR. Unfortunately, integration time cannot be increased indefinitely for further increases in SNR, owing to an additional constraint on pixel size known as *charge capacity* (symbol N_w, also known as the well depth, full well, or integration capacity). This is the number of electrons that can be generated during integration without saturating the detector. Saturation results in a loss of contrast between the high- and low-brightness levels in the scene as photoelectrons from the bright areas diffuse into the dark (see Fig. 6.43).

A plot of detector saturation is shown in Fig. 6.44. The figure indicates that, at low-to-moderate signals, the photocurrent is more or less linear with optical flux—with the slope of the straight line equal to the responsivity, $i_s = \mathcal{R}\Phi_s$. As the flux increases, there is a point at which the photocurrent starts to roll off and the response is no longer linear. There are various criteria for when the response is no longer considered linear—for example, a 5 percent drop-off from the best-fit straight line—but at *some* point the FPA ROIC (or single-pixel circuitry) has accumulated as many electrons as possible and is then said to be saturated.

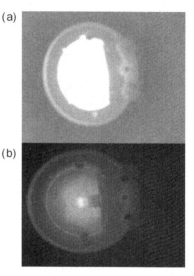

FIGURE 6.43 (a) Saturation of the FPA results in reduced contrast and MTF, often to the point of obscuring important details (b). (*Credit: O. Nesher et al., Proceedings of the SPIE, vol. 4820.*)

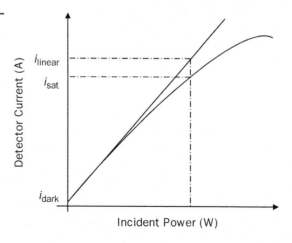

FIGURE 6.44 Detector saturation is a deviation from linearity that occurs when the incident power approaches the capacity of the charge well or detection circuitry.

For an FPA with charge collection and readout, the incident flux that saturates each pixel depends on two parameters. The first is the size of the pixel. Physically, pixel size controls the size of the capacitor (or "charge well"), which in turn determines the number of electrons (or "depth") that can be collected. Well depth also depends on the type of amplifier: for CMOS ROICs, source followers have the smallest depth and direct injection (DI) amplifiers the largest.[40] In any case, a larger pixel can have a bigger capacitor on the ROIC, and this allows the pixel to be exposed to a brighter source for a longer time before it saturates. As a result, charge-well capacity is often given in units of electrons per square micron of pixel area, clearly illustrating the capacity's dependence on pixel size. For example, an InGaAs FPA with CMOS ROIC and 25-µm square pixels may have a well capacity of 4000 e$^-$/µm^2 × (25 µm)^2 = 2.5 × 10^6 electrons (or 2.5 Me$^-$).

The second parameter that determines when incident flux leads to saturation is the voltage limit for the material system and design rules. For instance, silicon CMOS FPAs have an allowable voltage limit on the order of 3.3 V, which can easily be exceeded for strong signals or long integration times. As an illustration, we can calculate that a CMOS ROIC with an electron-to-voltage conversion sensitivity (or conversion gain) of 5 microvolts per electron (µV/e$^-$) collects only 660,000 electrons before exceeding 3.3 volts. Yet a different ROIC design, one with a conversion gain of 0.33 µV/e$^-$, can store as many as 10 million electrons before exceeding its well capacity during the integration time.

Well capacity N_w is thus bounded either by pixel size or a voltage limit; the capacity is determined by the smaller of the two (measured in units of electrons). Designs that do not use a larger aperture to compensate for smaller pixels must rely on a longer integration time to compensate for the drop in etendue. Unfortunately, smaller pixels have less well depth to compensate, which often pushes the design toward a smaller integration time. And this, according to Eq. (6.24), will increase the noise bandwidth Δf, resulting in more "noise" electrons collected by the electrical subsystem.

When the upper limit on collected charge is set by the well depth, the lower limit on the smallest irradiance that can be measured is set by the noise floor for a given integration time. This can be background noise collected over the integration time, read noise, dark current, or any of the other noises reviewed in Sec. 6.3.2. The ratio of the upper to the lower limit is known as the dynamic range (DR), which is an important spec for optical systems designed for large variations in brightness across a scene or from frame to frame.

For example, the dynamic range for CCDs used in low-light-level astronomy is set by the well depth (the upper limit) and read noise (the lower limit), so in this case we have DR = N_w/σ_r. Given a well depth of $N_w = 10^5$ electrons and a read noise of $\sigma_r = 10$ electrons, the dynamic range is equal to 10^4 [or 20 log$_{10}$ 10^4 = 80 dB (decibels)] in this particular case. When a bandwidth-dependent noise—for example, background or dark-current shot noise—sets the noise floor, reducing the bandwidth is one approach to increasing the dynamic range. Typical values for FPA dynamic range are 60 to 80 dB (or $10^{60/20}$ to $10^{80/20}$ = 1000 to 10,000 in real numbers that have not been converted to decibels). The use of saturation values—rather than the full dynamic range of the well depth—can also be used to specify the "useful" dynamic range if output nonlinearities cannot be corrected.

Dynamic range is sometimes translated into the number of bits (or "bit depth") that the ADC uses to convert analog values into digital numbers [DNs, or analog-to-digital units (ADUs), or least-significant bits (LSBs)]. This is possible only if the number of bins into which the ADC divides the signal is sufficient to cover the dynamic range. Thus, for a dynamic range of 4096 (72 dB), the ADC must be able to resolve 1 part in 4096; this requires 12 bits of ADC resolution. Given the discrete nature of ADC bits, the math typically does not work out so neatly; hence the number of bits should exceed the dynamic range requirement by at least 1. For example, the number of bits required for a DR = 5000 (74 dB) is 13—even though this gives the ADC the capacity to resolve 1 part in 2^{13} (for a dynamic range of 78 dB). In any case, the dynamic range to use in design calculations should be based on the well depth and noise floor, not on the number of ADC bits.

Applying all the concepts just discussed to detector subsystem design yields the following list of points to consider:

- Saturation can be avoided by using a detector with a lower responsivity, thereby increasing the dynamic range over which the SNR exceeds a certain value. The cost of doing so is an overall drop in SNR.

- Detector output is generally linear up to about 90 percent of well capacity. Beyond that, saturation starts to set in and the relation between output current and incident power is no longer a straight line.

- Specifications on integration time are often set for 50 percent of well capacity, the *half-well* condition. This allows for variations in the scene that produce many more and many fewer photons than the half-well level, resulting in a wide dynamic range.

- When background subtraction is necessary, the well depth must be large enough to collect the entire background for averaging and not only the background variance. Because of its higher thermal background, for example, an LWIR detector typically needs more well capacity than a SWIR system. These situations generally require a large dynamic range, because a high level of background flux can use up most of the well capacity and some must be available to measure the signal.

- In addition to the limitations on well size already described, deep wells have more dark current if the well depth is determined by pixel size. For a well depth based on either pixel size or voltage limit, deep wells also have more quantization noise (for a given number of ADC bits).

6.4.7 Uniformity

Detector specs such as responsivity, QE, integration time, linearity, and so forth apply to both single-pixel detectors and FPAs. Of the detector specs reviewed so far, well capacity, frame rate, and array size are FPA-specific. Two other specs that are unique to FPAs are uniformity and operability.

The concept of uniformity and its correction were covered in Sec. 6.2.3; here we discuss additional details needed for specifying FPAs. The first of these considerations is that just about any FPA property—responsivity, D^*, NEDT, and so forth—can vary from pixel to pixel. Spec sheets for UV-to-SWIR FPAs usually note the nonuniformity of responsivity, whereas specs for MWIR and LWIR arrays most often include the nonuniformity of NEDT. For low-signal situations, variations in dark current are also important.

In any case, the extent of nonuniformity is determined from a histogram of the FPA property of interest. Figure 6.45 is a sensitivity (NEDT) histogram of the nonuniform image pictured in Fig. 6.22; here the number of pixels with specific values of sensitivity is shown for a 300-K background temperature. The mean sensitivity is approximately 48 mK, and the nonuniformity is quantified as the ratio of the standard deviation to the mean, or 12.5 percent in this example ($\sigma/\mu = 6$ mK ÷ 48 mK = 0.125). Another common criterion is the ratio of the standard deviation to the dynamic range; in terms of

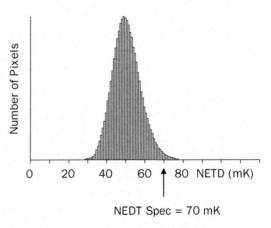

Figure 6.45 Histogram of number of pixels with given sensitivity values (measured as noise-equivalent temperature difference, NETD) for an IR focal plane. The image associated with this histogram is that in Fig. 6.22. (*Credit: J. L. Tissot et al., Proceedings of the SPIE, vol. 6395.*)

Detectors and Focal Plane Arrays

this metric, uncorrected nonuniformities can be as high as 5 to 10 percent for InGaAs and LWIR MCT arrays. Such values are high enough for the resulting spatial noise, if left uncorrected, to be the limiting factor on SNR in high-signal situations [see Eq. (6.14)].

To increase the uniformity and SNR, a nonuniformity correction (or NUC, pronounced "nuke") is used. These corrections involve adjustments of the ROIC amplifier gain affecting measured responsivity; they equalize, as much as possible the pixel-to-pixel variations. Equalizing typically employs a two-point correction for each pixel, where the points are two different temperatures or flux levels; this procedure is illustrated in Fig. 6.46 for five pixels. The FPA is exposed to a highly uniform scene at these two levels using a blackbody or integrating sphere as a source to ensure that the same radiance levels are incident on each pixel.

The NUC shown in Fig. 6.46 is known as an *offset* correction, which measures the pixel output at a specific flux (for quantum detectors) or background temperature (for thermal detectors) and then adjusts ("offsets") each pixel's photocurrent until all are equal at that flux. The offset correction narrows the variance in Fig. 6.45; an additional measurement at a second flux level results in "gain correction" (not shown), which modifies the slopes of these lines until they are equal. In reality the curves are not linear, so a cubic or quadratic fit may be required to minimize the variance at different fluxes. In addition, the required correction must be within the dynamic range of the ROIC, likely forcing the downward-sloping pixel in Fig. 6.46 to be *inoperable* (Sec. 6.4.8).

Corrected (or "residual") nonuniformities can be on the order of 0.05 percent or lower, depending on the material, array size, and optics. Figure 6.47 shows the primary limitation of the NUC process: the corrected uniformity applies only at the two points where

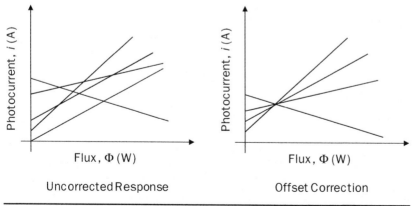

FIGURE 6.46 To obtain the same response from all pixels in an FPA, an offset correction (vertical shift) is first required to give the pixels the same output photocurrent at a specific level of flux.

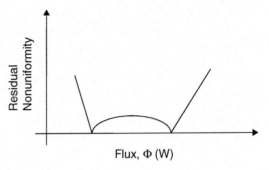

Figure 6.47 A limitation of the two-point NUC process illustrated in Fig. 6.46 is that the uniformity is corrected only at the two flux levels where measurements were made and may still vary widely at other points.

measurements were made, and the residual nonuniformity can be much worse at different flux or radiance levels—resulting in what is known as the "W-curve." For this reason, three- and four-point corrections are used for scenes with highly varying radiance. For MWIR and longer-wavelength systems, the NUC is further limited by the temperature uniformity of the optics; this may require a new correction if the optics or environment are changed.

In addition to the NUC methods just described, FPA nonuniformity may be modulated by temperature variations, array size, and detector material. The temperature profile across the FPA primarily affects dark current, which has a strong thermal dependence. Uniformity is generally better for smaller arrays because the fabrication tolerances in composition, doping, and layer thickness that create nonuniformities are smaller for smaller arrays. In addition, uniformity is another factor to consider in material selection—for example, InSb and InGaAs are more uniform than HgCdTe. Uniformity is especially sensitive to compositional variations for LWIR HgCdTe, and this is one of the primary motivations behind development of quantum-well infrared photodetectors. The LWIR QWIPs have their own disadvantages—in particular, responsivity and operating temperature—but solving these problems is an ongoing area of research and new results are published every year.

6.4.8 Operability

Even after correction for nonuniformity, there are almost always pixels that still do not meet the criterion of "good" performance. For example, Fig. 6.48 shows an FPA with hot pixels, dead pixels, clusters, and dead columns. Hot pixels are those with a large dark current such that the pixel seems to be responding to incident photons even when there are none; at the other extreme, the low responsivity of dead pixels renders them virtually nonfunctional. In neither case do the pixels meet specifications, which naturally creates problems with image interpretation, target tracking, and so on if there are too many of them.

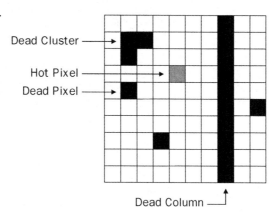

FIGURE 6.48
Operability is reduced in this FPA by hot pixels, dead pixels, dead clusters, and dead columns; dead rows are also possible. The operability of this array is only 83 percent; a value exceeding 99 percent is more common. (See also color insert.)

The percentage of out-of-spec pixels defines an array's *operability*. To determine if a pixel should be counted as having a "large" dark current or "low" responsivity, a cutoff level for these specs is compared with the histogram generated for uniformity. This is shown in Fig. 6.45: the allowable NETD is 0.070 K (70 mK), and all pixels with a higher NETD are considered to be inoperable. The cutoff is usually determined from system requirements; operability is then quantified as the percentage of pixels that meet the cutoff spec. Typical values for responsivity operability are on the order of 98 to 99.9 percent for large-format arrays. Specifying an operability of 100 percent should be avoided for all but the smallest arrays, since the manufacturing yield on such "cherry picked" FPAs will be quite low and the resulting cost extremely high.

It is also common to specify the percentage of out-of-spec pixels using a deviation from the mean. For example, a pixel with a responsivity that varies from the mean by more than 33 percent may be considered inoperable. Observe that a larger-than-normal responsivity also decreases the operability with this criterion, given the potential for saturation and uniformity issues. The pixel with the negative slope in Fig. 6.46 would likely exceed the ability of the ROIC to compensate and would therefore be considered inoperable.

More detailed specs also include additional requirements on the operability of adjacent pixels ("no two adjacent pixels shall be inoperable"), subarrays in the center 100 × 100 pixels (for example) of the FPA, smaller subarrays ("clusters") anywhere on the FPA, and entire columns or rows. Note also that pixels can change from being operable to inoperable over time (as responsivity decreases and dark current increases), and for this reason it is wise to obtain reliability and end-of-life data from the vendor.

6.4.9 Power Consumption

For single-pixel detectors and FPAs both, electrical power consumption can determine whether the detector subsystem "makes" or "breaks" the electrical budget. By themselves, detectors and FPAs usually don't consume much power; instead it is the peripheral equipment—coolers, amplifiers, ADCs, and so forth—that consume the most electrical power. In fact, the main motive for the development of microbolometers is their ability to operate without a cooler (or with a small cooler used only to ensure temperature uniformity across the array). A 1024 × 1024 MWIR InSb array, to take another example, may use only 200 mW of electrical power, but the electronics and cooling equipment required to maintain an 80-K operating temperature can easily consume 50 to 100 times as much.

That being said, reductions in detector consumption directly reduce the requirements on cooling power. For FPAs, that consumption depends on the FPA type (CCD versus CMOS), array size, and frame rate. CMOS arrays use significantly less power than a CCD and thus are always the first choice for low-end consumer electronics that operate on small batteries. Larger arrays have more detectors to cool, and power consumption is nearly proportional to the number of pixels. Faster frame rates also increase the cooling requirements, a result of transferring more electrons off the FPA per second.

There are four basic types of coolers. Two of them require significant electrical power: the Stirling-cycle cooler and the thermoelectric cooler (TEC). Electrical-to-cooling conversion efficiency depends on the cooling temperature. Stirling-cycle coolers require 25 to 30 watts of electrical power to remove a single watt of thermal power to reach an operating temperature of 80 K, which translates into a cooldown efficiency of 3 to 5 percent; however, the steady-state efficiency after cooldown is about 4 times higher. Thermoelectric coolers can't reach 80 K, although four-stage TECs can cool down to approximately 200 K—at which point they use 8–10 W of electrical power to remove 80–100 mW of heat from the detector plus surroundings (see Fig. 6.49). These four-stage coolers are sometimes used for specialized silicon CCD FPAs that require low levels of dark current. Single- and two-stage TECs are also used for scientific CCDs that require lower levels of read noise and dark-current noise.

To operate the cooler electronics, both Stirling-cycle and thermoelectric coolers require electrical source power that must be included in the electrical power budget. The other cooler types—Joule–Thomson gas-expansion coolers and cryogenic liquids (liquid N_2, LHe, etc.)—generally require only smaller amounts of electrical power. However, they have other disadvantages such as maintenance requirements and limited lifetime.

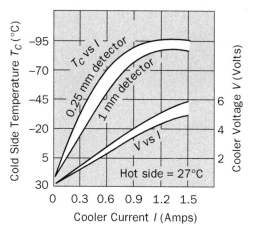

FIGURE 6.49 Current and voltage requirements for a four-stage TEC to reach 200 K with a cooling capacity of 80–100 mW. (*Credit: Teledyne Judson Technologies.*)

The ADCs used with FPAs also consume more power when operating at faster data rates: ROIC-integrated 12-bit ADCs use about 2 mW per megahertz, 14-bit devices use approximately 4 mW/MHz, and 16-bit ADCs use about 30 mW/MHz.[40] When combined with the pre-amps, additional amplifiers and electronics, and so on, the size, weight, cost, and power consumption of the detector subsystem can be significantly greater than that of the detector itself. This is an important aspect of the detector selection process, which is reviewed in Sec. 6.5.

6.4.10 Size and Weight

Size, weight, and power consumption (or "SWaP") are important properties of any optical system. After all, weight is expensive to launch into space as well as annoying to lug around on one's neck. Large, cumbersome systems are not only difficult to carry but also prevent additional instruments or features from being included in the launch or product offering. The factor determining SWaP in many optical systems is whether the FPA requires cooling.

For cameras imaging in the visible, the largest available FPAs are currently 60-megapixel uncooled versions with dimensions of about 50 mm by 50 mm (in the plane of the FPA) by 5 mm thick. The size and weight of such a camera is not dominated by the FPA itself, as most of the size and weight is in the optics, electronics, and packaging.

In contrast, the SWaP of cryogenically cooled FPAs for MWIR and LWIR imaging *can* be dominated by the FPA and its associated cooling, thus yielding a component known as the *integrated detector–cooler assembly* (IDCA). Stirling-cycle coolers for an 80-K operating temperature are most commonly used in such assemblies, where 0.5-W coolers have the dimensions shown in Fig. 6.50. The *vacuum*

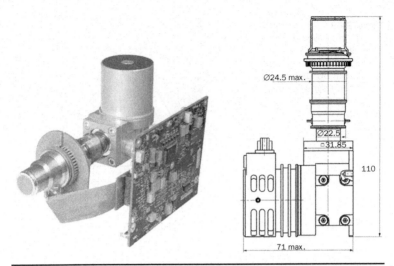

FIGURE 6.50 Typical dimensions of an IDCA required for an FPA temperature of 80 K. Light enters the dewar through the window at the lower left of the photo. (*Credit: AIM Infrarot-Module GmbH.*)

dewar that contains the FPA is considered to be part of the IDCA. Optics and additional packaging increase the SWaP beyond that of the IDCA shown here.

Uncooled microbolometers are also common for LWIR imaging. These sometimes require a TEC not for cooling per se but rather for temperature stability; the addition of a relatively small cooler does not dramatically increase the SWaP. As we have seen, the disadvantage of this setup is the poorer sensitivity compared with cooled IR FPAs.

6.4.11 Cost

As with optical sources, the cost of detectors varies widely. The largest cost difference is between single-pixel detectors and FPAs. For FPAs, integrated detector–cooler assemblies for MWIR and LWIR sensors are commercially available for approximately $30,000 (U.S.), a price that reflects the relatively high cost of cryogenic coolers and dewars. The operating and maintenance costs of IR systems are also high when one considers their electrical power consumption and limited lifetime.

Uncooled LWIR FPAs, are significantly less expensive: on the order of $3000 to $5000. For the visible FPA market, silicon arrays are by far the least expensive, a result of the huge volumes developed for consumer cell phones, cameras, and so on. Large-format (640 × 480) CMOS FPAs are available for these markets at volume prices less than $5.

6.5 Detector Selection

In light of the wide range of details that must be considered in detector specifications—that is, responsivity, bandwidth, pixel size, linearity, saturation, uniformity, operability, and so on—it is no surprise that detector selection is one of the most challenging aspects of optical systems engineering. As seen in the Sec. 6.4, a critical part of the process is understanding the information on the manufacturers' spec sheets. This information is necessary but by no means sufficient, and there are a few additional guidelines to follow that make the selection process a bit easier. In conjunction with the precautions reviewed in Chap. 5—normalized numbers, spec conditions, and missing information—the steps spelled out here constitute only one approach; there are other paths that lead to the same result.

As with the selection of optical sources, detector selection can be a time-consuming process. It requires a throughput budget to determine system sensitivity, a noise budget to calculate SNR, and a comparison of detector specs from different vendors of which none has exactly what's needed. The process may require an iteration or three if the detector has too much noise or isn't available in a material that matches the source wavelengths or if the system requirements are such that it's not possible to lower the bandwidth to reduce some of the noise components, and so on.

We have seen that there are many ways to classify detectors: by type (photoconductor, photodiode, thermal, and others not covered in this book), by geometry (single-pixel versus FPA), by FPA technology (CCD, CMOS, or hybrid), and by material (silicon, InGaAs, HgCdTe, InSb, etc.). However, the first step in detector selection is to compare the detector materials listed in Table 6.1 with the source wavelengths. When the source spectrum is unknown or still to be determined, it may be useful first to consider the detector requirements and then work backwards to source selection. In any case, detector selection follows the guidelines given in Sec. 6.5.1. Section 6.5.2 offers some additional comments on the various system-level trade-offs involved.

6.5.1 Detector Comparisons

For a known source spectrum, Table 6.1 lists the most common detector materials for matching the detector to the source emission. A limitation of this table is that it lists only the common materials for each wavelength range and none for specialized applications. Nevertheless, the table does indicate that silicon detectors are the most likely choice for visible applications. If the specs call instead for an MWIR imager, then InSb or HgCdTe FPAs are currently the best candidates.

Table 6.1 also shows that there are numerous options for an LWIR system: HgCdTe photoconductors and photodiodes, VOx microbolometers, arsenic-doped silicon, and QWIPs. A second step must therefore be included in the selection process—namely, a comparison of the different detector types based on four or five of the most critical specifications. These are usually wavelength coverage, detector geometry, sensitivity (using a metric such as peak D^* for the LWIR), cooling requirements, and total cost (including cooler); other known specs should also be included in the comparison. The following list summarizes the most important specifications for each detector technology:

- *Photoconductors*—Although there are exceptions, these are usually low-cost, single-pixel detectors for the visible and NIR; compared with photodiodes, they are relatively slow and noisy. The high-cost exceptions are specialized detectors for LWIR and VLWIR astronomy.
- *Photodiodes* ($V_b = 0$)—Unbiased photodiodes have little dark current, which places them among the most sensitive detectors; however, they are also relatively slow. These detectors are also often used for direct current (DC) applications such as conversion of solar energy to electricity.
- *Photodiodes* ($V_b < 0$)—A reverse-biased p-n junction photodiode has dark current but is still quieter than a photoconductor. Depending on type, these detectors can be very fast and extremely sensitive; they are available in large-format arrays with small pixels.
- *Microbolometers*—Although slower than photodiodes and photoconductors, this type of thermal detector is available uncooled. Focal plane arrays of moderate sensitivity are available at relatively low cost.

As with source types, detector evaluation is best performed with the aid of a comparison matrix. The one shown as Table 6.6 illustrates *the specific example of LWIR detectors.*

The properties listed here must be compared with the system requirements, which are often included as an additional column in the comparison matrix. One possibility is that cost is an extremely important requirement. If so, then VOx microbolometers are the best option. In focusing on cost, however, it is important that other requirements not be ignored. For example, a fast response time may also be important, in which case no obvious conclusion can be drawn as to which is the most appropriate detector. Cost requirements are always important, and discussions must be held between the engineering and marketing departments concerning why cost (or speed, for example) is the most important consideration. This

Property	Photoconductor	Photodiode	Bolometer
Detector mechanism	Photon	Photon	Thermal
Detector material	HgCdTe, Si:As, QWIP	HgCdTe	VOx
Wavelength coverage	LWIR	LWIR	LWIR
Detector geometry	Single-pixel, FPA	Single-pixel, FPA	Single-pixel, FPA
Cooling requirements	80 K (HgCdTe) 5 K (Si:As) 40–60 K (QWIP)	80 K	280–300 K (FPAs only)
Cost	High	High	Low to moderate
External responsivity	5–6 A/W (HgCdTe) 2 A/W (Si:As) 1 A/W (QWIP)	5–6 A/W (HgCdTe)	5–10 A/W
Dark current	High	Low	Not applicable
Pixel bandwidth	Moderate (~ 100 MHz)	Fast (PIN ~ 10 GHz)	Slow (~ 100 Hz)
Frame rate (typ)	30 Hz (QWIP)	30–100 Hz	30 Hz
Detector size (typ)	~ 25 μm	~ 20 μm per pixel	17 μm per pixel
Fill factor		100%	100%
Array size (typ)	512 × 256 (MCT, Si:As) 1024 × 1024 (QWIPs)	640 × 512	640 × 512
Well depth	Low	ROIC dependent	ROIC dependent
Dynamic range	Moderate	High	Low
Uniformity (uncorrected)	~ 10% (HgCdTe) ~ 1% (Si:As, QWIP)	~ 10%	High w/ TEC
Operability	> 99% (QWIP)	> 98%	> 99%
Power consumption	High	High	Low to moderate

TABLE 6.6 Comparison of specifications for the LWIR detectors listed in Table 6.1. An analogous table for other wavelength bands (VIS, NIR, SWIR, etc.) will have entries for different detector types and materials.

back-and-forth is typical of the detector selection process, though it is often three or more specs that are being "juggled" at the same time.

To continue with our example, suppose that the engineering department has decided that detector speed is the most important factor to meet the system requirements and sets the required minimum bandwidth at 1 MHz. (It's usually not possible to obtain these numbers from marketing or business development colleagues, so the engineering department will likely lead such technical decisions.) Given this decision, Table 6.6 shows that the bandwidth requirement excludes thermal detectors and thus leaves only photoconductors and photodiodes as viable options.

At this point, the "usually" and "generally" descriptions of photoconductor properties at the beginning of this section are no longer accurate. Of the remaining detectors, photoconductive Si:As is an extremely expensive option—not only the FPA itself but also its 4-K operating temperature. This makes HgCdTe and QWIP FPAs less expensive options. Comparing these latter two FPAs reveals that MCT—either photoconductor or photodiode—doesn't have the operability of a QWIP but does have better responsivity, a higher operating temperature, and less dark current. As a result, the lower noise associated with MCT implies that these FPAs are better suited for low-light applications whereas QWIPs may be better suited for situations involving a high level of incident photons if the high cost of cooling a QWIP to 60 K is acceptable—generally not a likely conclusion. Additional details on the comparison of LWIR HgCdTe and QWIPs can be found in Rogalski.[41]

Once the requirements point to a preference for MCT or QWIPs, the next step is to make a table similar to Table 6.6 that uses different vendors for the columns. Even though all the information needed may not be available, an SNR estimate for different irradiance levels will then point to the most promising manufacturers. This is also an appropriate place in the selection process to examine in more detail the various trade-offs allowed by the options still in play, including different pixel sizes, array formats, and the like (see Sec. 6.5.2).

In sum, focusing on critical differences between detectors allows us to reduce the number of options to a few promising candidates. The specs for the ideal candidate often must be changed and design compromises made. Our example of LWIR FPAs illustrates how to "down select" from a list of alternatives by first matching the detector response to the source spectrum and then excluding options that clearly don't meet other critical requirements such as speed or cost. The list of properties summarized in Table 6.6 will be completely different for selecting a LWIR detector than for selecting a UV, visible, NIR, or any other type of non-LWIR detector. For example, single-element visible-wavelength photoconductors are very inexpensive—unlike the HgCdTe, Si:As, or QWIP FPAs required for operating in the LWIR.

6.5.2 System Trades

In addition to the trade-offs between detector types and materials, a second aspect of detector selection is the many system-level trades that require design compromises between competing requirements such as sensitivity and resolution. These compromises have been illustrated throughout this chapter; in this section, the most important are summarized.

Aperture versus Image Quality

We saw in Chap. 3 that the angular resolution of a diffraction-limited optical system is given by $\beta = 2.44\lambda/D$. At the same time, the pixel IFOV given by Eq. (2.3) may or may not match the system resolution. For example, an imager that collects visible light through a 100-mm aperture has a diffraction-limited resolution $\beta = (2.44 \times 0.5 \text{ μm})/(0.1 \text{ m}) \approx 12$ μrad. If an imager has an FPA with a pixel IFOV or ground sample distance (GSD) of 6 μrad (for example), then any object in the scene that is less than two pixels in size will be imaged onto the FPA as a two-pixel "point."

The best-attainable resolution—or ground *resolved* distance (GRD)—in this case is thus two pixels, which does not match the resolution capabilities of the aperture to the FPA. If the f/# allows it, a better option from the perspective of spatial resolution seems to be doubling the aperture to 200 mm; this would decrease the telescope's resolving power to 6 μrad, *possibly* giving the best possible image quality for this system. The pixel size could also be doubled—or the focal length halved—in order to increase the IFOV to match that of the optical system. Yet even though either of these two options would not change the aperture size and optical resolution, the change in IFOV would reduce the image quality considerably with the sampling of the scene by the FPA now no better than 12 μrad.

In many designs, the image blur is intentionally larger than the pixel IFOV so that the telescope will be smaller and weigh less. This does not dramatically reduce the image quality, since the National Image Interpretability and Rating Scale (NIIRS) for subjective quality depends more on GSD than on blur-to-pixel ratio.[42] Reductions in image quality due to aperture reduction (larger f/#) may thus be somewhat compensated for by a smaller IFOV (smaller pixels). The larger f/# does increase the blur size; however, until the blur is about five pixels in size, such multipixel sampling reduces aliasing—at the expense of an increase in integration time and a reduction in SNR, optical MTF, and NIIRS-rated image quality. Depending on the system requirements, these reductions may still be within the bounds of acceptable.

Going in the other direction, it is oftentimes necessary to use a larger aperture—to collect more photons, for example—that yields a blur size of subpixel dimensions. There are also many optical systems for which the best attainable resolution is not relevant; radiometers are the most common example.

For imagers, these design options can be summarized using the Q-parameter. In Sec. 6.2.1 it was shown that this parameter—also known as "$\lambda f/D$"[19]—is sometimes used as a metric for image quality, where aliased artifacts are possible with $Q < 2$. Given that Q depends on the ratio of blur size to pixel pitch, it can be decreased by either: (1) decreasing the blur with a larger aperture; or (2) using a larger pixel pitch. Figure 6.18 illustrated the effects on Q for an increase in aperture, with Q decreasing from 2 to 0.2.

The consequences of a larger aperture on system MTF ($\text{MTF}_{opt} \times \text{MTF}_{det}$) are shown in Fig. 6.51, where it is seen that the MTF extends well beyond the detector Nyquist frequency ($f_N = 1/2x_p = 100$ lp/mm) for the larger aperture with $Q = 0.2$ (f/2, aperture diameter $D = 500$ mm). The large MTF at $f > f_N$ allows for the possibility of aliasing, but if there is none then the larger aperture yields much better image quality than the smaller-aperture design (f/20, $D = 50$ mm) with $Q = 2$. Because scenes with periodic features may result in aliasing, a common design compromise is to use $Q \approx 1$ and thus balance MTF and image quality due to aperture size against that due to aliasing—subject to the constraint of minimum f/#.[42] A more conservative design rule was proposed by Schade, who suggested that the total MTF at the detector Nyquist be *less than* 0.15 or so in order to avoid the subjective perception of aliasing, which is objectionable to human observers.[18] In either case, the ratio of blur size to pixel dimension (B/x_p) for these design rules is approximately 2.5, for which the 100-mm aperture used in the example at the beginning of this section is in fact appropriate.

Figure 6.51 implies that a smaller Q corresponds to a larger MTF. However, a larger aperture is only half the story because Q can also be increased by using smaller pixels. In that case, we expect a larger Q to be associated with higher MTF and image quality. This is illustrated in Fig. 6.52, where the pixel pitch is varied while the aperture is held constant at $D = 50$ mm (f/20). It is now seen that the smaller (5-μm) pixels with $Q = 2$ have a better MTF at any spatial frequency, which is consistent with physical expectations. Although the $Q = 0.2$ design has a larger MTF at its Nyquist frequency ($f_N = 10$ lp/mm) than the $Q = 2$ design does at its Nyquist ($f_N = 100$ lp/mm, where the MTF = 0), the $Q = 2$ design *resolves a much wider range of frequencies*, over which the MTF is larger than the $Q = 0.2$ design. The $Q = 2$ design *with smaller pixels* is thus better able to resolve higher spatial frequencies and also avoid aliasing.

In summary, then, Q is a metric for the absence of aliasing but is not a complete metric for image quality. A complete metric is the system MTF given by Eq. (6.5), which contains within it system-level trades between optical MTF and detector MTF. As we have seen

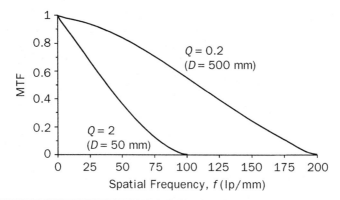

FIGURE 6.51 Increasing aperture size can yield a Q-parameter small enough to introduce aliasing of spatial frequencies $f > f_N$. For this figure, the pixel pitch $x_p = 5$ μm and the wavelength $\lambda = 0.5$ μm.

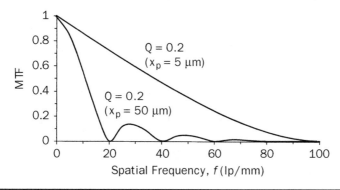

FIGURE 6.52 If the aperture remains constant, then smaller pixels have better system MTF and image quality despite the larger Q. Observe that the detector Nyquist changes from $f_N = 100$ lp/mm for $x_p = 5$ μm to $f_N = 10$ lp/mm for the pixel with $x_p = 50$ μm.

when changing aperture size, the smaller Q associated with larger apertures gives better results until $Q \approx 1$ (Fig. 6.51); when changing pixel size, the larger Q associated with smaller pixels is the better design (Fig. 6.52). Smaller pixels and large apertures always increase MTF. But image quality depends also on the presence of aliasing and so, given the potential for aliasing with $Q < 2$, it may be counterproductive to make the aperture as large as possible when attempting to obtain the best possible resolution.[42] Aperture reduction also reduces size and weight, a significant benefit in almost all optical systems.

Resolution versus Wavelength

The equation for angular resolving power shows that both larger apertures and shorter wavelengths improve the resolution. Varying wavelength can be a complicated trade, depending on the wavelengths; in fact, if the source spectrum is fixed then this trade is not even an option. When it is an option, the primary advantage of a shorter wavelength is that it does not require as large an aperture to attain the same resolution, which saves on size and weight. The trade-off is that the source at the shorter wavelength may not have enough emittance for a reasonable SNR, as occurs when trying to replace an LWIR or MWIR thermal imager with a SWIR wavelength (for example). Atmospheric transmission also plays a role in these trades, with more details given in Holst[19] for the LWIR versus MWIR comparison.

Sensitivity versus Aperture

It is surprising how often a small aperture with a million-dollar detector is perceived as "better" than a larger telescope with a five-dollar detector. These are extreme numbers, of course, that are used here to emphasize that a slightly larger aperture may result in a significantly less sensitive (and less expensive) detector. How sensitive the detector should be depends on the source power, the integration time, and the etendue collected by the telescope (which in turn depends on the telescope aperture and the pixel solid angle). In addition to excess cost, a detector that's overly sensitive will probably not have enough dynamic range to avoid saturating.

The trade between detector sensitivity and optical aperture is sometimes overlooked. The reason is that detector selection is often considered independently of the optical system engineering and instead focuses on evaluating high-sensitivity detectors without first estimating the radiometric power budget or SNR. The performance metric which allows for a meaningful comparison is the total cost required to obtain the SNR needed (quantified as cost per unit SNR), and not detector sensitivity.

A brighter source can also result in a less expensive detector. In considering the combined effects of throughput, integration time, and pixel solid angle, detector selection entails more than just detector specs in isolation: it strongly depends on the overall design of the source-plus-optics-plus-detector system.

Pixel Size

The many trade-offs that involve pixel size have been reviewed in the previous sections. These trades may be summarized as follows.

- *Etendue.* A smaller pixel has a smaller IFOV for the same focal length optics, thus reducing the etendue and system sensitivity. The focal length can be reduced—or the aperture increased—to keep the etendue the same with a faster lens (smaller f/#), where the lower limit on the focal length and aperture is set by the f/# and acceptable aberrations. If the relative lack of object motion allows it, the integration time can also be increased to compensate—although this can lead to shutter lag and image blur.

- *Resolution.* Improvements in spatial resolution (IFOV and GSD) are one of the motivations for smaller pixels. Unfortunately, the reduction in etendue for area sources also degrades the system sensitivity, with larger apertures and/or longer integration times required to maintain the same SNR.
- *Saturation.* Smaller pixels have smaller charge wells, which saturate faster than larger wells for the same aperture and IFOV.
- *Noise.* Smaller pixels have less dark current and quantization noise. However, they also have a shorter integration time before the charge well is full; this may require a larger electrical-system bandwidth, which would increase any of the Δf-dependent noise terms described in Sec. 6.3.2.
- *Size and weight.* Smaller pixels can be used to keep the same FOV, yielding a shorter focal length and a smaller aperture (to maintain the same f/#). This combination results in extremely small and compact optics, as are found in cell-phone cameras. Even with the same f/#, the smaller pixel size results in a reduction in etendue and so increases the integration time required to collect photons. This is evidenced by the shutter lag found with these cameras.

Despite the number of possible trade-offs, pixel size cannot be changed arbitrarily and is usually limited by what's available with commercial FPAs. There are technological limitations on minimum pixel size, such as the so-called bump-bonding process for hybrid arrays; pixels on the order of 3–6 µm are typical for visible arrays that do not require bump bonding, whereas pixels of about 15 µm are the normal minimum for IR FPAs that do. If the pixel size doesn't meet the requirements then the problem will have to be looked at a second and possibly third time by reworking the trades and reexamining the assumptions (wavelength, cooling, bandwidth, etc.). Specialized detectors and FPAs are quite expensive and should be avoided if at all possible.

Noise Balancing

We saw in Example 6.3 that two or three noise terms usually dominate the SNR. Ideally, these dominant terms should be about equal so that, if one grows larger than the others over time, the effect on total noise is minimal [a result of the RSS summation in Eq. (6.8)]. Given the complexity of system design, it's difficult in practice to balance the different noise sources in this way.

Operating Temperature

Cooling a detector to a lower temperature is almost always necessary to reduce dark current in IR devices, whereas control of the temperature *distribution* is part of ensuring FPA uniformity. Reducing the cooling requirements can have a dramatic impact on system cost. This means that reexamining the need for cooling and temperature control is an important system-level trade.

For example, microbolometers often use thermoelectric coolers not for temperature reduction but for temperature uniformity, thus reducing spatial noise. Yet there are many applications of uncooled arrays where spatial noise is not critical, which eliminates the need for a TEC. Similarly, dark current may not need to be reduced in many applications where another noise (e.g., background shot noise) is dominant; in such cases, cooling is either unnecessary or is traded against a larger aperture or brighter source such that overall system requirements are still met.

Problems

6.1 Why is photon energy lower at longer wavelengths? Is there a physical explanation based on the wave properties of photons? *Hint:* What happens to the wave frequency at longer wavelengths?

6.2 How much solar energy is at a wavelength that is longer than the bandgap of silicon? What fraction is this of the total solar output? If all of the solar energy with a bandgap shorter than silicon's could be converted to electricity, what would be the highest possible efficiency of silicon solar cells?

6.3 Show that Eq. (6.2) leads to the equation for open-circuit voltage: $V_{oc} = (kT/q) \ln[1 + (i_s/i_{dc})]$.

6.4 Does the responsivity of a PV system depend on the concentration ratio? *Hint:* What happens to responsivity as a detector becomes hotter?

6.5 For a photovoltaic system pointing straight up with a concentration ratio of 100 : 1, does the conversion efficiency change as the summer Sun moves across the sky? Is it possible to design the optics to cover the entire sky with the same efficiency over the entire day? If not, then how could a PV system be designed to maintain its efficiency as the Sun moves?

6.6 Why are microbolometer sensitivities specified at a particular f/#, such as f/1?

6.7 For a given f/#, the sensitivity of a microbolometer improves with bigger pixels. Discuss the design trade-offs involved when changing the integration time, given that a longer integration time can also result in better sensitivity. Is there a limit to the length of the integration time?

6.8 Even though the MTF is zero at the detector cutoff, is image irradiance at this spatial frequency still being measured?

6.9 Are there any effects on image MTF when there is no obvious periodic pattern in the image? Does your answer depend on whether the MTF is measured above or below Nyquist?

6.10 What happens to image aliasing as pixel size is reduced and the f/# is kept the same? Does your answer depend on the ratio of blur size to pixel dimension (B/x_p)?

6.11 Example 6.2 shows that the blur size for a ground-based telescope is limited by atmospheric distortion ("seeing"), so a larger aperture increases the number of photons collected but does not reduce the blur diameter. What about a space-based telescope? How wide would the aperture have to be in order to match the Nyquist IFOV of the FPA?

6.12 Add the optical and detector MTF curves to Fig. 6.51 and Fig. 6.52 for the case where $Q = 1$ (i.e., when $2f_N = f_c$).

6.13 A LWIR imager has a pixel IFOV of 500 μrad. Is the diffraction-limited blur for a 50-mm aperture oversampled or undersampled at a wavelength of 8 μm? At a wavelength of 12 μm?

6.14 What are the effects of wavelength on the Q-parameter for an optics-plus-detector system? For a visible imager that is required to be Nyquist sampled across the visible band, should the pixel size be selected for the longer or shorter wavelength in the band?

6.15 If a "large" Q is required for a high-resolution imager and a "small" Q for a high-precision radiometer, then what value of Q is required for an imaging radiometer? A radiometric imager?

6.16 Look through some manufacturers' specs to determine which has better uncorrected uniformity: silicon, InGaAs, or HgCdTe FPAs.

6.17 What type of source should be used to determine the correction for FPA nonuniformities?

6.18 What determines whether a source is shot-noise limited? Does it depend on the detector used to measure it?

6.19 At what wavelength does the photon energy equal the electron thermal energy at room temperature?

6.20 Does Johnson noise depend on whether or not any photons are incident on the detector?

6.21 Equation (6.12) gives the expression for Johnson noise *current*. What is the expression for Johnson noise *voltage*?

6.22 Which would be expected to have a lower room-temperature dark current: HgCdTe optimized for a peak responsivity at 10 μm or at 1 μm? How does the dark current for HgCdTe at 1 μm compare with InGaAs?

6.23 Is it possible to measure dark current for a LWIR detector sensitive to 8–12-μm photons? Is it possible to make the background "dark" at this wavelength?

6.24 For a detector whose dark current is greater than other types of noise and varies erratically from frame-to-frame, is it possible to subtract out its average and then use its standard deviation as the noise level? If not, then why not?

6.25 For a BLIP situation, does it make sense to try to increase SNR by reducing the pixel size? Does it matter whether we are imaging a point source or an area source? *Hint:* What are the effects on the signal?

6.26 At low light levels, read noise may dominate a CCD. Is there a simple way to reduce the noise in this case? *Hint:* See Sec. 7.3.

6.27 Smaller pixels restrict the solid angle from which the detector collects signal photons, thus increasing the NEP. Are there similar effects with respect to f/#? *Hint:* What is the expression for etendue in terms of f/#?

6.28 An InSb data sheet gives a value for D^* of 10^{11} cm·Hz$^{1/2}$/W. Is it safe to assume that this is D^*_{BLIP}? Why or why not?

6.29 With both the SP35 and SP28 filters at the same temperature in Fig. 6.42, why does the higher-transmission cold filter (SP28) yield a greater improvement in D^*_{BLIP}?

6.30 Is it possible to extend silicon's response out to 1064 nm (a common laser wavelength), by using a thicker absorption layer to compensate for the reduced absorption coefficient α_m? How thick would the layer need to be to give a responsivity of 0.1 A/W?

6.31 Do solar cells have dark current? If so, does it matter—or is *any* current that a solar cell generates useful?

6.32 List the detector types from noisiest to quietest, including photodiodes (both $V = 0$ and $V < 0$), photoconductors, and VOx microbolometers.

6.33 Make a separate detector comparison matrix for each of the UV, VIS, NIR, SWIR, and MWIR bands.

6.34 Use the detector selection process described in Sec. 6.5 to choose the appropriate detector for a laser radar system operating at a wavelength of 1.55 μm. Start by listing the requirements for your system.

Notes and References

1. M. Bass, E. W. Van Stryland, D. R. Williams, and W. L. Wolfe (Eds.), *Handbook of Optics*, 2nd ed., vol. 1, McGraw-Hill (www.mcgraw-hill.com), 1995, chaps. 15–23.
2. E. L. Dereniak and G. D. Boreman, *Infrared Detectors and Systems*, Wiley (www.wiley.com), 1996.
3. Silvano Donati, *Photodetectors*, Prentice Hall (www.phptr.com), 2000.
4. G. H. Rieke, *Detection of Light From the Ultraviolet to the Submillimeter*, 2nd ed., Cambridge University Press (www.cambridge.org), 2002.

5. Philip C. D. Hobbs, *Building Electro-Optical Systems*, Wiley (www.wiley.com), 2000.
6. Paul Horowicz and Winfield Hill, *The Art of Electronics*, 2nd ed., Cambridge University Press (www.cambridge.org), 1989.
7. Michael A. Kinch, *Fundamentals of Infrared Detector Materials*, SPIE Press (www.spie.org), 2007.
8. Jozef F. Piotrowski, *High-Operating-Temperature Infrared Photodetectors*, SPIE Press (www.spie.org), 2007.
9. P. W. Kruse, *Uncooled Thermal Imaging Arrays, Systems, and Applications*, SPIE Press (www.spie.org), 2001.
10. Ed Friedman and John Lester Miller, *Photonics Rules of Thumb*, 2nd ed., SPIE Press (www.spie.org), 2004.
11. Depending on the material, the same process can occur with electrons; in that case, it is extra holes that are injected into the photoconductive layer to conserve charge.
12. Paul L. Richards and Craig R. McCreight, "Infrared Detectors for Astrophysics," *Physics Today*, February 2005, pp. 41–47.
13. Stephen B. Alexander, *Optical Communication Receiver Design*, SPIE Press (www.spie.org), 1997.
14. Sarah Kurtz and Daniel Friedman, "Photovoltaics: Lighting the Way to a Brighter Future," *Optics and Photonic News*, 16(6), 2005, pp. 30–35.
15. Unfortunately, p-n junction and PIN photodiodes are often called photovoltaic detectors even when they are reverse-biased.
16. Avalanche photodiodes have extra factors for excess noise and gain (M^2F), where most APD materials have an excess noise factor $F > 2$; for details, see Dereniak and Boreman[2] as well as Donati.[3]
17. In the MWIR band, for example, InSb photodiodes are better than the available thermal detectors at detecting small temperature differences.
18. R. Vollmerhausen and R. Driggers, *Analysis of Sampled Imaging Systems*, SPIE Press (www.spie.org), 2000.
19. Gerald C. Holst, *Electro-Optical Imaging System Performance*, 5th ed., SPIE Press (www.spie.org), 2008.
20. R. E. Fischer, B. Tadic-Galeb, and P. R. Yoder, *Optical System Design*, 2nd ed., McGraw-Hill (www.mcgraw-hill.com), 2008.
21. Steve B. Howell, *Handbook of CCD Astronomy*, 2nd ed., Cambridge University Press (www.cambridge.org), 2006.
22. Frame transfer and interline transfer are two common methods of reading out data from CCDs. A recent innovation is the orthogonal-transfer CCD, which is useful for many applications including adaptive-optics technology that corrects for atmospheric distortions.
23. James R. Janesick, *Scientific Charge-Coupled Devices*, SPIE Press (www.spie.org), 2001.
24. The MOS capacitor is also known as a metal-insulator-semiconductor (MIS) capacitor.
25. Gerald C. Holst and Terrence S. Lomheim, *CMOS/CCD Sensors and Camera Systems*, SPIE Press (www.spie.org), 2007.
26. B. R. Frieden, *Probability, Statistical Optics, and Data Testing*, 2nd ed., Springer-Verlag (www.springer.com), 1991.

27. Bahaa E. A. Saleh and Malvin Carl Teich, *Fundamentals of Photonics*, 2nd ed., Wiley (www.wiley.com), 2007.
28. In CCD FPAs, a large component of read noise is due to the temperature T of the charge-transfer capacitance C; hence read noise is sometimes known as kTC noise (where k is Boltzman's constant).
29. H. J. Christian, R. J. Blakeslee, and S. J. Goodman, "The Detection of Lightning from Geostationary Orbit," *Journal of Geophysical Research*, 94(D11), 1989, pp. 13,329–13,337.
30. David Kryskowski and Gwynn H. Suits, "Natural Sources," in J. S. Acetta and D. L. Shoemaker (Eds.), *The Infrared and Electro-Optical Systems Handbook*, vol. 1, SPIE Press (www.spie.org) chap. 3, table 3–17.
31. This is also known as adding "in quadrature" because the noises are independent, just as the legs of a hypotenuse that are oriented perpendicular with respect to each other (at 90 degrees, oriented in quadrature).
32. Burle Electron Tubes, *Electro-Optics Handbook*, Burle (www.burle.com), 1974.
33. For example, the background temperature at which the detector becomes BLIP must be defined—"the temperature at which the background-generated current is 10 times the dark current," for example—and the f/#, integration time, and optical transmission of the instrument must also be specified.
34. Dereniak and Boreman[2] also include results for photoconductors, which have twice as much noise due to recombination.
35. Responsivity includes the number of electrons generated that can be measured; if these electrons recombine before being collected at the contacts, then they do not contribute to the photocurrent. In the extreme case, we can imagine a signal that creates many electrons all of which recombine before being detected and so the responsivity is close to zero. This can occur at very high electrical frequencies when the carrier lifetime is short—see page 193 of Dereniak and Boreman.[2]
36. Arnold Daniels, *Field Guide to Infrared Systems*, SPIE Press (www.spie.org), 2007.
37. For large angles, the most accurate definition of relative aperture is given by $f/\# = 1/2NA = 1/\{2\sin[\tan^{-1}(D/2f)]\}$.
38. J. F. James, *A Student's Guide to Fourier Transforms*, Cambridge University Press (www.cambridge.org), 1995.
39. Paul R. Norton, "Infrared Image Sensors," *Optical Engineering*, 30(11), 1991, pp. 1649–1663.
40. Thomas Sprafke and James W. Beletic, "High-Performance Infrared Focal Plane Arrays for Space Applications," *Optics and Photonic News*, 19(6), 2008, pp. 23–27.
41. Antoni Rogalski, "Third-Generation Infrared Photon Detectors," *Optical Engineering*, 42(12), 2003, pp. 3498–3516.
42. Robert D. Fiete, "Image Quality and $\lambda FN/p$ for Remote Sensing Systems," *Optical Engineering*, 38(7), 1999, pp. 1229–1240.
43. Philip C. D. Hobbs, "Photodiode Front Ends: The Real Story," *Optics and Photonics News*, April 2001, pp. 44–47.

CHAPTER 7
Optomechanical Design

Despite the "lessons learned" from the Hubble Space Telescope, decisions made during the development of optical systems continue to be misguided. A more recent example was covered in depth in the *New York Times* (Philip Taubman, "Death of a Spy Satellite Program," 11 November 2007), where it was disclosed that a major spy satellite program was unable to deliver the performance promised by the aerospace contractor. After $5 billion was spent, the optical system was still not ready for launch. After more delays, what *was* eventually launched was an investigation into the root cause of the problem; an aerospace executive concluded: "There were a lot of bright young people involved in developing the concept, but they hadn't been involved in manufacturing sophisticated optical systems. It soon became clear the system could not be built."

Aside from the obvious question of where the bright *experienced* people were when the concept was developed, it was reasonable to expect that the young optical engineers had some basic knowledge of optical system manufacturing—preferably with direct, hands-on experience. It seems, however, that their universities lacked such a course; this does not justify the failure so much as it puts the onus on management to recognize what skills and educational background their personnel do, in fact, possess. Absent such recognition, managers and technical leads may prove ill-suited to the task, a gamble that neither the company nor the taxpayers can ever win.

The purpose of this chapter is to provide some background for determining whether or not a particular optical system can be built. Optomechanical engineering has traditionally focused on the mechanical aspect of optical systems, such as lens mounts and alignment mechanisms.[1-6] Although both optical and mechanical engineering skills are required to build hardware, the emphasis in this chapter is on the "opto" part of optomechanical design.

This aspect of optomechanical engineering starts with the fabrication of the lenses, mirrors, and so forth that make up the system (see Fig. 7.1). These building blocks are collectively known as

FIGURE 7.1 Fabrication of one of the 8.4-meter primary mirrors used in the Large Binocular Telescope. (*Credit: David Anderson and Jim Burge, "Optical Fabrication," in* Handbook of Optical Engineering, *Marcel Dekker, www.dekker.com, 2001.*)

"elements" or "components," and an inability to manufacture them with the required quality results in a system that will not meet specs. For example, a diffraction-limited, f/1 mirror of 2-meter diameter is not straightforward to build, and fabrication technology may be the limiting factor in its optical performance. Yet if the mirror needs only to collect photons and need not generate good image quality, then a fast and large-diameter optic becomes more feasible. The ability to meet specs thus depends in part on how well the image quality requirements of the system match up with the optical quality available from the fabrication process.

Fabrication starts with an optical prescription provided by the lens designer (Table 3.6), from which optomechanical drawings are created for each component. These drawings are then used by an optical shop to grind and polish the lens into the shape specified on the drawing. Fabrication quality is measured by a number of metrics. For imagers, wavefront error and MTF are the most important, and a poorly fabricated lens will increase the blur size to well beyond the diffraction limit. As reviewed in Sec. 7.1, other important parameters that affect optical performance include variations in the refractive index, optical transmission, surface roughness, and dimensional control.

After fabrication, optical elements are assembled into an instrument or system; this requires that they be accurately aligned with respect to each other, again to prevent the blur from increasing.

The distance between the primary and secondary mirrors in the Hubble, for example, was approximately 5 meters—to a tolerance of ±1.5 microns, or 1 part in 1.5 million! Typical alignment problems for an optical system include: (1) the tilt angle with respect to either of the two axes perpendicular to the optical axis; (2) decenter, or displacements along either of those two axes; (3) despace, or the incorrect placement of one element with respect to another as measured along the optical axis; and (4) defocus, a specific type of despace pertaining to the placement of the detector or FPA with respect to the imaging optics as measured along the optical axis. These assembly alignments and their effects on optical performance are reviewed in Sec. 7.2.

Once aligned, an optical system will usually remain so on a vibration-isolation table in a climate-controlled building; however, the system may become useless once it is moved out into the field, where environmental effects—including temperature changes, shock, vibration, humidity, ocean spray, vacuum, and radiation—become important. The two most common problems are temperature and vibration, either of which can quickly destroy the instrument's ability to meet a diffraction-limited WFE spec. Even systems that are precisely fabricated and accurately aligned can fail to meet blur-size requirements when, for example, the temperature drops below freezing or the instrument has been bouncing around in the back of a truck during a bumpy off-road trek through the Colorado mountains.

The effects of temperature on blur size are reviewed in Sec. 7.3, which describes the change with temperature in element geometry, refractive index, element-to-element spacing, and alignment. Of particular importance are design techniques for minimizing the effects of such temperature changes, including thermal management (Sec. 7.3.1), the use of materials with small thermal expansion (Sec. 7.3.2), and athermalized lenses (Sec. 7.3.3).

Section 7.4 reviews the effects of mechanical forces ("loads") on blur size. Loads can be classified as static (due to gravity, e.g.) or dynamic (due to shock and vibrations); in either case, surfaces can distort and structures bend, changing not only the shape and WFE of each element but also the element-to-element alignments. Structural design aims to control both effects through the use of appropriate shapes, sizes, materials selection, and so forth.

Given that blur size is affected by fabrication processes, alignments, and thermal and structural design, software has been developed that identifies when optical, thermal, and structural parameters can be traded. These integrated design tools, which are usually classified under the label of "structural, thermal, and optical (STOP) analysis," are reviewed in Sec. 7.5. An important output from this analysis is the system WFE budget, which summarizes the effects of fabrication, element alignment, and other factors on system

WFE. Even more important are the possible cost savings that can be derived from the use of such collaborative engineering on complex projects.

The most critical factors determining whether or not an optical system can be built are the fabrication and assembly *tolerances*. These are the acceptable variations of the fabrication and assembly parameters; an f/2 optic is relatively straightforward to fabricate, for example, whereas an f/2.0756 design represents an unacceptably "tight" specification.

Tolerances are determined by how sensitive performance is to small changes in system-level parameters such as f/#. This sensitivity flows down to component-level tolerances through the dependence of f/# (for instance) on aperture and effective focal length; EFL, in turn, depends on the refractive indices, surface curvatures, and thickness and spacings of the lenses in the system. An EFL that is extremely sensitive to changes in index, for example, is considered difficult to fabricate; similarly, a lens-to-lens spacing that is highly sensitive to misalignments is difficult to assemble. Moreover, tight assembly tolerances imply that the optics are also extremely sensitive to vibration and changes in temperature.

Optical tolerances are calculated by the lens designer, whose responsibility it is to find a design that does not entail tight component tolerances.[7] The tolerances must be determined in conjunction with the optical system engineer, mechanical engineer, thermal analyst, structural analyst, and fabrication engineer to determine whether or not the system can even be built (is "producible"). Tolerances, which are quantified in Secs. 7.1 and 7.2, are typically classified as standard ("loose"), precision, and high-precision ("tight"). Volume quantities of low-cost, commercial off-the-shelf (COTS) lenses are fabricated to standard tolerances; at the other extreme, "one-off" state-of-the-art designs usually involve high-precision (and thus high-cost) tolerances. Looser tolerances are generally more producible up to the manufacturer's standard tolerance level, at which point looser tolerances are no longer useful.

In short, better *design* performance does not necessarily translate into a better optical system. Many other factors affect overall performance, including fabrication, alignment, and environmental influences such as temperature and vibration. The *sensitivity* to fabrication tolerances and misalignments is the key determinant of cost and producibility. The best design is therefore one that has loose tolerances in all areas; those for component manufacturing and fabrication are reviewed first.

7.1 Fabrication

The polishing error that led to the spherical aberration in the Hubble Space Telescope was only 2 microns on the rim of the 2.4-m-diameter primary mirror. By conventional mechanical engineering standards, this is an extremely small error and illustrates the wavelength-level precision to which optical systems must be manufactured. This includes the fabrication of all types of optical elements (lenses, mirrors, beam splitters, windows, prisms, optical filters, and polarization components) over a range of sizes from about 0.1 mm up to the 8.4-m mirror fabricated at the University of Arizona for the Large Binocular Telescope (Fig. 7.1).

The fabrication of standard-size optics (10–100 mm in diameter) begins with the rough shaping of a block of glass (BK7, SF11, etc.) or crystalline material (silicon, germanium, etc.). After machining ("generating") the surfaces roughly to shape, the resulting blank is mounted ("blocked") with other blanks for additional grinding with finer and finer abrasives in preparation for final polishing.[8–10] Along with material quality, this polishing is a critical factor in determining the image quality produced by the lens.

Lenses can be polished to spherical, cylindrical, or aspheric shapes. Beam splitters, windows, prisms, filters, and polarization components usually have flat surfaces. Mirrors can be flat, spherical, or aspheric—including parabolic, elliptical, hyperbolic, or polynomial. The focus of this section is on the final polishing of standard-size optics with spherical and flat surfaces. The asymmetry of aspheres makes them difficult to polish, although diamond turning of such surfaces is common for MWIR and LWIR materials. For further details on aspheres, see Anderson and Burge,[10] Goodwin and Wyant,[11] Malacara,[12,13] and Karow.[14]

In its most basic form, the fabrication process must be able to produce a lens of the required focal length and wavefront error. As we have seen, the requirements are based on the optical prescription; for example, the focal length depends on the index of refraction, center thickness, and surface radii. Unfortunately, it is easy to fabricate a lens whose refractive index, thickness, and radii are not *exactly* what the prescription specifies.

It is therefore necessary for the optical designer to specify an acceptable *range* of parameters, and this range is known as a *tolerance*. Standard-size COTS lenses, for example, typically have a tolerance on focal length of ±0.5 percent; for a nominal focal length of 100 mm this is equivalent to a tolerance of ±0.5 mm, so the lens received from the vendor will, at least 99 percent of the time, have a focal length ranging anywhere from 99.5 mm to 100.5 mm. A larger ("looser") tolerance is possible, and it is the responsibility of the optical engineer to determine whether that would be acceptable.

If the lens is not a COTS component but rather a custom part being fabricated under contract, then the optical shop needs to know what tolerances are acceptable. This sometimes takes the form of a tolerance on focal length but can also flow down into specifications on the index, thickness, and radii that determine the focal length.

These factors are "toleranced" by the lens designer using a tool known as a sensitivity (or "change") table,[7] which is developed into a tolerance budget (see Example 7.1) and communicated to the shop on component drawings for each system element (Fig. 7.2).

The most critical factors determining whether or not an optical system can be built at reasonable cost are the fabrication and assembly tolerances. A design that requires high-precision tolerances for every component will be much more expensive than one requiring only standard tolerances. Systems requiring tolerances that are tighter even than high precision may be possible for a hand-crafted, "one-off" device but are probably not producible in quantity. Designs with still tighter tolerances cannot be built with current, state-of-the-art technology; they require the development of new fabrication techniques.

Figure 7.2 shows that there are requirements on other parameters—in addition to those on index, center thickness, and surface radii—over which the shop has control. As reviewed in Sec. 7.1.1, these include a number of refractive-index properties that affect image quality. Of particular importance are how closely the polished surfaces match a sphere (Sec. 7.1.4) and the angular error (wedge) between surfaces (Sec. 7.1.8). Good surface finish and quality also affect the image SNR through control of scatter (Secs. 7.1.5–7.1.6).

This section reviews the tolerances expected for each of these parameters for standard, precision, and high-precision fabrication processes. There can be large cost differences between them, so it is

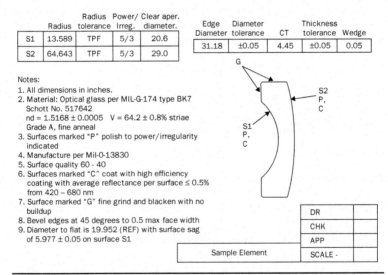

Figure 7.2 Conventional component drawing used by optical vendors to fabricate lenses. Drawings using the ISO 10110 standard are also common. (Credit: R. E. Fischer, B. Tadic-Galeb, and P. R. Yoder, Optical System Design, McGraw-Hill, www.mcgraw-hill.com, 2008.)

perfectly acceptable to question the lens designer's choice of index grade (for example), which is sometimes selected without considering cost or availability; revisiting the design with looser tolerances can result in significant cost savings. That being said, a specialized, high-precision system with challenging requirements is not the place to nickel-and-dime on tolerances. It is when the potential cost savings are considerable—as they often are after another look at the overall design—that such redesigns are worthwhile.

The fabrication tolerances reviewed in the following sections are organized according to (1) material properties of index of refraction and internal transmission; (2) surface properties that determine how well a component surface matches the radius of curvature specified in the optical prescription, its shape in comparison with spherical, the surface finish, and surface quality; and (3) the overall geometrical properties of center thickness, wedge, and clear aperture.

7.1.1 Index of Refraction

Except for metallic mirrors, optical components start as a block of glass or crystalline material whose refractive index is part of what determines its performance (in terms of focal length and wavefront error). The glass can be BK7, SF11, or one of the many others available from vendors such as Schott, Heraeus, Hoya, and O'Hara; the crystals include germanium, silicon, zinc selenide, and others. The key factor in a producible design at this stage is material availability; many materials are listed in vendor catalogs, but only a fraction are available within a reasonable time frame.

With any material, the index of refraction is not an exact number and has a tolerance associated with it. For example, the average index for Schott product number 517642 at a wavelength of $\lambda = 587.6$ nm is $n_d = 1.5168 \pm 0.001$ for standard-grade glass. Table 7.1 shows that

Property	Standard	Precision	High-Precision
Index of refraction	±0.001	±0.0005 (Grade 3)	±0.0002 (Grade 1)
Homogeneity	±10^{-4}	±5 × 10^{-6} (H2)	±10^{-6} (H4)
V-number	±0.8%	±0.5% (Grade 3)	±0.2% (Grade 1)
Birefringence	10 nm/cm	6 nm/cm	4 nm/cm
Striae	30 nm (Grade C)	15 nm (Grade B)	< 15 nm
Bubbles/Inclusions	0.5 mm² (Class B3)	0.1 mm² (Class B1)	0.03 mm² (Class B0)

TABLE 7.1 Standard, precision, and high-precision tolerances associated with various index-of-refraction properties for glasses.[10] For high-index glasses (i.e., with $n_d > 1.83$), the tolerances on refractive index are doubled.[30]

higher-precision grades are also available for glasses, with tolerances that are 2× and 5× smaller than the standard grade.

In addition to the tolerance on average index, there are also small index variations throughout a volume of optical material. Known as *index inhomogeneities*, these spatial variations result in wavefront error across a lens that are similar to the atmospheric distortions described in Sec. 3.3. Variations throughout the volume of large blocks of glass that come out of a melt furnace are $\delta n \approx \pm 10^{-4}$. Measured in radians, the resulting wavefront (phase) error is $2\pi\, \delta n(t/\lambda)$; to obtain the WFE in cycles or waves, this value is divided by 2π radians per cycle; thus,

$$\text{WFE} = \delta n \frac{t}{\lambda} \quad \text{[waves PV]} \tag{7.1}$$

where PV denotes "peak to valley". At a wavelength of 0.5 μm, an index variation $\delta n = 2 \times 10^{-4}$ (PV), and an element thickness $t = 10$ mm, the resulting WFE = 4 waves (4λ PV), a very large fabrication error for precision optics designed to λ/4 or better. Fortunately, the inhomogeneity of smaller pieces of glass that are cut out of the melt to fabricate standard-size optics are on the order of 10^{-5} to 10^{-6}, which results in a WFE that is smaller by one or two orders of magnitude.

Achromats rely on dispersion of the refractive index to produce a color-corrected optic. As reviewed in Chap. 3, this dispersion is governed by the V-number, which also has a tolerance associated with it. The standard tolerance is ±0.8 percent, which gives Schott product number 517642 a V/# of 64.2 ± 0.51. Should this tolerance result in performance variations outside the acceptable range, smaller tolerances are also available.

Birefringence ("double refraction") occurs naturally in some materials; in others, it is seen when the material is stressed. It is observed as a focal length that's different for each of the two principal polarizations, a result of their different refractive indices (the "ordinary" and "extraordinary"). Many materials are not birefringent; however, calcite and crystalline quartz have inherent birefringence and so are not usually used except for polarization components. Plastics are highly birefringent when stressed, and a few materials (e.g., SF57) have a small stress-induced birefringence.

Stress can result from external forces, but there are also internal stresses that are "locked in" to the material during the cooling process in manufacture. It is these internal stresses that are specified to the material supplier or fabrication facility. The spec is given as the stress-induced index difference δn_b, which is measured in units of nm/cm. A typical value for δn_b is 10^{-6}, which can be written as $(10 \times 10^{-9}\text{ m})/(10^{-2}\text{ m}) = 10$ nm per cm (nm/cm); this is an appropriate spec for visual instruments such as photographic and microscopic instruments. The highest precision is used for components such as polarizers and interferometers that require very low birefringence. Glasses with $\delta n_b \approx 4$ nm/cm are available that have been precision annealed to remove the locked-in stresses for such applications.

Another index-of-refraction property, known as *striae*, must also be specified. Striae (striations) are small, localized regions of index inhomogeneities that may be caused by temperature gradients that occur as the molten glass from the melt furnace cools off. There are various grades for classifying striae, depending on their visibility and orientation with respect to the surface of the element; striae that are parallel to the direction of propagation impart the most WFE (Fig. 7.3). For a material that meets the high-precision grade, the effects on WFE are small: parallel striae values of $\delta n = 10^{-6}$ produce a WFE of $t\delta n = 0.01$ m × $10^{-6} = 10$ nm ($\lambda/50$ at $\lambda = 0.5$ μm) in a 10-mm thick optic. Thinner elements have proportionally less WFE, which allows for the use of a less expensive, coarser-grade striae spec.

Finally, optical materials can also have highly localized regions, such as air bubbles and inclusions, with large index differences. These can sometimes be seen as shadows when found in optics such as field flatteners that are located near an image (see Fig. 3.25). Table 7.1 lists the acceptable size of these imperfections for various material classes, where it understood that the total allowable area of the bubbles and inclusions (in units of mm²) is per 100 cm³ of material volume. Only smaller imperfections are acceptable for applications requiring very low scatter and for high-power laser systems in which absorption at metallic inclusions creates heating and fracture of the optic.

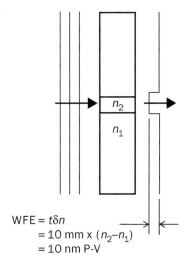

FIGURE 7.3 Effects of high-index striae on the wavefront error for an optic with a high-precision striae spec. The index difference may not extend through the entire optic, in which case the WFE is smaller.

WFE = $t\delta n$
= 10 mm × $(n_2 - n_1)$
= 10 nm P-V

7.1.2 Internal Transmission

As seen in Chap. 4, the internal transmission (aka transmittance) is an inherent material property that depends on the material's purity. The transmittance is usually not specified on a component drawing unless the optic is an absorption filter or other optical component with critical transmission properties. High-power laser systems also specify the minimum transmission for all optics in the system, where excess absorption can result in heating and fracture of the components.

When not specifically given ("called out"), the internal transmittance is assumed to meet the specs listed in a vendor's material data sheet, such as that shown in Fig. 3.31. When specified, the internal transmission should be given at all wavelengths of interest for which the vendor has data and for a given material thickness—typically 10 mm or 25 mm. Standard transmission values can sometimes be improved upon by specifying high-transmission (HT) and HHT grades when these are available for a specific material. The overall transmission, including surface reflections, is then specified separately if optical coatings are used.

7.1.3 Surface Curvature

Starting with known index and transmission properties, a material can be fabricated into lenses, prisms, and other components. The process entails shaping the element surfaces with the flat, spherical curvature, or aspheric shape given in the optical prescription. For flat and spherical surfaces, the curvature determines the refractive power of the surface; for aspheric surfaces, the surface power is not constant and varies across the aperture. In either case, a surface cannot be polished into a perfect shape; there are tolerances associated with the curvature that affect the optic's refractive (or reflective) power and hence its focal length.

The traditional method for tolerancing the power of spherical surfaces is by comparison with a precise reference optic known as a *test plate*; more recently, interferometers are also used to make final measurements.[11] The transparent test plates are fabricated to very high precision with opposite curvature—for example, a concave test plate is used to measure a convex surface—so that a correctly fabricated element fits into the test plate with only a small error (Fig. 7.4). The test-plate fit (TPF) thus measures how closely the radius of curvature of the fabricated surface matches the design ideal when the test plate radius was used to create the optical prescription.

The error in curvature is measured as the gap between the surface being polished and the test plate. For spherical surfaces, it is observed as contour fringes (Newton's rings) when looking at how well the surface fits the test plate (Fig. 7.5); in reflection, each bright or dark ring represents a distance of one-quarter of a wavelength (peak-to-valley)

Optomechanical Design 351

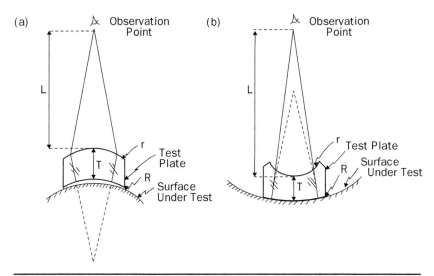

FIGURE 7.4 The curvature of an optical surface is commonly measured by comparison with a known reference called a test plate. (*Credit: D. Malacara*, Optical Shop Testing, *3rd ed. Wiley, 2007.*)

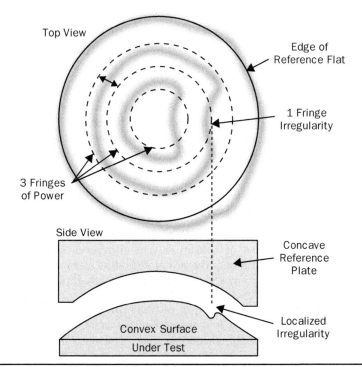

FIGURE 7.5 Interference pattern produced by an imperfect spherical surface when measured with a test plate. (*Credit: Courtesy of Edmund Optics, Inc.*)

between the surface and the test plate.[11] For example, an incorrectly polished surface may be perfectly spherical yet still deviate by $\varepsilon = 1$ μm (for instance) from the *required* curvature; this deviation is measured by the test plate as $N = 2\varepsilon/\lambda = 4$ fringes at a wavelength of 0.5 μm. These are observed as four bright and four dark rings—or lines for a flat optic—across the aperture in an alternating sequence indicative of the constructive and destructive interference that occurs as the optical path difference (OPD) between the surfaces increases or decreases in half-wavelength increments.

The wavelength of the measurement must therefore be specified, since a test-plate fit which produces 4 fringes at a wavelength of 0.5 μm produces only 2 fringes at a test wavelength of 1 μm. A typical test wavelength for optics that transmit in the visible is 632.8 nm, the red color of the helium-neon (HeNe) laser. Thus, a spec may read: "Test plate fit over the clear aperture shall be no more than 4 fringes of power at a wavelength of 633 nm." It's also necessary to relate the measurement to the *design* wavelength, because ultimately it's the optical system performance, not the measurement, that is important.

A precise mapping of the surface curvature to the test plate is often not critical; typical specs for the refractive power of a surface are on the order of 4–10 fringes (2–5 waves), a seemingly large amount for an optic with a $\lambda/10$ WFE spec (for example). As shown in Sec. 7.1.4, these smaller wavefront errors are characterized by using what is known as a *surface figure* spec; a surface curvature (or "power") spec of 4–10 fringes for the focal length is adequate for many applications.

Using a thin plano-convex lens with a focal length of 100 mm as an illustration, the nominal radius of curvature of the convex side is $R = 50$ mm for $n = 1.5$ [see Eq. (2.1)]. Variations from this radius due to fabrication tolerances have a direct impact on the focal length $[\Delta f = \Delta R/(n - 1)]$; by equating the gap for mismatched radii to the OPD $(=N\lambda)$ in terms of the number of fringes N that deviate from circular, we find that[8]

$$\Delta R \approx N\lambda \left(\frac{2R}{D}\right)^2 \quad [\text{mm}] \tag{7.2}$$

Here D is the diameter over which the fringes are counted and R is the test plate's radius of curvature. The difference between the fabricated radius and that of the test plate is thus measured directly from the number of fringes. Equation (7.2) shows that $\Delta R \approx 10$ μm for the conditions just described, so $\Delta f = 20$ μm for an f/2 lens with 4 fringes of power at a wavelength of 0.633 μm. This is not a large fabrication error, though it is magnified by index variations and errors in fabricating the second surface; however, the resulting variation of the focal length can be compensated for by focus adjustments at assembly. Figure 7.6 illustrates the fabrication tolerance

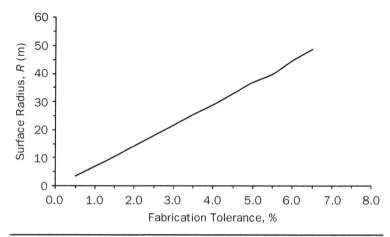

FIGURE 7.6 Larger radii are more difficult to polish accurately and so have a larger tolerance associated with their fabrication; Eq. (7.2) shows that the tolerance $\Delta R/R \propto R$. *(Credit: Data courtesy of CVI Melles-Griot.)*

$\Delta R/R$ for lenses manufactured by a volume manufacturer of optical components.

The fabrication of beam splitters, windows, mirror substrates, prisms, optical filters, and polarization components may involve the polishing of flat surfaces. In general, the fit of a planar surface on a planar reference will not produce circular fringes, depending on how close the polished surface is to being "flat." Polishing can result in a wide variety of surface shapes and fringe patterns; these include rings, straight lines, and hyperbolas as well as combinations of these (and more); see Fig. 7.7.

When there is no reference surface against which to measure fringes of power, as in the fabrication of the test plates themselves, the radius and tolerance of the surface are specified. This may be necessary when the lens designer has included a surface in the prescription that does not match the test plates that the fabrication shop has in stock. Using available plates is always the preferred approach; however, new test plates can usually be made within a few weeks for any spherical radius and to a precision of ±0.1 percent.

7.1.4 Surface Figure

The tolerance on surface *curvature* determines how closely a spherical or flat surface fits that of a test plate; another property known as the *surface figure* specifies the magnitude of small-scale surface irregularities such as those shown in Fig. 7.5. These irregularities from the polishing process may be found in the fabrication of any surface, including flats, cylinders, spheres, and aspheres. Deviations from the ideal surface induce wavefront error (see Fig. 7.8) and

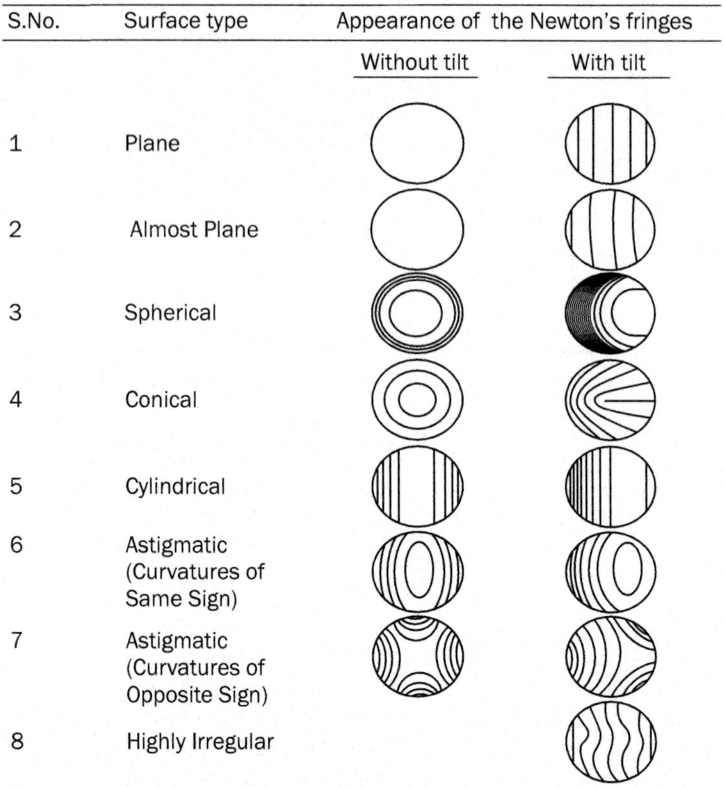

354 Chapter Seven

S.No.	Surface type	Appearance of the Newton's fringes	
		Without tilt	With tilt
1	Plane		
2	Almost Plane		
3	Spherical		
4	Conical		
5	Cylindrical		
6	Astigmatic (Curvatures of Same Sign)		
7	Astigmatic (Curvatures of Opposite Sign)		
8	Highly Irregular		

Figure 7.7 Comparison of a flat test plate with different surface types produces the fringe patterns shown. (*Credit: Daniel Malacara,* Optical Shop Testing, *3rd ed. Wiley, 2007.*)

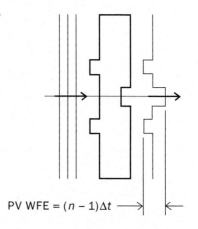

Figure 7.8 Wavefront error created by surface irregularities in a flat plate. The surface figure error (SFE) is the height or depth (Δt) of each irregularity; the PV WFE = $(n - 1)\Delta t$ for an optic in air.

PV WFE = $(n - 1)\Delta t$

associated aberrations (usually astigmatism). The surface figure WFE is in addition to the effects of design residuals and refractive-index variations.

Test-plate fits and interferometers are also used to evaluate surface figure, where the deviation (or "irregularity") from circular fringes measures the variation from a spherical surface;[11] this is shown in Fig. 7.5 as one fringe of deviation from the circular rings defining power. Note that it is difficult to measure irregularities using test plates that are a small fraction of the power spec, and a ratio of 4–5 fringes of power for every fringe of irregularity is a common measurement limitation for production optics. Interferometers have no such a limitation and are used for high-precision optics. Independent of the fringe ratio, the resulting wavefront error is WFE = $(n-1)\Delta t/\lambda$ due to a surface figure error $\Delta t/\lambda$ in units of wavelengths; in terms of fringes, the WFE is given by[8]

$$\text{WFE} = \frac{(n-1)N}{2} \quad \text{[waves PV]} \tag{7.3}$$

for a lens measured in air; the factor of 2 in the denominator is a result of using test plates, as they double the optical path difference due to surface mismatch. For $N = 1$ fringe of irregularity and a material with $n = 1.5$, this corresponds to 0.25 waves ($\lambda/4$ PV) of WFE—equal to that required for a diffraction-limited lens. A lens has two surfaces, so a typical spec on surface figure for a diffraction-limited lens in the visible might read as follows: "Test plate fit over the clear aperture shall be no more than 1/2 fringe of irregularity at a wavelength of 633 nm."

Unlike those shown in Fig. 7.8, the WFE errors on the surfaces of different lenses are generated randomly with respect to each other; hence the errors add as a root sum-of-squares (RSS) for many surfaces in a system. Specifying one-half fringe of irregularity then seems a bit conservative for volume manufacturing, given that 0.707 fringes for each of two surfaces will RSS to produce 1 fringe of WFE. However, this fails to take several factors into account: the design and refractive-index contributions to WFE, the possible existence of multiple elements in a lens assembly, and the alignment and environmental effects reviewed in Secs. 7.2–7.4. It is thus common to use irregularity specs as low as 0.05 waves ($\lambda/20$) for high-precision applications, such as interferometers or photolithography optics, and to use elements with a small enough aspect ratio (e.g., ≤ 6 : 1, diameter to thickness) that they will not be deformed during polishing to such tight tolerances.

The fabrication of mirrors with low WFE is particularly difficult because the effects of surface figure are amplified for a reflected beam. This phenomenon is analyzed by using a refractive index of $n = -1$ in Eq. (7.3) for a mirror; this gives a WFE = 1 wave for 1 fringe of irregularity, larger by a factor of 4 than that for a refractive surface with $n = 1.5$. The aspect ratio is thus critical for mirrors, as is using the correct substrate material. Zerodur is stiff enough for a $\lambda/20$ irregularity spec, but BK7 and fused silica are not (see Sec. 7.4).

Wavelength also plays a major role in the smallest WFE that can be obtained. A lens with 1 fringe of irregularity in the visible, for example, has a surface figure error on the order of 0.25 µm, and the surface must be accurately polished to within this value. In contrast, a LWIR lens with the same refractive index at a wavelength of 10 µm can have a SFE approximately of 5 µm (20 times larger) before the WFE reaches 1 fringe, a much easier tolerance to maintain during fabrication.

An exact comparison between wavelengths must also take into account the use of materials that have a different refractive index. Physically, each fringe corresponds to one-half of a wavelength of fabrication error from the ideal spherical or flat surface; Eq. (7.3) therefore implies that the WFE induced by the surface figure error Δt is $(n - 1)\Delta t/\lambda$ waves PV. A visible lens with an index $n = 1.5$ thus has an SFE of 0.25 µm associated with 1 fringe (0.25 waves) of irregularity WFE at $\lambda = 0.5$ µm; a high-index LWIR lens with $n = 3.5$ (silicon) polished to the same number of fringes has an SFE of $\Delta t = $ WFE $\times \lambda/(n - 1) = (0.25 \times 10 \text{ µm})/2.5 = 1$ µm, or 4 times larger than the visible lens. This increased tolerance at longer wavelengths allows the fabrication of faster lenses in the LWIR, where the highly curved surfaces required for f/1 optics are much easier to polish (to within 1 µm of spherical) than are visible lenses (which require 0.25-µm accuracy).

Both mirrors and lenses are also specified as meeting an irregularity spec before optical coatings are applied. Depending on the aspect ratio, coatings can distort the optical surface as they absorb moisture or change temperature. Temperature changes are particularly important: the difference in coefficients of thermal expansion between the optic and the coating can strain the optic into a "potato chip" shape, potentially increasing the SFE and WFE well beyond the system requirements.

Finally, note that off-the-shelf optics are often specified for irregularity over the entire aperture even though many elements in a system do not require this level of performance. In particular, only those lenses that are at or near a pupil (entrance, exit, or intermediate) must meet the irregularity spec over the full diameter of the lens. Other lenses, such as the eyelens shown in the afocal system in Fig. 2.26, intercept the image cone from the entrance pupil (defined by the objective) at a smaller size.

The diameter of the smaller cone on the optic has fewer fringes of irregularity than the larger pupil size. As a result, commercial optics often perform better than expected when used away from a pupil, and the cost of fabricated optics can be significantly reduced by specifying the irregularity only over the required cone size (approximately 10 mm for a 5× afocal with a 50-mm entrance pupil, for example). The spec for a fabricated optic might therefore read: "Test plate fit shall be no more than 1/2 fringe of irregularity at a wavelength of 633 nm over any 10-mm diameter over the clear aperture." Here 10 mm has been used to illustrate the concept, not as an example of how all component specs should be written.

7.1.5 Surface Finish

As the final polishing of surface figure reduces the WFE to acceptable levels, yet another surface property must simultaneously be maintained. Known as *surface finish* (or surface roughness), it is a measure of the RMS height variations across a surface (Fig. 7.9). State-of-the-art fabrication and characterization methods are able to provide extremely fine, molecular-level finishes as smooth as 2 angstroms (Å). Surface finish thus occurs on a much smaller scale than surface irregularities, so it has no direct effect on surface WFE.

However, we saw in Chap. 4 how surface finish increases the scattering of a surface, which is measured as angular scattering (bidirectional scatter distribution function) or as the sum over all angles (total integrated scattering). Most applications do not require a super-polished surface with only 2-Å microroughness; more typical specs which uniquely determined by the stray-light requirements for each system are 10 to 50 angstroms. Metals such as beryllium and

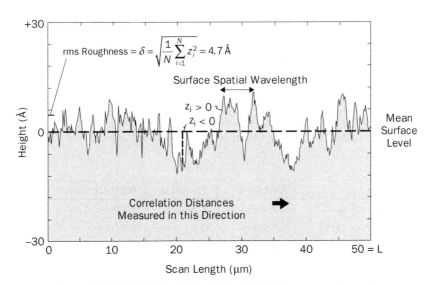

Figure 7.9 Surface finish of a polished or diamond-turned optic is measured as an RMS surface roughness. (*Credit: J. M. Bennett and L. Mattsson*, Introduction to Surface Roughness and Scattering, *Optical Society of America, www.osa.org, 1989.*)

aluminum typically cannot be polished to better than about 50 Å, so they are coated with electroless nickel when used as mirror substrates requiring better surface finish. In contrast, fused silica and ULE (ultra-low expansion) glass can be polished to very fine finishes—down to 5 Å if required.

Along with surface figure, surface finish is a major cost component because meeting these specs involves a time-consuming process of fine polishing. It is expensive to simultaneously obtain high-precision figure and finish, since the additional polishing required to give good finish often affects the figure. A recent fabrication method that avoids this trade is magnetorheological finishing (MRF), which uses a magnetic material (iron) in the polishing fluid to control figure. The method is useful for surface figures down to $\lambda/20$ as well as surface finishes to about 10 angstroms RMS.[15]

7.1.6 Surface Quality

The polishing process can leave marks on a surface that are too large for the finishing process to remove. These are the types of scratches often seen on a windshield or pair of eyeglasses, but they also include larger pits and divots known as "digs." They are mostly cosmetic defects but can be seen as shadows when found on surfaces—e.g., field flatteners and field lenses—that are located near an image. They can also be a concern in systems where stray light can reduce the SNR as well as in high-power laser systems, where enhanced electric field strength at these small-area imperfections can destroy the optic.

Scratch and dig are specified in terms of allowable sizes, spacings, and densities. In its simplest form, the scratch is measured in units of length and specified as 10 times larger than the scratch width in microns.[16] A scratch specification of 60, for example, corresponds to surface scratches that are no more than 6 μm in width.

The dig spec is also in units of length but is given as 10 times smaller than the actual diameter in microns; a dig spec of 40 thus refers to a dig with a 400-μm diameter. Both scratch and dig are usually examined visually by an experienced inspector. Table 7.2

Property	Standard	Precision	High-Precision
Surface radius (PV)	5 fringes	3 fringes	1 fringe
Surface figure (PV)	2 fringes	0.5 fringe	0.1 fringe
Surface finish (RMS)	50 Å	20 Å	10 Å
Surface quality	80/50	60/40	20/10

TABLE 7.2 Standard, precision, and high-precision tolerances associated with surface properties.[8-10] Surface radius is specified as "power"; surface figure is specified as "irregularity."

summarizes the tolerances for the surface quality (scratch/dig), as well as the surface properties of surface radius (power), surface figure (irregularity), and surface finish for standard, precision, and high-precision tolerances. Not included are midspatial frequency properties such as "quilting" or "ripple." As with the index-of-refraction tolerances in Table 7.1, the cost of high-precision tolerances is significantly higher than that of standard tolerances.

7.1.7 Center Thickness

It's possible to fabricate a lens with exactly the correct index and radii as called out in the prescription but still end up with the wrong focal length. This outcome is due to the lens thickness—shown in Eq. (2.1) as the variable CT, which is set to zero for thin lenses. For thick lenses, the distance traveled in the lens affects the height at which the rays strike the second surface, changing the focal length and WFE of the optic. Clearly, the steeper the rays and the thicker the optic, the larger the effects of thickness variations.

The thickness must therefore be toleranced, but the tolerance may refer to the center thickness, the edge thickness (ET), or somewhere in between. A common standard is the CT (Fig. 7.10), although this is unfortunately a difficult parameter to control. Typical thickness tolerances are on the order of ±200 μm for standard optics in volume production and ±50 μm for high-precision elements. This precision is not possible for softer materials, for which the normal variations in grinding pressure allow less control over the material removed during polishing.

7.1.8 Tilt/Wedge

Yet another fabrication tolerance refers to the angular relationship (tilt) between surfaces. A plane-parallel plate, for example, is specified as having surfaces that are both flat (e.g., a surface figure of $\lambda/10$) and parallel to some tolerance. If one surface is tilted with respect to the other (Fig. 7.11)—as all surfaces are, to some degree—then rays will

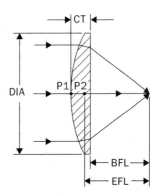

FIGURE 7.10 The center thickness (CT) is commonly used to specify a lens thickness. (*Credit: Janos Technology LLC.*)

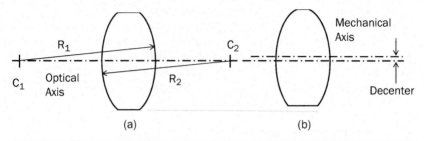

FIGURE 7.11 Lens tilt is created by an offset ("decentration") between the mechanical axis determined by the edge grinding and the optical axis defined by the centers of curvature of the surfaces. For the lens (a), the optical and mechanical axes are coincident.

deviate from their expected path. Different wavelengths will also be slightly dispersed, as even a small amount of tilt acts as a weak prism. In addition, astigmatism and coma will be imparted to converging wavefronts, a result of the asymmetric ray bending induced in the plane of the plate as compared with the bending in the perpendicular plane. It is therefore necessary to quantify the degree to which surfaces must be parallel. For elements that are flat on both sides, this tolerance is known as *wedge* or *surface tilt*; for lenses, it may also be called *centration*.

Looking first at planar elements, we see that the two sides cannot be polished perfectly parallel; hence parallel rays that pass through the element will change direction. This is the basis of a thin prism, where the refractive index and wedge angle determine the angular deviation δ:[8]

$$\delta = (n-1)\alpha \quad [\text{rad}] \tag{7.4}$$

The wedge angle α between the two surfaces is specified as a maximum for flats, where typical numbers for standard optics are on the order of 5 arcminutes. (An arcminute—sometimes written simply as a "minute"—is a very old unit still used by opticians and astronomers; it is equivalent to 1/60 of a degree, or approximately 291 µrad.) For a material with a refractive index $n = 1.5$, a wedge of 5 minutes produces a deviation $\delta = (n-1)\alpha = 0.5 \times 1.455$ mrad ≈ 0.73 mrad—enough to produce a linear displacement of $d = \delta \times L = 730$ µm when propagated over a distance $L = 1$ meter.

For lenses, the optical axis defined by the center of curvature of the surfaces will be offset from the mechanical axis defined by the edge grinding so that the outer diameter is not concentric with the diameter defined by the optical axis. This is shown in Fig. 7.11 for a biconvex lens, where the offset is equivalent to inserting a wedge between the curved and flat surfaces (Fig. 7.12); for small offsets, the

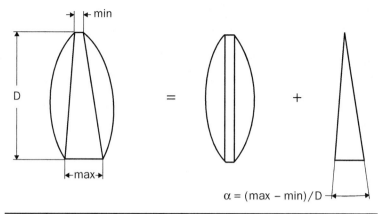

FIGURE 7.12 Element tilt deviates rays in the same manner as a prism. (Credit: Jim Burge, University of Arizona, College of Optical Sciences, *www.optics.arizona.edu*.)

Property	Standard	Precision	High-Precision
Center thickness	±200 μm (±0.008″)	±100 μm (±0.004″)	±50 μm (±0.002″)
Lens tilt/Wedge	5 arcmin	1 arcmin	0.15 arcmin
Centration	200 μm	50 μm	10 μm
Clear aperture	90% of OD	90% of OD	95% of OD

TABLE 7.3 Standard, precision, and high-precision tolerances associated with geometrical fabrication properties for standard-size optics.[8-10] (For reference, 25.4 μm is approximately equal to a very old unit of length known as a "mil," or 0.001 inches; OD denotes "outer diameter.")

wedge angle α is approximately equal to the edge thickness difference (ETD) divided by the lens diameter. Table 7.3 shows typical offsets ("decentration") for various levels of lens precision, although the acceptable level must be determined not by rule of thumb but rather by including the effects of wedge in the tolerancing change table as a tilt angle.

Wedge and centration specs take on different forms in a component drawing, depending on the type and use of the component. A centration spec, for example, is generally not required for inexpensive lenses such as those used in cell-phone cameras. In the design of prisms, however, specific (and precise) values of wedge are intentional. Similarly, for flat elements used in laser systems, a small amount of tilt is often introduced in order to prevent Fresnel reflections from the two surfaces from interfering.[17]

In addition to offset-induced surface tilt, the optical axis will also be tilted with respect to the mechanical axis. These tilts are difficult to quantify on a heuristic basis but, when combined with decentration, are critical specs in light of Kingslake's comment that "[surface] tilt does more damage to an image than any other manufacturing error."[18] In addition, the assembly of lens elements into mechanical housings with tolerances on their diameters results in *assembly tilt*, a topic discussed in Sec. 7.2.1.

7.1.9 Clear Aperture

Not all of the surface and element tolerances reviewed so far need to be satisfied over the entire size of the optic. We have seen, for example, that power and irregularity specs are given over a "clear aperture" (CA). Figure 7.13 shows that the CA is smaller than the mechanical diameter; for standard size optics—that is, of about 5–100 mm in diameter—a typical value for the CA is 90 percent of the element outer diameter (OD). One area for potential cost savings is that the CA can be different on each side of a lens (see Fig. 7.2), depending on how the chief and marginal rays work their way through the element. Table 7.3 summarizes the CA and other tolerances associated with geometrical fabrication properties.

Example 7.1. In Chap. 3 we saw that a diffraction-limited lens has no more than $\lambda/4$ of peak-to-valley WFE. Example 3.8 illustrates that obtaining such a lens requires less than quarter-wavelength performance from the design given the

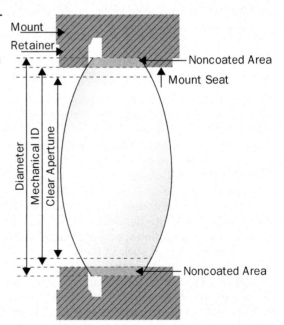

Figure 7.13 The clear aperture of an optical element, over which fabrication and coating quality are specified, is typically about 90 percent of its outer diameter. (*Credit: Courtesy of Edmund Optics, Inc.*)

other contributors to WFE (e.g., fabrication, alignment, and thermal stability). In this example we look at the tolerance levels required for only the design and fabrication contributors to a diffraction-limited singlet.

The starting point for such an analysis is a prescription sensitivity (change) table, which summarizes the effects of fabrication tolerances on such performance metrics as off-axis aberrations.[8] The table is then used to generate a tolerance budget, which summarizes the net effects of these changes on the WFE. An example is shown in Table 7.4, where the design residual has been determined by the lens designer to be 0.2 waves (PV) at the edge of the FOV, which is possible for a high-f/#, narrow-field singlet. Because the design and fabrication wavefront errors RSS to obtain the total [i.e., $\lambda/4 = (WFE_d^2 + WFE_f^2)^{1/2}$], the fabrication contribution to the total WFE must be less than $[(\lambda/4)^2 - (\lambda/5)^2]^{1/2} = 0.15$ waves PV (approximately $\lambda/7$) to obtain a diffraction-limited lens. Is this possible with the standard tolerances given in this chapter?

These tolerances are listed in the second column in Table 7.4: 5 fringes of power, 1 fringe of irregularity, an index variation of ±0.001, surface tilts of 5 arcmin, and a thickness variation of ±0.2 mm. The effects of the irregularity tolerance can be calculated from Eq. (7.3), where $N = 1$ fringe corresponds to 0.25 waves for a lens with an index $n = 1.5$; the effects of the other tolerances on WFE must be calculated by the lens designer for each specific design.

The results in Table 7.4 show that there are four terms in the second column (irregularity and tilt of each surface) whose WFE are each $\lambda/4$ and so together produce a WFE of $\lambda/2$. Hence we can see by inspection that the standard tolerances will not meet the diffraction-limited requirement, even if the design WFE could somehow be reduced to zero.

Property	Standard Tolerance	Standard WFE (PV)	Precision Tolerance	Precision WFE (PV)	Comment
Design WFE	—	0.20λ	—	0.20λ	f/10, FOV = 1°
Radius 1	5 fringes	0.05	1 fringe	0.01	Power
Radius 2	5 fringes	0.05	1 fringe	0.01	Power
Surface 1	1 fringe	0.25	0.2 fringe	0.05	Irregularity
Surface 2	1 fringe	0.25	0.2 fringe	0.05	Irregularity
Index	±0.001	0.01	±0.001	0.01	
Surface 1	5 arcmin	0.25	1 arcmin	0.05	Surface tilt
Surface 2	5 arcmin	0.25	1 arcmin	0.05	Surface tilt
Thickness	0.2 mm	0.10	0.2 mm	0.10	
RSS Total		0.55λ		0.246λ	

TABLE 7.4 Tolerance budget for a narrow-field singlet based on standard and precision surface tolerances. To illustrate the influence of the largest contributors, the index and thickness tolerances have not been changed for the precision tolerance. For designs with multiple lenses, the power, irregularity, and tilt tolerances would each be merged into row entries summarizing the RSS WFE for all of the elements.

Meeting the requirement thus depends on the ability to reduce these four dominant terms, which are shown in the fourth column in Table 7.4 for precision tolerances. The power tolerance is also modified in order to maintain a 5 : 1 ratio between the number of power and irregularity fringes; the index and thickness are kept the same to illustrate the dependence of WFE on the dominant errors. The RSS sum of 0.246 waves is just within the $\lambda/4$ requirement, with little margin for mistakes, though it may be possible to reduce the index and thickness tolerances to increase that margin.

In summary, the ability to manufacture an element (its "producibility") depends on the requirements and the fabrication tolerances needed to meet them. If a design has an inherently large WFE—as may a fast, wide-angle singlet, for example—then the requirements for a diffraction-limited lens cannot be met no matter how good the fabrication quality and how tight the tolerances. Example 7.1 shows that it is possible to manufacture a diffraction-limited singlet without using high-precision tolerances, but this is no guarantee that the system performance will also be diffraction limited. The next step along that path, described in the following section, takes into account the assembly tolerances that result in tilt, decenter, and despace of the elements.

7.2 Alignment

Even when lenses are designed and fabricated with diffraction-limited tolerances or are purchased as off-the-shelf components of sufficient quality, their use can still result in a system that is not diffraction limited. The image degradation can be due to misalignments of the individual elements at assembly, temperature changes, and/or mechanical motion due to structural loads. These effects all increase an optic's WFE, which limits the performance (e.g., blur size) that can be obtained.

In addition to fabrication tolerances, the most critical factors in whether or not an optical system can be built at reasonable cost are the alignment tolerances. When assembling an optical system consisting of multiple lens elements, unavoidable misalignments between elements include angular or linear errors resulting from poor assembly, temperature changes, or structural vibrations. The angular errors are known as alignment tilt; linear errors are classified as either "in the plane of the element" (decenter), or "along the optical axis" (defocus and despace).

As with the fabrication of optical elements, there are tolerances on each of these misalignments and the tolerances determine the design's producibility. In this case, producibility refers to the ability of a machine shop to fabricate *mechanical* parts that maintain the required tilt, decenter, and defocus/despace. The allowable tolerances are determined by the lens designer and are based on system requirements such as blur size, MTF, ensquared energy, and so on. In conjunction with the shop, the optomechanical engineer must then determine whether parts such as lens housings and spacers can be fabricated and assembled to within these tolerances; if not, then alignment adjustments will be required during assembly. The balance of this section describes assembly alignment in more detail. The effects of thermal and structural loads will be reviewed in Secs. 7.3 and 7.4, respectively.

7.2.1 Tilt

Example 3.3 illustrates that a tilted, plane-parallel plate in a converging or diverging beam creates astigmatism in the image, a result of the asymmetric ray-bending induced in the plane of the plate as compared with the bending in the perpendicular plane.[8] The same phenomenon occurs when lens and mirror elements are tilted, and the tolerances on such tilt must be established by the lens designer using criteria such as the modulation transfer function (MTF) or ensquared energy. Tolerances on tilt angle between lenses are, in general, fairly tight; this is because assembly tilt can be the largest contributor to alignment WFE,[18] just as surface tilt is often the primary contributor to fabrication error.

Figure 7.14 shows a tilted and decentered lens, where the tilt is about the axis in and out of the page; for tilt about both axes, the term *tip/tilt* is sometimes used. In either case, we have seen that fabrication tilt results from a misalignment of the optical axis with the mechanical axis of the element (Sec. 7.1.8); assembly misalignment due to excess clearance between the lens diameter and its housing, when coupled with wedge in the spacers between elements, results in element tilt. A spacer that is machined to be thicker on one side than the other by 0.127 mm (0.005 inches) over a 25-mm diameter, for example, has a tilt angle of 5 milliradians; this is a very large tilt, approximately 3.5 times larger than the standard surface tolerance given in Table 7.3. Reducing assembly tilt thus requires the optomechanical engineer to work closely with the lens designer to ensure that the tilt specs are met—not only on the mechanical drawings released to the machine shop but also on the parts that are received.

Physically, the incidence angles on the lens change when tilt is introduced, and this affects both the geometrical and aberration properties of the lens. The surfaces that intercept rays with the largest angle of incidence, such as those found in high-speed (low-f/#) or wide-field lenses, are the most sensitive. Geometrically, tilt moves one part of a lens toward an object and the other part away from it. Even with a single-element lens, the part of a finite-conjugate object that's closer to the lens (has a smaller object distance) is imaged slightly farther away, as shown in Fig. 7.15. Known as the Scheimpflug condition, this angular relation between finite-conjugate object and image can result in a blurred image (Sec. 7.2.3) unless the focal plane

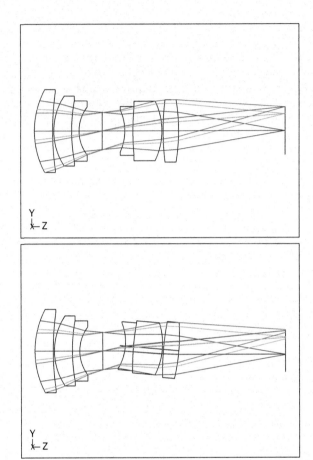

Figure 7.14 Tilt and decenter of the final three elements in a double Gauss lens, illustrating the large blur size for off-axis rays (bottom) in comparison with a lens whose elements have not been misaligned (top).

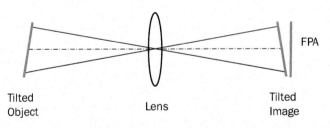

Figure 7.15 A finite-conjugate object that is tilted with respect to the lens results in a tilted image; alternatively, one can view the lens as being tilted with respect to the object. In either case, the image may not meet focus requirements over most of the FPA.

is also tilted. This is not a practical solution because the direction of tilt is usually unknown, so the tilted lens results in a tilted image.

Tilt between elements also results in astigmatism and coma, increasing the blur size to larger than diffraction limited.[18] As we saw with plane-parallel plates, thicker elements are more sensitive to tilt because of the longer distance that rays travel through the element. Example 3.3 also reveals an index dependence to the astigmatism that is evidenced by a reduced tilt sensitivity for plane-parallel plates with a larger index. Mirrors do not have thickness or chromatic sensitivities; however, they are twice as sensitive to tilt because a tilt angle θ changes the reflected rays by a factor of 2θ.[8]

Tilt is sometimes introduced intentionally—for example, in the Mersenne telescope design shown in Fig. 4.30, where the off-axis secondary mirror is used to prevent obscuration and to reduce off-axis scattering into the FOV. The tilt tolerances for such designs are typically tighter than for an on-axis design, given the larger angles of incidence on each mirror surface.

There are also situations where it is not possible to keep the WFE under control with alignment tolerances or to maintain the required alignment under thermal and structural loads. In such cases, active tilt-adjustment mechanisms are required. For instance, the segmented mirrors of the 6.6-m primary mirror in the James Webb Space Telescope require tilt adjustments to achieve interferometrically accurate alignment of the segments. In addition to angular tilt adjustments about all three axes, each segment also has three linear adjustments to ensure the correct phasing: two adjustments for decenter and a third for focus.

7.2.2 Decenter

We have seen that the fabrication error known as wedge can result from the decenter between the optical and mechanical axes of a lens. In addition to being a fabrication tolerance, decenter is also an alignment tolerance; in this case it describes the parallel offset between the optical axes of different elements when they are assembled. As illustrated in Fig. 7.11, these axes are defined by the centers of curvature of each lens. The offset between axes can be a result of element wedge or of small clearances between the lenses and the mounts that retain them. The offset can be parallel to both the x and y axes; thus decenter is also known as xy alignment with reference to the relative placement of the optic in the xy plane perpendicular to the nominal optical axis z.

As with tilt, the asymmetry induced by decenter results in coma and astigmatism in the image; on-axis coma is a typical symptom.[19] The acceptable decenter is determined by the lens designer, where the steeper ray angles associated with fast, wide-field optics again having tighter tolerances. An important exception is the decenter for flat surfaces, which has no effect on WFE and image quality because offsets do not affect the incident angles on these surfaces. In contrast, curved surfaces are affected by decenter, with off-axis elements being more so owing to the initial asymmetry of the ray-intercept angles. Tolerances for both on- and off-axis elements are generally not as tight as they are for tilt; nonetheless, they must be specified by the lens designer in order to meet such imaging requirements as blur size, MTF, Strehl ratio, and so forth.

Alignment decenter determines the allowable mechanical tolerances when fabricating lens and housing diameters. When the clearance between the lens diameter and the inner diameter of the housing is too large, for example, the resulting decenter will degrade image quality; a standard tolerance (and thus decenter spec) for this clearance is on the order of 0.1 mm (~ 0.004 inches). Another tolerance that creates decenter is the lack of concentricity between different lens assemblies, which results when the subassemblies are attached together with flanges or threads; this tolerance can be alleviated by mounting the assemblies (or *lens cells*) in the same housing (Fig. 7.16). The tolerance "stack-up"—of each lens in its lens cell and the lens cells in the housing—still contribute to decenter, though this is minimized by the use of a compliant bond that allows decenter adjustments during assembly and by the ability to hold tighter tolerances on the diameter of the lens cell than on the lens itself.

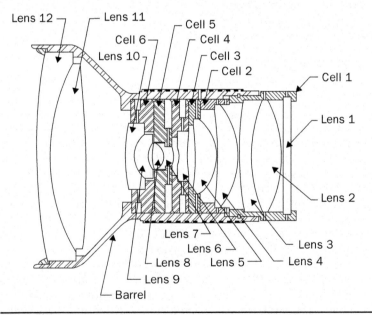

FIGURE 7.16 Decenter can be minimized in complex assemblies by using lenses bonded in lens cells with compliant bonding, and press fit in a common housing. (*Credit: Robert E. Fischer,* Proceedings of the SPIE, *vol. 1533.*)

7.2.3 Defocus and Despace

At some point, rays propagating through an optical system will exit the last lens and come to focus on a single-pixel detector or FPA. If the detector is not placed at the correct distance from the lens then images will be blurred, and all the effort that has gone into designing and fabricating a diffraction-limited lens will not produce the crisp image expected. Fortunately, the detector need not be located an exact distance from the lens, and the tolerance with respect to this distance is known as depth of focus (DOF) or allowable *defocus* (i.e., the distance over which the blur size remains "acceptable"). There are two criteria for depth of focus: one based on geometrical blur and one based on diffraction.

For defocus based on the geometrical blur of a non–diffraction-limited system, similar triangles (the dashed lines in Fig. 7.17) for the focused beam show that $\pm \Delta z / B = f / D$, where Δz is the defocus and B is the allowable blur size for a well-corrected wavefront. The defocus in this case can therefore be written as

$$\Delta z = \pm B \times f/\# \quad [\mu m] \quad (7.5)$$

where the narrower cone angle for a bigger f/# allows for a longer defocus distance. The allowable blur may be based on the size of a pixel, to which the blur size may be matched. Using this criterion for a visible-wavelength FPA with 7-μm pixels, the defocus $\Delta z = \pm 7\ \mu m \times 2 = \pm 14\ \mu m$ for an f/2 lens; one benefit of an f/4 lens would

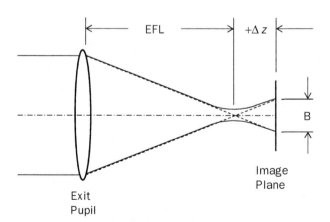

Figure 7.17 The depth of focus for the distance between the exit pupil and image plane can be based on either geometrical optics (dashed lines) or wavefront diffraction (solid lines). The figure illustrates Eq. (7.5) with the "+" sign; with the "−" sign, the image plane would be to the left of the focus by a distance Δz.

be an increase in the geometrical DOF to ±28 µm for the same pixel size.

However, a focused beam does not result in a geometrical point but instead creates a diffraction- or aberration-limited blur. For defocus based on the Rayleigh criterion of λ/4 PV wavefront error, the DOF Δz depends on the wavelength—which is expected for a diffraction-based criterion. As with the geometrical defocus, it also depends on the f/#:[1]

$$\Delta z = \pm 2\lambda (f/\#)^2 \quad [\mu m] \tag{7.6}$$

This DOF is based on the Rayleigh limit, where quarter-wavelength peak-to-valley WFE produces a focus shift of Δz. This means that design and fabrication efforts will be wasted if the lens isn't focused properly or if the FPA itself isn't sufficiently flat. The distance from the lens to the FPA can sometimes be controlled with a focus mechanism; the value obtained from Eq. (7.6) indicates how difficult it would be for such a mechanism to work. Typical numbers based on this criterion for an f/2 lens in the visible are Δz = ±2 × 0.5 µm × 4 = ±4 µm, a value that increases linearly with wavelength to ±80 µm for LWIR optics at a wavelength of 10 µm.

The field curvature reviewed in Chap. 3 may provide yet another restriction on depth of focus, but the choice between Eqs. (7.5) and (7.6) depends on imaging requirements such as blur and pixel size. Although Eqs. (7.5) and (7.6) are useful rules of thumb for first-order calculations, in practice a more exact determination requires through-focus plots of spot size such as those shown in Figs. 3.13 and 3.20.

Defocus is the allowable tolerance on lens-to-detector distance, but equally important is the spacing between surfaces in multiple-element lenses such as photographic objectives. As we saw in the ray-trace equations [Eqs. (2.9) and (2.10)], such spacing determines the height and angle at which rays intercept the following surface and thereby affect aberrations and WFE. Fast lenses (e.g., high-NA microscope objectives) are particularly sensitive to "despace" distance because of the marginal ray's large slope for these lenses. As with tilt and decenter, despace tolerances are determined by the lens designer and are based on comparing system requirements with criteria such as aberration WFE, ensquared energy, Strehl ratio, and the ability to fabricate or align parts to the appropriate tolerances.

7.3 Thermal Design

In a more recent case study of telescope performance not meeting its requirements due to testing errors, a scientific instrument was found to have a large defocus. In this case, it was determined by the manufacturer that the error was not due to an incorrectly polished mirror (as it was for the Hubble); instead, it was found that the cryogenic temperature of the test chamber caused the test optics to warp. As a result, a test mirror that was thought to be flat had reflective power, and the focus of the imager was thus set incorrectly.

Thermal effects such as mechanical distortion are only one of many that occur in optical systems. An optical system may stay aligned in a temperature-controlled building on a vibration-isolation table, for example, but not be able to meet requirements once it is moved out to the field where environmental effects become important. Changes in the thermal environment may lead to the following effects:

- Optical elements can exceed their recommended operating temperature, causing optical glasses to either melt or flow.
- Optical elements can shatter if their temperature changes too quickly.
- The transmission of some optical materials—typically semiconductors—decreases significantly as they become hotter.
- The blackbody emission of optical and structural materials increases with temperature.
- Both optical and structural materials expand as they get hotter, potentially affecting surface power and figure as well as tilt, decenter, despace, and defocus.
- Differential expansion between an FPA and its ROIC can cause the array to warp, decreasing its flatness to the point where the allowable defocus may be exceeded.
- Refractive indices change with temperature, resulting in WFE beyond that caused by design, fabrication, and assembly.
- As reviewed in Chap. 4, the spectral band-pass of thin-film filters shifts to longer wavelengths at higher temperatures.
- As reviewed in Chap. 5, the output wavelength of LEDs shifts with temperature.
- As reviewed in Chap. 6, detector dark current and SNR depend strongly on temperature.

Figure 5.7 shows a typical optical system where thermal design must be considered. The figure shows a lens imaging an on-axis point source at infinity onto a cryogenically cooled LWIR FPA. The lens is not cooled, but the cold stop and cold filter are cooled by mounting them on the cryogen tank. Optically, the system is identical to many seen in previous chapters; thermally, photons absorbed in the dewar are analyzed as heat rather than light. Figure 7.18 shows another example, where heat is generated by the wall-plug *in*efficiency of an LED and then transferred to the surroundings at ambient temperature.

The distinction between radiation as heat or light is based on material absorption, which converts light to heat. Heat can also be transferred directly to optical and structural elements either from other elements or from heaters, detectors and FPAs, electronic and mechanical components that dissipate power, and electromechanical components (electrical amplifiers, motors, etc.). The term *heat* can mean power [W], power density [W/m^2], or energy [J]; the term *light* is used in the same ways, with the units and context making clear which meaning is intended. This book uses the "power" definition.

Heat thus depends in part on the power collected and the resulting material absorption. In Chap. 4 we saw how to calculate the collected power from sources such as the Sun or an arc lamp by using the radiometric concepts of incident power (flux) Φ, irradiance E, radiance L, and their spectral equivalents; after subtracting out the reflected light, the heat load Q on each optical element is then equal to its absorption multiplied by the incident flux: $Q = (1 - R)(1 - T)\Phi$. It is important to note that the flux not collected by the optics can still be intercepted by the surrounding structure, where it can be absorbed

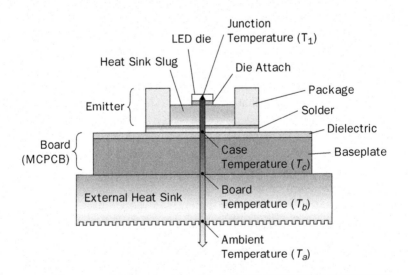

Figure 7.18 LEDs create heat because they are not 100 percent efficient at emitting photons. In order to avoid excessive LED temperatures, the heat must be efficiently transferred to the surroundings at the ambient temperature. (*Credit: U.S. Department of Energy, Pacific Northwest National Labs, www.pnl.gov.*)

and converted to heat and then transferred to the optics. The thermal "aperture" is thus the entire optical system, not just the radiometric aperture.

Temperatures change in response to incident heat flux, although variations in environmental ("ambient") temperature also change component temperatures. Ambient changes are largely out of the designer's control, but the system's response to heat and temperature differences is not. In this section we review the methods needed to reduce the effects of heat on an optical system. How heat is transferred to, from, and within an optical system is discussed first; this topic is referred to as *thermal management* (Sec. 7.3.1). The two most important thermo-optic parameters are reviewed in the following sections: thermal expansion (Sec. 7.3.2) and the change in refractive index with temperature (Sec. 7.3.3).

7.3.1 Thermal Management

The throughput budgets developed in Chap. 4 provide more information for the system designer than just a radiometric accounting of the flux incident on the detector; they also indicate how much flux is absorbed by each component, which increases their temperature. These heat loads on an optical system can be significant, and there are many consequences. We saw in Chap. 5, for example, that arc lamps have an output power on the order of 1 kW; this leads to the optics absorbing as much as 100 watts of heat. In this situation, large increases in WFE—and even fracturing of refractive elements—are common in the absence of proper temperature control. At lower power levels, the heat-induced increase in temperature increases the size and index of refractive optics; these effects will be reviewed in Secs. 7.3.2 and 7.3.3, respectively.

In addition to directly affecting the optics, thermal loads also affect the mechanical structure that maintains the tilt, decenter, despace, and defocus tolerances. Asymmetric heating and cooling—the norm in almost every system—causes differences in expansions and contractions of the structure, which in extreme cases may warp the structure well beyond the tolerance requirements of the optics. Such heat loads may be due to an optical source such as an arc lamp or the Sun, but they can just as easily result from nearby electrical subsystems that dissipate heat. In any case, maintaining the temperature of optics at a known, stable temperature requires appropriate thermal management and control.

A number of technologies are available for thermal control, including paints and coatings, conductive epoxies, fans, thermoelectric coolers, Stirling-cycle coolers, Joule–Thomson coolers, cryogenic liquids, heat pipes, and even simple fins. Regardless of the technology used, the principle of efficient heat transfer remains the same. This principle is illustrated in Fig. 7.19, where heat is transferred

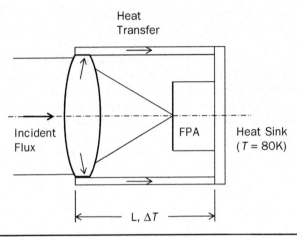

FIGURE 7.19 Absorbed flux increases the temperature of an optical element until its temperature is hot enough to transfer the heat away to the heat sinks via Eq. (7.7) and Eq. (7.9).

from higher temperatures to lower as the kinetic energy of atomic vibration is imparted to colder, slower-moving atoms.

An increase in component temperature depends on more than the heat load; it also depends on how well the heat is removed. A system with large "barriers" to the removal of heat becomes relatively hot; a system with small barriers (i.e., with low thermal resistance) stays significantly cooler. Conservation of energy requires that the heat load Q conducted away from an optic must equal the heat that it absorbs: $Q = (1 - R)(1 - T)\Phi + Q_e + Q_t$ for an incident flux Φ and for thermal loads Q_e from electrical dissipation and Q_t from absorption by nonoptical surfaces. Physically, the temperature of the optic increases until the absorbed heat equals that transferred, and the temperature increase ΔT is given by[20]

$$\Delta T = T_{op} - T_s = QR_t \quad [K] \tag{7.7}$$

Here R_t is the thermal resistance to heat flow, T_s is the (lower) temperature of the thermal sink to which the heat is transferring, and T_{op} is the resulting operating temperature—that is, the temperature the component must reach to remove the heat load Q, given the barrier to removal R_t.

Determining the heat load is not always straightforward. For the simple case of a lens in an optical system collecting 1 kW/m² of solar irradiance E, the nominal load $Q = (1 - R)(1 - T)EA = 0.995 \times 0.1 \times 1$ kW/m² $\times 0.25\pi(0.025 \text{ m})^2 = 48.8$ mW for an AR-coated lens ($R = 0.5$ percent) with a 25-mm aperture and an average optical transmission of $T = 90$ percent over the solar spectrum. As shown in Fig. 7.20,

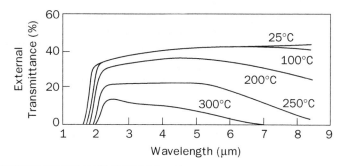

FIGURE 7.20 The transmittance of lens materials—such as the semiconductor germanium, graphed here—drops dramatically as its temperature increases. (*Credit: William L. Wolfe, "Optical Materials," in J. S. Acetta and D. L. Shoemaker (Eds.), The Infrared and Electro-Optical Systems Handbook, vol. 3, SPIE Press, www.spie.org, chap. 1.*)

however, the transmission of some materials depends strongly on temperature. This means that the assumption of a constant transmission T is not always correct; instead, the initial absorption increases the temperature, which in turn results in more absorption. The process feeds back on itself, stopping only when the absorbed heat is balanced by the heat transferred via a combination of conduction, convection, and radiation.

The thermal resistance R_t for conduction depends on the material and geometry of the path to the heat sink. For the linear geometry shown in Fig. 7.19, which consists of heat conducting along a metering tube with cross-sectional area A, the thermal resistance is given by[20]

$$R_t = \frac{L}{kA} \quad [\text{K/W}] \tag{7.8}$$

This is a direct measure of the temperature difference created per watt of absorbed power. The material component of the thermal resistance is the thermal conductivity k (units of W/m·K; see Table 7.5); note that some materials are more efficient than others at transferring heat, which means they have greater conductivity. The geometric contributor is the ratio of length L to area A. Physically, a longer length exhibits more thermal resistance—a sort of "inefficiency" as heat is transferred from atom to atom—as does a smaller cross sectional area A, a result of fewer atoms being available to transfer heat. For example, a common method for thermally isolating cryogenically cooled FPAs from room temperature is to use an evacuated dewar whose vacuum has an extremely high resistance to conductive heat flow; the FPA is mounted directly on the cryogen tank, which is in turn mounted on a long, thin, fiberglass tube with low conductivity and a high L/A ratio.

Material	CTE α_t [10^{-6}/K]	k [W/m·K]	α_t/k [10^{-6} m/W]	α_t/D [sec/m²·K]
Glass Substrates				
Borosilicate (Pyrex)	3.25	1.13	2.88	4.66
ULE	0.03	1.3	0.023	0.04
Zerodur	0.0 ± 0.02 (Class 0)	1.46	0.014	0.03
Refractives				
Fused silica	0.5	1.4	0.36	0.59
Germanium	6.1	58.6	0.10	
N-BK7	7.1	1.11	6.40	
N-SF6	9.03	0.96	9.41	
N-SF11	8.52	0.95	8.97	
Silicon	2.62	140	0.019	0.03
ZnSe	7.1	18	0.39	
ZnS	6.6	16.7	0.39	
Structural Materials				
Aluminum (6061-T6)	23.6	167	0.14	0.35
Beryllium (I-70H)	11.3	216	0.052	0.20
Copper (OFHC)	16.5	391	0.042	0.14
Graphite epoxy	−1 to +1	3.5	−0.3 to +0.3	0.05
Invar 36	1.0	10	0.10	0.38
Silicon carbide (Si-SiC)	2.6	155	0.017	0.03
Stainless steel (304)	14.7	16	0.92	3.68
Titanium (6Al-4V)	8.6	7	1.23	3.03

TABLE 7.5 Coefficient of thermal expansion α_t, thermal conductivity k, thermal distortion α_t/k, and specific thermal diffusivity α_t/D for various materials measured at room temperature; the CTE is an average value measured over a temperature range of −30°C to +70°C. Data are from *Optical System Design*,[1] Paquin,[28] Weber,[29] and *The Crystran Handbook of Infra-Red and Ultra-Violet Optical Materials*.[30]

As illustrated in Fig. 7.21, interfaces are also sources of thermal resistance; a standard one-dimensional thermal model shows these resistances for heat transfer from an LED.

Equation (7.7) ignores convection transfer to the heat sink temperature, a particular form of conduction with complex expressions for the thermal resistance that depend on whether the convection is natural or forced. Convection is absent in the vacuum of a dewar, yet the heat load on the dewar can still be significant. This

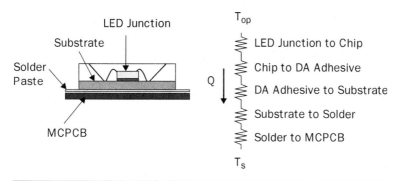

Figure 7.21 The thermal resistance in Eq. (7.8) for an LED is often modeled as one-dimensional heat flow using resistances in series; the model takes into account interface resistances for LED to adhesive, adhesive to substrate, and so on. (*Credit: Avago Technologies, www.avagotech.com.*)

is due in part to conduction but is also a result of a third heat-transfer mechanism: radiation.

This mechanism was mentioned in Chap. 5; there it was shown that all objects emit electromagnetic radiation based on their temperature, thus converting heat to light. As the conversion implies, one may view this radiation as consisting of either emitted photons or of heat; the distinction between photons and radiative heat transfer is based on any subsequent absorption that converts the electromagnetic field into heat. The irradiance from the Sun, for example, is carried as electromagnetic radiation through the vacuum of space; some of this radiation is converted to heat when absorbed by the atmosphere, the ground, the oceans, or an optical system.

If radiation is the only mechanism available to transfer this absorbed heat, then conservation of energy again requires that the radiated heat flux Q must equal that absorbed. In this case, the temperature rise depends on the heat load as follows:[20]

$$Q = \varepsilon\sigma A(T_{op}^4 - T_s^4) \quad [\text{W}] \tag{7.9}$$

an expression whose form is similar to that of Eq. (5.1). In reality, a combination of conduction, convection, and radiation remove the absorbed heat. Radiation typically starts to dominate at a temperature of approximately 400 K for thermal sinks at room-temperature, depending on the object's conductive resistance and radiative emissivity. Observe that Eq. (7.9) ignores the geometrical "view factor" (solid angle), a factor that is critical to radiative thermal transfer between surfaces[21] and must be included in any complete thermal analysis.

Equation (7.9) does not explicitly include a term for thermal resistance to heat flow. This resistance in the absence of conduction or convection is clearly not infinite, so Eq. (7.9) can be put in a form similar to that of Eq. (7.7) and Eq. (7.8). The result is $R_t = 1/[\varepsilon \sigma A (T_{op} + T_s)(T_{op}^2 + T_s^2)]$ such that $T_{op} - T_s = QR_t$; this shows that, in addition to its dependence on temperature, the resistance to radiative heat transfer increases in response to decreases in emissivity ε and surface area A.

Coatings and spectrally selective paints are often used for radiative thermal control. Here the absorbed radiation from a source depends on the coated surface's absorption over the source spectrum; for example, an aluminum mirror absorbs little of the solar spectrum. At the same time, the barrier to removing this heat via radiation depends on spectral emissivity, with the spectrum of interest now at the emission wavelengths (peak $\lambda \approx 10$ μm for a graybody at 300 K).

A temperature-control coating thus relies on the absorption-to-emission $[\alpha(\lambda_1)/\varepsilon(\lambda_2)]$ ratio as a figure of merit, where a small ratio *at the wavelengths of interest* is key for effective thermal management.[20] For instance, white paints with a high solar reflectivity (low solar absorption α_s) are available that are also "black" (have high emissivity) for the LWIR wavelengths at which the radiative heat transfer from a room-temperature (300-K) graybody is significant [see Eq. (5.3)]. Room-temperature surfaces with these coatings are found to be significantly cooler when exposed to solar radiation than are uncoated surfaces; paints that are reflective and emissive at other wavelengths (as output from arc lamps, for example) are also available. In all cases, the emissivity and absorption change with time as the coating ages or becomes dirty. Therefore beginning-of-life (BOL) and end-of-life (EOL) values must be carefully distinguished and performance expectations adjusted as the instrument nears retirement.

Even when appropriate coatings and paints are used, variations in the flux incident on a surface can create warping and distortion of the instrument. These variations can occur on a curved surface such as a cylinder, for instance, part of which is pointing directly at the source while other parts are angled away from it. The flux on the angled portions is smaller, so the varying heat load results in the source-pointing part of the cylinder becoming hotter than the other parts. Such temperature differences (or *gradients*) cause the hotter part to expand more than the cooler, thus distorting the structure; nonuniform absorption and emissivity of the surface have the same effect.

Even when the heat load is uniform over an entire surface, the temperature profile may not necessarily be the same. Conductive temperature distributions vary with geometry; a common example of this phenomenon is the solar flux incident on the lens shown in Fig. 7.19, whose conductive path for heat removal is around the circumference of the metering tube. Such an arrangement results in a hot spot in the center of the lens with cooler temperatures toward the outer diameter (Fig. 7.22). As we shall see in Sec. 7.3.3, the resulting temperature profile results in thermal WFE that causes distortion[22] in addition to that caused by design, fabrication, and assembly.

Example 7.2. The thermal WFE depends in part on the temperature change induced by heat loads. This example shows how to use the conservation of energy to determine the temperature change due to absorption and conduction. The goal is to keep the temperature increase at the center of the cryogenic lens shown in Fig. 7.19 as small as possible, reducing the effects of thermal WFE.

The lens is nominally held at 77 K by a LN_2 cryogen; it receives radiation from the 300-K background of the dewar window as well as solar flux, from outside the detector FOV, that illuminates the lens with 1 kW/m² of irradiance over all wavelengths. The thermal resistance between the lens and the cryogen tank determines how much the lens temperature rises above 77 K as a function of the radiation heat load and absorption in the lens itself. Conservation of energy for the lens shows that its temperature increases until the incident heat is dissipated by the conduction losses to the cold sink temperature of $T_s = 77$ K. Hence $Q_r + Q_s = Q_c$, where Q_r is the radiation heat load from the 300-K background, Q_s is the solar power absorbed by the lens, and Q_c is the conduction transfer to the cryogen tank.

The conduction loss $Q_c = (T_L - T_s)/R_t = (T_L - T_s)kA/L$ [using Eqs. (7.7) and (7.8)], while the absorbed power $Q_s = (1 - R)(1 - T)\Phi$ for an incident solar flux of $\Phi = EA = 1$ kW/m² $\times 0.25\pi (0.05$ m$)^2 = 1.96$ W on a lens of 50-mm diameter. With $R \approx 0$ and with 10 percent absorption, 0.196 W of power will be transferred to the cryogenic cooler; since radiation from the low-temperature lens is relatively low, conduction is the primary heat-transfer mechanism. Radiation absorbed by the lens from the 300-K environment is also relatively small because the lens has been coated with a high-reflectivity IR filter ($\alpha_s \approx 0$). As a result, the absorbed

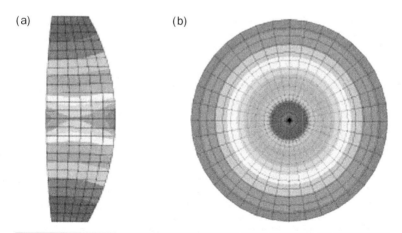

FIGURE 7.22 Temperature distribution (a) and average WFE (b) of a lens like the one shown in Fig. 7.19; the center is hotter owing to heat transfer out toward the heat sink at the edges. (*Credit: Sigmadyne, Inc., www.sigmadyne.com.*) (*See also color insert.*)

solar flux is the dominant heat load that must be dissipated by conduction ($Q_s = Q_c$).

For a given lens absorption, the only design parameter that can be controlled is the thermal resistance between the lens and the cryogen. To keep the lens as cool as possible, the thermal resistance must be kept as small as possible; this requires a lens mount of short length, large cross-sectional area, and high thermal conductivity [see Eq. (7.8)]. The 100-mm focal length constrains the mount length to $L \approx 100$ mm, while the conductivity requirement indicates that copper ($k = 391$ W/m·K) or aluminum ($k = 167$ W/m·K) are good choices for the material. The heat-transfer area of the mount is $A = 0.25\pi(0.051^2 - 0.045^2) = 4.52 \times 10^{-4}$ m^2. Substituting into $Q_s = Q_c$, we find that $\Delta T = 0.26$ K; this gives a lens temperature $T_L \approx 77.26$ K for aluminum. The use of an aluminum mount is thus appropriate for thermal management; however, as shown in Sec. 7.3.2, its thermal expansion is unfortunately excessive for most applications.

The use of a high-conductivity mount with a large cross-sectional area keeps the average temperature of the lens relatively close to that of its cryogenic heat sink. In order to minimize the temperature change across the lens itself, we apply the same principles: a large thermal conductivity and thickness-to-diameter ratio ensures the smallest ΔT across the lens and, all else being equal, the lowest possible thermal wavefront error.

7.3.2 Thermal Expansion and Contraction

Despite the problems with polishing the primary mirror of the Hubble Space Telescope to the correct asphericity, a number of extraordinary technological advances were made in maintaining the despace tolerance between the primary and secondary. This tolerance is approximately 1.5 μm over a center-to-center distance of about 5 meters, and a key challenge in obtaining this performance was the requirement to operate over an average temperature change of ±15°C. What makes this temperature range challenging is that, with few exceptions, an increase in temperature expands a given length of material; a consequence of this phenomenon is a change in spacing between mirrors, which makes small ("tight") despace tolerances extremely difficult requirements to meet.

The dimensional change is characterized by the coefficient of thermal expansion (CTE) α_t, defined as the fractional change in length $\Delta L/L$ that results from a given change in temperature ΔT. Quantitatively, $\alpha_t = (\Delta L/L)/\Delta T$ [1/K]. As illustrated in Fig. 7.23, the dimensional change produced by an average change in temperature is therefore given by[33]

$$\Delta L = \alpha_t L \Delta T \quad [\text{mm}] \qquad (7.10)$$

Equation (7.10) shows that ΔL depends on the temperature change, with positive values of α_t resulting in material expansion for an increase in temperature. Physically, each atom or molecule in the material expands a small amount when heated; longer lengths have

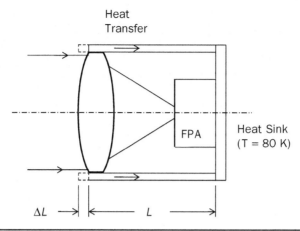

FIGURE 7.23 Heat transferred from a metering tube contracts the tube both radially and axially (along the optical axis). The axial contraction ΔL may or may not result in defocus (see Sec. 7.3.3).

more atoms or molecules and thus expanding more. In other words, two effects made the despace tolerancing of the Hubble mirrors such a difficult task: a large center-to-center distance L and a wide range of operating temperatures ΔT.

Table 7.5 shows typical values of the CTE for both optical and structural materials. Typical values are on the order of 1×10^{-6}/K to 25×10^{-6}/K, where units for the fractional change are sometimes given as parts per million per Kelvin (ppm/K). The CTE and the thermal conductivity k depend on temperature, but they are usually constant over a particular temperature range; for example, the linear data in Table 7.5 is valid over the range $-30°C$ to $+70°C$. If α_t is not constant but instead varies over a larger temperature range, then the change in length is found numerically by integrating $\alpha_t(T)$ with respect to T; in this case, Eq. (7.10) is generalized as $\Delta L = \int \alpha_t(T) L \, dT$. The figure of merit α_t/k indicates that small values (corresponding to low expansion and high thermal conductivity) are preferred for reducing steady-state distortions; a small α_t/D (i.e., low expansion and high thermal diffusivity D) is preferred for increasing transient heat transfer.

The range of CTEs found in Table 7.5 varies from about 24×10^{-6}/K for aluminum to almost 0 ppm/K for graphite-epoxy (GrE) composites. For the lens shown in Fig. 7.23 with a 100-mm focal length, an aluminum structure will create a defocus of $\Delta L = \alpha_t L \Delta T = 23.6 \times 10^{-6}$/K $\times$ 100 mm $\times$ 220 K = 0.52 mm for a uniform, 220-K decrease in cooling from room temperature to the cryogenic (LN$_2$) "ambient." For GrE, the change in length can be many times smaller—well within the diffraction-limited depth of focus even for small-f/#

systems in the visible. Ultra-low-expansion glasses such as ULE (Corning) and Zerodur (Schott) are useful as low-distortion mirror substrates, but they do not have the optical quality needed for fabrication into refractive elements. For lenses, windows, prisms, and the like, fused silica—also known as fused quartz—has the lowest CTE of commonly used refractive materials.

In addition to affecting despace between elements, there are two further consequences of the CTE in optical systems: the change in refractive power of individual elements, and the possible distortion of surface radii and figure. If the index stays the same then the power decreases owing to the change in focal length with temperature, which Eq. (7.10) shows to be $\Delta f = \alpha_L f \Delta T$; for a positive α_L, the magnitude of the focal length thus increases with temperature for both positive and negative lenses.

Physically, the increase in focal length results from the expansion of every lens dimension. This is illustrated in Fig. 7.24, where both the thickness and outer diameter of the lens have been thermally expanded. The figure shows that this expansion results in a larger surface radius for a positive lens. Equation (2.1) indicates that the longer radius for the positive lens increases the focal length of a positive lens by Δf; since $\Delta f < 0$ for a negative lens, a longer radius makes its focal length $(f + \Delta f)$ more negative. A pedagogical analogy often used for *unconstrained* elements is that the thermal expansion is similar to a photocopy machine, whereby every dimension increases (or decreases) according to the magnification ($\Delta L/L$) of the copy.

In addition to inducing changes in refractive power, thermal expansion can also distort ("potato chip") or otherwise warp the surface figure of a lens or mirror. The mechanism is explained by Eq. (7.10) in that CTE differences (ΔCTE) or thermal gradients across a surface result in mismatches in expansion. No material has perfectly uniform properties; there are always axial and radial

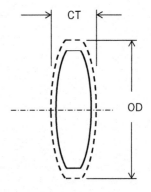

FIGURE 7.24 Both the outer diameter (OD) and the center thickness (CT) of an unconstrained element expand as its temperature increases. An expansion constrained at the OD will increase the CT more than the OD and may thus change a spherical surface to an asphere.

variations (tolerances) in properties, including the CTE. Likewise, in the absence of an isothermal environment, small temperature differences across a surface are likely; this creates a *change* in ΔT across an element.

Regardless of the cause, if some areas of a surface expand more than others then surface distortion and changes in figure will inevitably result. Such CTE mismatches can also occur with layered surfaces such as thin-film filters with multilayer coating CTEs that match neither the filter substrate nor each other. Silicon ROICs bump-bonded to InGaAs FPAs are another example, and differential contraction of these dissimilar materials at cryogenic temperatures is sometimes enough to separate a large-format array from its ROIC.

The changes in CTE and in ΔT also result in tilt and decenter misalignments between elements. A completely homogeneous material in an isothermal environment will cause despace and defocus in a circularly symmetric system, as indicated by Eq. (7.10). Note that the *asymmetries* in CTE and ΔT variation also produce linear displacements (decenter) with respect to the optical axis in addition to tilts about the axis.

Minimizing these misalignments was yet another challenge faced by the Hubble primary-to-secondary thermal design. The material that maintains ("meters") the mirror spacing in the Hubble is graphite epoxy, whose *average* CTE was only 0.02×10^{-6}/K. The thermal expansion found from Eq. (7.10) is thus $\Delta L = \alpha_t L \Delta T = 0.02 \times 10^{-6}$/K $\times$ 5 m $\times$ 30 K = 3 µm, which overshoots the axial requirement by a factor of 2.

Unfortunately, GrE has significant CTE variations from part-to-part and within each part, variations that could lead to excessive tilt and decenter. Electrical heaters can sometimes compensate for these gradients by changing the local ΔT to balance the ΔCTE; in a spaced-based telescope, however, this may not be possible given the limited number of watts available for heaters. The solution adopted by the Hubble designers was to measure the CTE of each of the metering struts and then select ("cherry pick") those that were within the tolerances dictated by the tilt, decenter, and despace requirements.[24] Of course, this procedure would be difficult to implement in a production environment, but was a cost-effective design strategy for a "one-off" telescope such as the Hubble.

The 1.5-µm despace tolerance for the Hubble is a derived requirement that incorporates thermal expansion of the primary and secondary mirrors. If the optical system also uses refractive elements, then the requirements analysis would have to take an additional thermal effect into account: the change in index with temperature. This consideration is reviewed in the following section.

7.3.3 Athermalization

For refractive elements, the index of refraction also changes with temperature. This effect, illustrated in Fig. 7.25, is characterized by the thermal slope dn/dT. Table 7.6 shows that the index for many materials increases as the element gets hotter; as a result, the refractive power of the element increases and thus shortens the focal length.

If we combine the effects of dn/dT and thermal expansion, then the change in focal length (Δf) when the temperature of a thin lens changes by ΔT is given by[25]

$$\Delta f = f\left[\alpha_L - \frac{dn/dT}{n-1}\right]\Delta T = -f\nu\Delta T \quad \text{[mm]} \qquad (7.11)$$

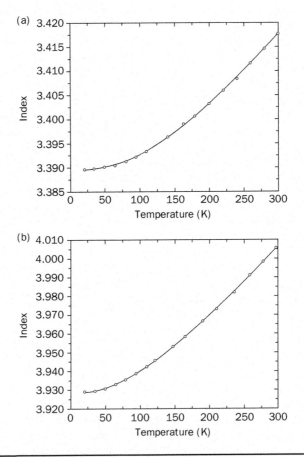

FIGURE 7.25 The refractive index for many optical materials increases with temperature, as shown here for (a) silicon and (b) germanium. (Credit: J. M. Hoffman and W. L. Wolfe, "Cryogenic Refractive Indices of ZnSe, Ge, and Si at 10.6 µm," Applied Optics, 30(28), 1991, pp. 4014–4016.)

Material	Wavelength	dn/dT [10^{-6}/K]	ν [10^{-6}/K]
AMTIR 1	LWIR	72	36
Germanium	LWIR	385	126
N-BK7	541.6 nm	3.0	1.3
N-SF6	541.6 nm	1.5	7.2
N-SF11	541.6 nm	2.4	5.5
Silicon	MWIR	162	63
Zinc selenide	LWIR	61	36
Zinc sulfide	LWIR	45	26

TABLE 7.6   Relative change in index with temperature (dn/dT) and thermo-optic coefficients (ν) for various refractive materials measured in air over the temperature range 20–40°C at the wavelengths indicated. Values supplied by different references vary, especially for IR materials. The uncorrected value of dn/dT should be used for optical systems that operate in a vacuum, where the change in the refractive index of air with temperature does not contribute to the total dn/dT. Data are from Smith[8] and from Rogers and Roberts.[25]

where a positive thermal slope results in lenses with a slightly shorter focal length as the temperature increases. The term in brackets defines the thermo-optic coefficient (also known as the T-value or glass constant γ), which is given as $\nu = (dn/dT)/(n-1) - \alpha_L$ [1/K]. The magnitude of ν varies weakly with temperature, since the CTE, index, and dn/dT all vary with temperature; observe that both n and dn/dT vary with wavelength also.

Equation (7.11) shows that a large temperature difference produces an appreciable focus change, depending on the relative magnitudes of the two terms that contribute to the thermo-optic coefficient. If the thermal expansion can be balanced against the dn/dT effects, then it is possible to design a temperature-insensitive ("athermal") singlet that does not change focus with temperature. The need for such athermalization is generally determined by the temperature changes and the depth-of-focus criteria given in Sec. 7.2.3. In practice, the thermal slope for many materials is much larger than the CTE; this makes athermal designs difficult and impractical for large temperature changes, particularly for MWIR and LWIR lenses made from materials such as silicon or germanium with large dn/dT.

Lenses are attached to mounts, of course, and including the thermal expansion of the mount brings the design of an athermal singlet somewhat closer to practicality; this process is illustrated in Example 7.3. Other approaches that are used to obtain some degree of athermalization include doublets and triplets with a negative lens,[8] reflective optics that do not exhibit dn/dT effects, reflective designs with dissimilar mounting materials chosen specifically to balance thermal expansion, unimaterial designs in which mirrors and mounts are fabricated from the same material and used in an isothermal environment, and active mechanisms for adjusting the thermal defocus. Active focus-adjustment mechanisms are usually expensive and prone to failure; they should be avoided if the DOF tolerance allows the limited range of a passive mechanism.

In all cases, unavoidable temperature differences throughout the lenses and their mounts—either axially or radially—produce geometry and index nonuniformities render performance less than that predicted by first-order analysis. One effect of these gradients is to modify the focus shift given by Eq. (7.11); as shown in Sec. 7.3.2, more important consequences are the WFE induced in the lenses and the misalignments in the mounts. We shall see in Sec. 7.5 that predicting the actual performance of the optical system requires an integrated structural-thermal-optical (STOP) analysis.

Example 7.3. The design of an athermal singlet that does not change focus with temperature must take into account the thermal expansion of both the lens and its mount. Given the materials that are commonly available, is it possible to passively athermalize a LWIR singlet without the use of a focusing mechanism? If not, then what range of operating temperatures allows for diffraction-limited performance? The aim of this example is to answer these questions.

Including the mount, the thermal defocus given by Eq. (7.11) becomes $\Delta f = -f(v + \alpha_m)\Delta T$; therefore, an athermal lens is possible if $(n - 1)(\alpha_L - \alpha_m) = dn/dT$. The common refractive materials for the LWIR band are AMTIR, germanium, zinc selenide, and zinc sulfide. Because the goal is to design an athermal lens, our first instinct is to choose a mount with a low CTE; for this, Invar ($\alpha_m = 1.0 \times 10^{-6}$/K) is selected (Table 7.5).

Table 7.7 compares the thermal slope with the $(n - 1)(\alpha_L - \alpha_m)$ term for the four materials considered; the table shows that an athermal LWIR singlet is not possible even with a low-expansion Invar mount. Looking closer at these results, we see that in all cases $dn/dT \gg (n - 1)(\alpha_L - \alpha_m)$; this implies that the large dn/dT of LWIR materials requires a mount material with a *negative* CTE. These values are shown in the table's fourth column, where graphite epoxy can be fabricated with a CTE < 0 ($\alpha_m = -1.0 \times 10^{-6}$/K). This improves our results slightly, but a completely athermal LWIR singlet remains infeasible—unless a passive mount made from dissimilar materials can be used to yield an effective CTE that is much less than 0.[1]

Lens material	dn/dT [10^{-6}/K]	Invar $(n - 1)(\alpha_L - \alpha_m)$ [10^{-6}/K]	GrE $(n - 1)(\alpha_L - \alpha_m)$ [10^{-6}/K]
AMTIR 1	72 (n = 2.5)	16.5	19.5
Germanium	385 (n = 4.0)	15.0	21.0
Zinc selenide	61 (n = 2.4)	8.60	11.3
Zinc sulfide	45 (n = 2.2)	4.32	6.72

TABLE 7.7 Comparison of dn/dT with $(n - 1)(\alpha_L - \alpha_m)$ for four commonly used LWIR lens materials.[26] The mount materials compared are Invar ($\alpha_m = 1.0 \times 10^{-6}$/K) and graphite-epoxy ($\alpha_m = -1.0 \times 10^{-6}$/K).

Although the LWIR singlet cannot be completely athermalized, there is still some range of operating temperatures over which the thermal defocus given by Eq. (7.11) is within the range of diffraction-limited tolerances. In this case, the allowable defocus is given by Eq. (7.6); for an f/2, LWIR lens with a 100-mm focal length and a center wavelength of 10 μm, the defocus $\Delta z = 2\lambda(f/\#)^2 = 2 \times 10$ μm $\times 4 = 80$ μm.

Comparing this tolerance with the thermal defocus obtained for a zinc sulfide lens with an Invar mount, Table 7.6 and the modified Eq. (7.11) show that $\Delta f = -f(v + \alpha_m)\Delta T = -100$ mm $\times [(26 + 1) \times 10^{-6}]/K \times 30$ K $= -81$ μm for $\Delta T = 30$ K, which would be adequate for many applications. For comparison, a germanium lens in an aluminum mount remains athermal over a temperature range of approximately ±5 K before exceeding the diffraction-limited defocus requirement.

7.4 Structural Design

Structural design is often thought of as belonging to the realm of civil engineers, builders of bridges and skyscrapers.[27] The transfer of these skills to optomechanical design requires a change in thinking—from designs that don't collapse to designs that don't distort. Breakage due to mechanical shock is always a concern, with calcium fluoride a good example of a structurally weak optical material. Of course, using cardboard and bubble gum as structural elements would result in misalignments, whereas using steel and titanium fasteners likely will not. But steel and titanium are relatively heavy, so how can optical alignments be maintained without exceeding weight requirements?

The design of lightweight structures for optical systems addresses two critical factors: (1) the distortion of optical elements due to applied forces, which increases the surface figure error (SFE); and (2) structural deformation that changes the tilt, decenter, despace, and defocus alignments between elements. The "structure" considered in Example 7.3, for example, is an Invar lens mount that expands thermally and also bends under load. Even the most elegant optical design will not work if an undistorted mirror is mounted at the end of a long, wobbly structure. Thus the structural design of optical systems—which are seldom exposed to forces sufficient to challenge strength—is concerned primarily with stiffness and deflection.

This section reviews the basic concepts needed to prevent excess component distortion and structural deflection. Addressing component distortion mainly concerns element size and shape, material, stiffness-to-weight ratio, and mounting techniques (Sec. 7.4.1). Alignment tolerances can be maintained by proper selection of structural geometry, materials, and natural frequency (Sec. 7.4.2). Natural frequency is related to dynamic deflection, whereby a structure experiences greater deflection when an applied force is changing than when it is not. Such variable forces are an inevitable part of all optical systems and they, too, affect alignment and component surface figure.

7.4.1 Component Distortion

All materials distort when a force is applied. For instance, the pressure difference ΔP across the optical window of an evacuated dewar is approximately 0.1 MPa (14.7 psi), causing the window to deflect toward the FPA. Because of this distortion, the window is not flat and instead becomes a weak negative lens that induces a small amount of defocus. It is possible to accommodate this defocus by adjusting the window-to-FPA distance before reaching the operating conditions of vacuum and cryogenic temperatures, although doing so involves a lengthy process of evacuation and cooldown after each adjustment. An alternative is to use a thicker window with enough resistance to bowing that its effects on focus are negligible.

Although such changes in surface curvature may be small, the resulting change in surface figure may not be. We saw in Sec. 7.1.4 that fabrication tolerances result in deviations from perfectly spherical surfaces. Applied forces can have the same effect, changing surfaces from nearly spherical with diffraction-limited performance to something highly nonspherical with large WFE (Fig. 7.26). This was also discussed in Sec. 7.3.2, where it was shown that thermal expansion and contraction—in addition to the other structural forces applied—can change the shape of a surface.

Applying a force (or "load") to the simplest possible geometry—a bar of length L and cross-sectional area A—illustrates how more complex shapes are distorted. When a load F pulls on the bar along its axis, the bar becomes slightly longer and a little bit thinner. The elongation (or distortion or deflection) ΔL that occurs is given by the ratio FL/AE, where E is the elastic (or Young's) modulus, a measure of the material's resistance ("stiffness") to deformation by pushing, pulling, or bending. The fractional deflection (or *strain*, symbol ε) is a

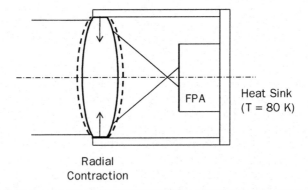

Figure 7.26 If the metering tube's CTE is larger than that of the lens, then differential thermal contraction produces a radial force that distorts the shape of the optical surfaces; this distortion leads to an optical path difference $(n - 1)\Delta t$, which in turn results in wavefront error.

Optomechanical Design

common metric for this deformation; it is given by $\varepsilon = \Delta L/L = F/AE$. Physically, the strain is smaller when there are more atoms (more area A) to resist the load and when there are greater bonding forces between atoms (larger E). Typical values of Young's modulus, in units of force per unit area (newtons per square meter or pascals, symbol Pa), are given in Table 7.8.

Material	E [GPa]	E/ρ [10^6 N·m/kg]	μ	dn/dσ [10^{-12}/Pa]
Glass Substrates				
Borosilicate (Pyrex)	64	28.3	0.2	3.9 [29]
ULE	67	30.5	0.17	4.0 [29]
Zerodur	93	36.9	0.24	3.0 [29]
Refractives				
AMTIR 1	22.0	5.0	0.266	
Fused silica	73.0	33.2	0.164	
Germanium	104	19.3	0.278	
N-BK7	82.0	32.7	0.206	2.77 [30]
N-SF6	93	27.6	0.262	2.82 [30]
N-SF11	92	28.6	0.257	2.94 [30]
Sapphire	335	84.4	0.25	
Silicon	131	56.2	0.279	
ZnS	74.5	18.3	0.29	
ZnSe	70.3	13.3	0.28	
Structural Materials				
Aluminum (6061-T6)	68.9	25.5	0.33	—
Beryllium (I-70H)	287	155	0.043	—
Copper (OFHC)	129.8	14.5	0.343	—
Graphite epoxy	250	135	—	—
Invar 36	144	17.8	0.259	—
Silicon carbide (Si-SiC)	330	114	0.24	—
Stainless steel (304)	193	24.1	0.27	—
Titanium (6Al-4V)	114	25.7	0.34	—

TABLE 7.8 Young's modulus E, specific stiffness E/ρ, Poisson's ratio μ, and the change in index with stress ($dn/d\sigma$) for various materials measured at room temperature. To convert MPa to the traditional units of pounds per square inch (psi), the approximate conversion factor of 145 psi/MPa is used. Data are from *Optical System Design*,[1] Paquin,[28] Weber,[29] and *The Crystran Handbook of Infra-Red and Ultra-Violet Optical Materials*.[30]

Three-dimensional load geometries and shapes that are more complex—such as windows, cylinders, cantilevers, and metering trusses—require a more sophisticated analysis but embody the same physical principles. For example, the "flat plate" shape of a dewar window deflects at the center to an extent that depends on the thickness, diameter, material, and circumferential ("edge") support:[23]

$$\delta = \frac{3\Delta PR^4(1-\mu^2)}{16Et^3} \quad [\text{mm}] \tag{7.12}$$

Here the resistance to bowing builds up very quickly with window thickness t, but it drops off even faster with radius R. A much smaller effect is that due to Poisson's ratio μ, which quantifies the material property of how much thinner a square bar gets when elongated by an axial load (generally about 0.2–0.3 times the axial strain). Using values for a sapphire window ($E = 335$ GPa and $\mu = 0.25$ from Table 7.8) with a 25-mm diameter ($R = 12.5$ mm) and a 2-mm thickness, we can calculate the center deflection as $\delta = 3 \times 10^5$ N/m² × (0.0125 m)⁴ × $(1 - 0.25^2)/[16 \times 335 \times 10^9$ N/m² × (0.002 m)³] = 0.16 μm, or approximately $\lambda/30$ for a MWIR imager at a wavelength of 5 μm. In this case, then, the effects of deflection δ on surface radius and WFE are small compared with the fabrication tolerances.

Equation (7.12) is included here not to be memorized as the formula for a specific geometry, but rather to illustrate important physical trends that are applicable to all geometries. For example, the aspect ratio (i.e., the diameter-to-thickness ratio $2R/t$) is usually specified to fall within the range of 6 : 1 to 12 : 1. The reason is clearly seen in Eq. (7.12), where a large $2R/t$ ratio increases the deflection of the dewar window. More generally, this ratio also influences the deflection of optical elements during polishing and mounting. In both situations, forces applied to the element increase its strain and thus its potential for surface figure error. An element that is thin compared with its diameter (a high aspect ratio) increases this deformation, yet specifying too small a ratio results in a component that is unnecessarily large and heavy. Although the optimum ratio for most materials is thus between 6 : 1 and 12 : 1, smaller ratios are sometimes used; for example, $\lambda/20$ Zerodur mirror blanks require an aspect ratio on the order of 2 : 1 to 5 : 1 in order to achieve extremely low SFE.[31]

Even with the appropriate aspect ratio, an element's SFE and WFE can be large if the optic is strained ("potato chipped") when mounted. Such surface deformations can be induced by improper mounting of a lens in its housing, of the housing to the structure, or the structure to an optical bench. As with thermal deflections, mirrors are particularly susceptible to WFE, given the 2× change in reflected angle as the local surface angle changes in response to distortion. In contrast, lenses usually require a *change* in thickness Δt to create wavefront error that is not defocus—see Fig. 7.26.

As reviewed by Yoder[1-4] and Vukobratovich,[5,6] there are a number of techniques available for low-strain mounting; these include semikinematic mounts to avoid overconstraining the optic, elastomeric supports, and integral threads for metallic mirrors. Surface figure error from mounting distortion is difficult to predict analytically. For this reason, the computational methods reviewed in Sec. 7.5 are used only when, as with space-based instruments, there is typically just one chance for success. For lower-budget, commercial projects, an empirical approach—combined with experience and back-of-the-envelope calculations—is the norm.

Low-strain mounting is especially difficult for large optics, where the deflection due to the element's own weight also contributes to SFE. In this case, the load used in Eq. (7.12) is not the pressure difference ΔP but instead the element weight $\rho g V$ [weight density ρg (N/m^3) multiplied by material volume V (m^3)] per unit area; the stiffness-to-density (E/ρ) ratio is thus a useful material figure of merit for resistance to self-weight SFE. Typical values are listed in Table 7.8, which indicates that beryllium and SiC are both very light and stiff whereas aluminum is heavier and more flexible. Although there are concerns about the low fracture strength of ceramics, SiC been used on a number of challenging projects—including the European Space Agency's development of a 3.5-meter mirror for the Herschel Space Telescope. Polishing of the porous SiC (sintered or reaction bonded) to a low surface finish is not possible. Therefore, a SiC substrate is clad with a nonporous layer via chemical vapor deposition (CVD), which can be polished to a surface finish of less than 10 angstroms RMS.[32]

Structural designs using trusses or honeycomb chambers are useful for reducing the total weight, but the best structural performance is still obtained with materials of greater specific stiffness E/ρ. However, any design decisions based on a single metric such as E/ρ will often conflict with other metrics such as thermal performance. Hence, as shown in Sec. 7.5, a complete structural-thermal-optical analysis is essential when evaluating design options.

An additional metric for evaluating refractive components is the change in index with strain. We have previously seen that the index can change with temperature; if a refractive element is strained for any reason—e.g., temperature gradients, nonkinematic mounting, self-weight—then its index also changes, as captured by the *strain optic* coefficient $dn/d\varepsilon$. The analysis of this effect can be complex, so the one-dimensional model for isotropic glasses (e.g., BK7) employs the *stress optic* coefficient $K = dn/d\sigma$.

Below a certain threshold stress, the linear relation between stress σ and strain ε (i.e., $\sigma = \varepsilon E$) determines the index difference due to strain. Using the entries in Table 7.8 as an example, $K = 2.77 \times 10^{-12}/$Pa for N-BK7;[30] thus, the birefringence in this case is $\Delta n = K \Delta \sigma = K E \Delta \varepsilon = 2.74 \times 10^{-12}/\text{Pa} \times 82 \times 10^9 \text{ Pa} \times 10^{-6} = 0.23 \times 10^{-6}$ for $\Delta \varepsilon = 1$ microstrain (or $\Delta\varepsilon = 1$ µm/m $= 10^{-6}$). The effect of strain on refractive index can therefore be small, but it still must be checked—especially for polarized optics using materials that are known to be *stress birefringent*.[39]

7.4.2 Structural Deflection

The performance of an optical system depends not only on the distortion of optical elements but also on the deflection of mechanical structures that hold the optics in alignment. The initial alignments established at assembly for tilt, decenter, despace, and defocus may not be maintained when loads are applied to the system. All structures deflect when loads are applied; these deflections may be due to rigid-body ("rocking") deflection or may occur because the structure itself bends or twists. A flexible structure allows a large degree of relative motion between the elements, resulting in both image motion and an increase in blur size (Fig. 7.14). The degree of flexibility that's acceptable flows down from the system requirements; it therefore determines the structural materials, geometry, and need for alignment mechanisms or rotationally insensitive optics such as pentaprisms.

Static Deflections

A common structural element for maintaining alignment is the optical bench on which an instrument is assembled. These can range from simple aluminum plates to honeycomb graphite-epoxy (GrE, also known as "carbon fiber") structures to meet a range of requirements pertaining to stiffness, weight, flatness, CTE, ΔCTE, jitter, and long-term stability (drift). An off-the-shelf honeycomb bench with low-CTE (Invar) facesheets is shown in Fig. 7.27. These are too heavy for space-based instruments; instead, GrE—or any of the other graphite-based materials, such as the graphite-cyanate shown in Fig. 1.4—are commonly used even though they have their own problems (e.g., "outgassing" and water loss or "desorption" that alters the flatness).

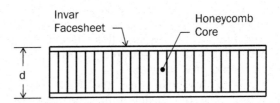

Figure 7.27 A common optical bench resists deflection via its thickness d and maintains low weight by using a honeycomb core; it therefore has a high stiffness-to-weight ratio.

Mounted on such benches is the optical instrument, which itself may incorporate another common structural element known as a *truss* for maintaining alignment between the primary and secondary mirrors of a reflective system (Fig. 7.28). For smaller refractive and catadioptric systems (~ 300 mm or smaller), a tube to maintain alignments between components is more common (Fig. 3.35); however, such tubes are too heavy for larger systems. A truss allows for portability of ground-based 'scopes and also minimizes self-weight deflection for extremely large designs, such as the Large Binocular Telescope shown in Fig. 7.1.

A simple tube structure is shown in Fig. 7.29, which illustrates that a downward load—due to self-weight and any additional applied force—creates tilt and decenter of the structure with respect to its end. The tilt and decenter depend on the load, the material, and the shape and size of the structure. For a cantilevered beam that is clamped on one end and has a lens weight or other force that is much

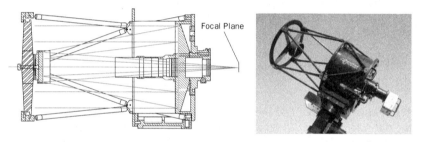

FIGURE 7.28 This telescope with a 430-mm aperture uses lightweight trusses to maintain portability. Compare this design with that seen in Fig. 3.35 for a telescope with a 300-mm aperture. (*Credit: Planewave Instruments, www.planewave.com.*)

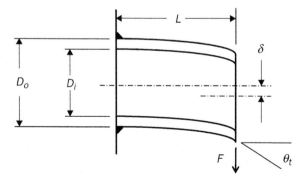

FIGURE 7.29 A cantilevered tube or truss deflects when pushed or pulled on by a force F; hence the centerline of the tube on one end is deflected by an amount δ with respect to the centerline on the clamped end.

larger than the tube weight on the other end, the decenter (or deflection) δ is given by[33,34]

$$\delta = \frac{FL^3}{3EI} \quad [\text{mm}] \tag{7.13}$$

This equation is valid when a heavy lens is located at the end of a much lighter tube. Although this is not the case for most designs of interest, the equation is not much different for cases that are of interest and serves to illustrate the physical principles.

These principles are that the resistance to decenter has both a material and geometric component. The material resistance is the elastic modulus E. The geometric component has a much larger impact on the deflection through the length L and the bending moment of inertia I (this bending moment is a different concept from the rotational moment of inertia for rigid-body motion). The bending moment of inertia is a measure of the structure's ability to resist bending and depends on the size and shape of the structural element; its value for the cylindrical tube that is typical of optical structures is given by[33,34]

$$I = \frac{\pi}{64}(D_o^4 - D_i^4) \quad [\text{m}^4] \tag{7.14}$$

This expression's dependence on outer diameter D_o usually controls the decenter, since D_o is raised to the fourth power. A tube designed to mount an f/2 lens with a 100-mm focal length, for example, resists bending with a moment $I = \pi[(0.054 \text{ m})^4 - (0.050 \text{ m})^4]/64 \approx 1.106 \times 10^{-7} \text{ m}^4$ for a wall thickness of 2 mm (D_o = 54 mm). The moment I also controls the angular tilt,[34] which is given by $\theta_t = FL^2/2EI$. Thus, for an aluminum tube (E = 71 GPa) holding a lens that weighs 3 newtons (~ 0.67 pounds), we have θ_t = 3 N × (0.1 m)²/[2 × 71 × 10⁹ N/m² × 1.106 × 10⁻⁷ m⁴] = 1.9 μrad.[35,36]

This value is significantly smaller than the high-precision tolerance (~ 72 μrad) for element tilt listed in Table 7.3; therefore, the structural tilt will create a slight image shift but will not significantly increase the blur size by inducing off-axis aberrations of coma and astigmatism. For comparison, a smaller f/2-lens mount with the same wall thickness and a 50-mm EFL (D_i = 25 mm) has a moment $I = \pi[(0.029 \text{ m})^4 - (0.025 \text{ m})^4]/64 \approx 0.015 \times 10^{-6} \text{ m}^4$ and a tilt angle θ_t = 3 N × (0.05 m)²/[2 × 71 × 10⁹ N/m² × 0.015 × 10⁻⁶ m⁴] = 3.5 μrad when holding a lens of the same weight. In any case, the consequences of tilt are typically predicted by exact ray tracing generated by software such as ZEMAX or CODE V.

If the load includes the self-weight of the tube then Eq. (7.13) is modified slightly and, as with the self-weight distortion of optical elements, illustrates the usefulness of a large stiffness-to-density ratio as a *material* figure of merit for structural deflection. The self-weight also includes the mass of the material, which can be reduced by cutting holes in the wall of the tube with minimal effect on the moment I. As the holes are made larger, the tube is transformed from a cylinder into a truss, which relies on the moment's strong dependence on size (D_o^4 for the tube) to resist deflection. In this way, the stiffness-to-weight ratio of the structure—and not just the stiffness-to-density ratio of the material—plays an important role in larger optical systems that require minimum weight and maximum stiffness.

Dynamic Deflections

The load F in Eq. (7.13) is classified as "static"—in other words, it does not change with time. Also typical are time-varying ("dynamic") forces, of which there are two types: sinusoidal and random (i.e., a collection of many different sinusoids of varying amplitude and frequency). For either type, the deflection for a tube can be larger or smaller than that predicted by Eq. (7.13) depending on how the vibration frequencies of the force compare with the resonant frequencies of the structure.

The so-called *resonance* frequencies of any structure depend on its mass, stiffness, and geometry. These are the frequencies at which structures such as tubes or guitar strings naturally vibrate (i.e., resonate or "ring") after an applied force is removed[42]—and are thus also known as *natural* frequencies. A structure of large mass oscillates at lower frequencies because there is more inertia associated with large masses. A stiff structure tends to vibrate at high frequencies because such structures have a larger spring force available to move the mass. For the textbook mass–spring system with spring stiffness k and mass m, the there is only one natural frequency $\omega_0 = (k/m)^{1/2}$; it is measured in units of radians per second.[37] It is also commonly expressed in units of cycles per second (Hz), in which case the natural frequency for the mass-spring system is given the symbol $f_0 = \omega_0/2\pi$. Note that the symbol f now denotes a temporal frequency, unlike the spatial frequency it represents in earlier chapters.

Structures such as tubes and trusses are more complex and therefore have multiple natural frequencies associated with the combination of shapes ("modes") into which the structure naturally distorts when vibrating. The smallest frequency is called the *fundamental* frequency, typically denoted with subscript "0"; for a cantilevered tube bolted stiffly on one end, a uniformly distributed self-weight, and a bending moment of inertia I for a cylinder or beam [given by Eq. (7.14)], this frequency is given by[37]

$$\omega_0 = 3.52 \sqrt{\frac{EI}{mL^3}} \propto \sqrt{\frac{k}{m}} \quad [\text{rad/sec}] \quad (7.15)$$

The structure's stiffness k is proportional to the EI/L^3 term, which controls static deflection [Eq. (7.13)]. A long structure oscillates at lower frequencies because of the increased time it takes for a longer cantilever to vibrate through a cycle of motion. The mass $m = \rho V$ also increases for a longer structure, indicating that the specific stiffness E/ρ is a useful figure of merit for both static and dynamic deflection. Physically, we also expect that more mass m (in the form of a lens at the end of the tube) will lower the natural frequency unless the stiffness $k \propto EI/L^3$ is increased to compensate.

These complex structures also have resonance frequencies that are higher than the fundamental frequency given by Eq. (7.15). For instance, the frequency of the second mode for the "clamped-free" cantilever is higher than the fundamental by a factor of 6.25, and the third-mode frequency is higher still (Fig. 7.30). Which of these modes are excited depends on the frequencies of the dynamic load. For very slow ("quasi-static") vibrations with an excitation frequency $\omega = 2\pi f$ much lower than the resonance frequency ($\omega \ll \omega_0$), the structure closely follows the applied load: the dynamic deflection δ_d is controlled by the static deflection δ [from Eq. (7.13), or its equivalent for different structures], which is a function of the applied force $F(t) = F_o \sin(\omega t)$ for a forcing frequency ω. In practice, the dynamic force $F(t)$ is described in terms of the number of "g's" of acceleration (5 g's, for example)—that is, the normalized acceleration a is measured in units of the gravitational acceleration (g). Hence the low-frequency deflection is $\delta_o = \delta \times a$, or 5δ for a 5-g acceleration, since the low-frequency dynamic weight is 5 times the static weight.

However, as the vibration frequency approaches a natural frequency such as the fundamental, the applied force becomes more "in sync" (in phase) with the velocity, causing the deflection near resonance to be *larger* than the low-frequency deflection. This phenomenon is illustrated in Fig. 7.31, a graph common to many engineering disciplines and known as a *Bode plot*. The plot shows how the deflection of an oscillating mass–spring system varies with sinusoidal forcing frequency and damping. Of particular interest is how the structure's vibration frequency matches that of the applied load, where the structure's deflection near a resonance can be greater than the low-frequency ($f/f_0 = \omega/\omega_0 \ll 1$) deflection δ_o. That is, the structure oscillates near resonance with a deflection that exceeds the low-frequency deflection such that $\delta(t) = \delta_d \sin(\omega t)$. The ratio of the structure's dynamic deflection δ_d to low-frequency amplitude δ_o is called the "amplification" or "gain," and it increases also in response to decreases in the structure's damping ξ.

Damping is evident in all structures as friction, air resistance, and inherent material damping, and it causes vibrations to "damp out" to zero when the force is removed. High-quality (high-Q) vibrations do not damp out quickly; they continue to resonate over many cycles, as one would expect in a well-designed bell, for example.

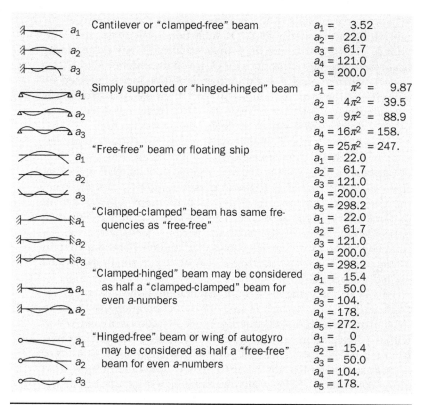

Figure 7.30 The mode shapes and associated natural frequencies of various beam supports. The resonant frequency $\omega_n = 2\pi f_n = a_n(EI/mL^3)^{1/2}$ in units of radians per second for f_n in units of Hertz (cycles per second). Note that the fundamental frequency in this figure is given the symbol ω_1, not ω_0. (Credit: J. P. Den Hartog, Mechanical Vibrations, Dover, 1985.)

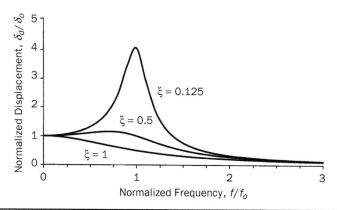

Figure 7.31 Depending on the value of the structural damping ξ, the displacement $\delta(t) = \delta_d \sin(2\pi f t)$ of a simple mass–spring system near its natural frequency f_0 can exceed that of the low-frequency displacement δ_0.

When a force is maintained, a structure's damping controls how much the near-resonance dynamic deflection δ_d can exceed the static, with $\delta_d = \delta_o \times Q_f$. Note that the transmissibility Q_f may exceed unity near resonance—typical values for Q_f are on the order of 10 to 100 for optical systems—and that this transmissibility, too, is affected by damping. If the level of damping is very high (i.e., if a structure is *critically damped* with $\xi \geq 1$), then damping limits the deflection at most frequencies to less than the low-frequency deflection.

In practice, dynamic loads have a number of different sources, none of which are pure sinusoids. Mechanisms moving on or near the optical bench, Stirling-cycle cryogenic coolers, buildings that shake as traffic moves on the road outside, aircraft vibrations from the engines or airflow—these forces (and many others) are all dynamic loads that must be considered in the structural design of an optical system. Because the forces are not pure sinusoids, it follows from Fourier analysis—which shows that random vibrations contain a range of frequencies—that various structural resonances may be excited at the same time.

A tool for analyzing these random vibrations is the power spectral density (PSD) curve, which is a plot of the acceleration produced by the source as a function of frequency. The acceleration is in comparison with the acceleration g due to gravity (9.81 m/sec²), squared, per unit bandwidth (Hz); this yields units of g²/Hz, analogous to the units of A²/Hz used for the detector PSD in Chap. 6. The physical interpretation of the mechanical PSD is that work done on a structure equals the dot product of the force and velocity ($F \cdot v$, or the component of the force that's moving in the same direction as the velocity) in units of kg·m²/sec³. For a unit mass of 1 kg, this work is equal to the acceleration squared (m²/sec⁴) per unit bandwidth (m²/sec³). The PSD is thus a measure of the work applied, per unit mass, to the structure at a specific frequency during vibration; thus it is also known as the acceleration spectral density (ASD).

A typical PSD plot is shown in Fig. 7.32 for a vibration source with constant PSD over the mechanical frequency band from 80 to 500 Hz. The net acceleration *applied to* the structure is found by integrating the PSD over the frequency band. Ignoring the ramp-up and ramp-down at the ends of the curve, the integration yields an acceleration $a = [0.04 \text{ g}^2/\text{Hz} \times (500 - 80) \text{ Hz}]^{1/2} = 4.1\text{g}$ RMS (within 1σ). For a Gaussian distribution of frequencies, the PV acceleration for 3 standard deviations is approximately 3 times the 1σ RMS,[38] or $a \approx 12.3\text{g}$ PV in this example ($a = 12.3 \times 9.81$ m/sec² = 120 m/sec² PV).

The extent to which the applied acceleration has any effect depends on the overlap of the PSD with the resonant frequencies of the structure. It is common for a structural resonance to occur within the frequency band of the PSD; when this occurs, the PSD drives any structural resonances and the specific value of the PSD *at a resonance*

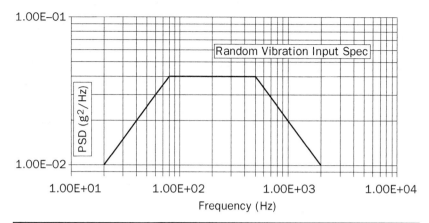

FIGURE 7.32 Typical PSD curve for acceptance-level testing of a satellite-launch vibration environment. (*Credit: NASA, www.nasa.gov.*)

frequency (e.g. PSD_0) determines the dynamic deflection. In other words, the RMS acceleration of the structure depends on the PSD_0 of the source and the transmissibility Q_f:[38]

$$\frac{a^2}{f_0} = \frac{\pi Q_f}{2} PSD_0 \quad [g^2/Hz] \tag{7.16}$$

The quantity on the left-hand side is the 1σ RMS acceleration (in units of g's) of the structure per unit natural frequency f_0. This equation is valid near a natural frequency where $Q_f \approx 1/2\xi$ (with Q_f greater than 10 or so) and if the PSD_0 is nearly constant around the natural frequency. For example, a structure with a fundamental frequency $f_0 = \omega_0/2\pi = 100$ Hz and a transmissibility $Q_f = 10$ at f_0 has an acceleration $a = [0.5\pi \times 10 \times 100 \text{ Hz} \times 0.04 \text{ g}^2/\text{Hz}]^{1/2} = 7.93$g RMS at 100 Hz for the PSD shown in Fig. 7.32.

Equation (7.16) shows that a structure with less damping (higher Q_f) accelerates more and thus has a greater near-resonant dynamic deflection $\delta_d = a/\omega_0^2$. As with the low-frequency case of a PSD spectrum that excites only the below-resonance frequencies, this deflection increases in proportion to the normalized acceleration a. Unlike the case of low-frequency vibrations, however, here the transmissibility $Q_f > 1$; this increases the dynamic force at the fundamental frequency f_0. Therefore, the equation for δ_d and Eq. (7.16) show that the near-resonant dynamic displacement is proportional to the square-root of the transmissibility. One difference between the low-frequency and near-resonant situation is thus the additional deflection due to a transmissibility $Q_f > 1$. The dependence of transmissibility on frequency is illustrated in Fig. 7.33 for the case of

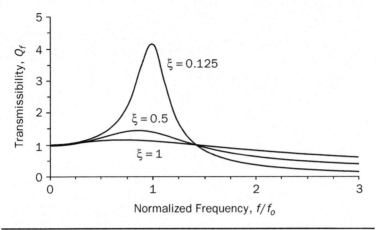

FIGURE 7.33 The ratio of dynamic to low-frequency force (or transmissibility Q_f) depends on frequency in a manner that is similar to the dynamic-to-low frequency displacement shown in Fig. 7.31. An important difference is that damping increases the dynamic force at frequencies greater than $1.4 f_0$.

a simple mass–spring–damper system. The resulting large deflections near a structural resonance can dramatically increase the tilt, decenter, and other misalignments of the optical system. For this reason, it is common practice to aim for the natural frequencies of a structure to be greater by a factor of at least 1.2 than the frequencies over which the applied PSD is large.

Equation (7.16) also seems to imply that a structure with a higher natural frequency has a larger dynamic deflection. But this is not the case: after all, we have just seen that the dynamic deflection δ_d depends also on the acceleration a and the transmissibility Q_f. Stiffer, lighter structures with a higher natural frequency f_0 also have a smaller static deflection in proportion to $1/f_0^2$ [use $F = mg = k\delta$ and compare Eq. (7.13) with Eq. (7.15)]. If Q_f is constant then the decrease in deflection for a stiffer structure dominates the increase in acceleration [$a \propto f_0^{1/2}$ from Eq. (7.16)], so a structure with a larger natural frequency has a smaller dynamic deflection $\delta_d \propto 1/f_0^{3/2}$ —emphasizing once again the importance of using a structure with a higher natural frequency than the frequencies found in the source PSD.

Optical systems actually contain more than a single mass–spring–damper structure; they contain multiple masses and springs coupled together to interact in ways not predicted by the 1-degree-of-freedom model on which Figs. 7.31 and 7.33 are based. These interactions can be extremely complex. For example, increasing the stiffness of one structure to move its natural frequency farther away from the PSD forcing frequencies may also increase the force transmitted by that vibration into other structures and thereby increase their dynamic deflection. A good rule of thumb for such cases is to design the different structures to have resonant frequencies that differ by a factor of 2 or more.[38]

Line-of-Sight Jitter

The tilt and decenter due to structural deflections have the effect of shifting the instrument's line of sight (LOS). For instance, a small downward tilt of the secondary mirror in Fig. 7.28 has the effect of shifting the image up on the FPA. This effect can result from either static or dynamic ("jitter" and "drift") displacements and misalignments in the pointing of the FOV. If the FPA is integrating over the time it takes for the LOS to shift dynamically, then there will be a larger blur on the FPA than predicted by the static equations. A structure with a fundamental frequency of 100 Hz, for example, takes 10 milliseconds to oscillate in its fundamental mode—shorter than the integration time of many systems with 30-Hz frame rates. These relatively short oscillation periods of structural vibration are known as LOS *jitter*, which results in tilt and decenter. This phenomenon is to be distinguished from long-term changes in LOS pointing, or *drift*, which can also be a problem with the long exposures—sometimes on the order of 10 to 20 hours for space-based telescopes—that are common when collecting enough photons for astronomical observations.

For structural tilt and decenter that are significantly smaller than the allowable element tolerances for off-axis aberrations such as coma and astigmatism, the optical blur θ_o and jitter blur θ_j are independent and thus can be added as an RSS to find the total image blur θ:

$$\theta = (\theta_o^2 + \theta_j^2 + \theta_a^2)^{1/2} \quad [\text{rad}] \tag{7.17}$$

(here a third term for the atmospheric blur, θ_a, has been added; cf. Sec. 3.3). To keep the effects of jitter small, typical values are $\theta_j \approx 0.1$–0.5 pixels (RMS). Thus, for a space-based imager with $\theta_a = 0$ and a diffraction-limited blur imaged onto one pixel, we have $\theta_o = 2.44\lambda/D = 1$ pixel and a total LOS jitter $\theta_j = 0.5$ pixels to cover both dimensions (or $0.5/\sqrt{2} = 0.35$ pixels in each of the x and y directions). Hence the total blur size $\theta = (1^2 + 0.5^2)^{1/2} = 1.12$ pixels, which is not much larger than the optical blur itself.

Ignoring the atmospheric blur, Eq. (7.17) shows that the optical design can trade the optical blur θ_o against the structural jitter θ_j—at least within the limits of the equation's validity. This is illustrated in Fig. 7.34, where the total blur θ is plotted as the radius of a circle whose value depends on the system requirements. The extreme cases thus allocate the entire burden of meeting these requirements either to the optical design (horizontal axis) or the structural design (vertical

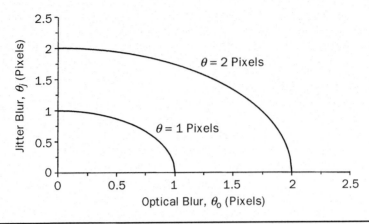

FIGURE 7.34 This plot of Eq. (7.17) illustrates how the blur size due to optical design and LOS jitter trade against each other for a total blur requirement of θ.

axis). A compromise between the two, where both components contribute roughly equal amounts to the blur, leaves the most margin for error against design uncertainties and long-term LOS stability. If the resulting jitter proves to be unacceptable, then it may be necessary to stiffen the structure, reduce its mass, increase the damping, or use a fine-scan mirror (see Fig. 2.25) or some other mechanism to compensate (e.g., the FPA motions or lens decenter intentionally used in commercial cameras to compensate for hand jitter).

An alternative to Eq. (7.17) is the use of an MTF analysis that describes how the modulation transfer function varies with spatial frequency for a "caffeinated" system. Jitter MTF depends strongly on the LOS motion θ_j, which randomly moves the blur spot around on the FPA at a rate faster than the frame rate; thus,[43]

$$\mathrm{MTF}_j = \exp\{-2(\pi\theta_j f)^2\} \tag{7.18}$$

where f is the spatial frequency in units of cycles per radian. This equation assumes that the motion occurs quickly during the integration time so that blurring of a point source occurs—rather than the visible smearing or "dancing" that results when the motion occurs more slowly over multiple frames. Stated differently: the frequency of motion is at a much higher rate than the frame rate (e.g., 30 Hz) and so the frame rate is unable to "freeze" the motion, resulting in the larger blur given by Eq. (7.17). Bessel functions are also used to approximate the MTF for this situation.[39]

The benefit of including the jitter MTF in the analysis is that it clearly illustrates the limiting contributor to image quality. The jitter MTF may be the limiting MTF in the system, in which case θ_j must be reduced to a smaller fraction of the pixel IFOV or optical blur—depending on the relative contributions of the FPA and optical MTF. Ideally, the relative contributions of the optical, FPA, and jitter MTF will be approximately equal at the FPA Nyquist frequency. In practice, this entails trading off the cost and risk of changing each component.

All element tilts and decenters have some WFE associated with them, and it is possible for initial designs to have tilt and decenter that increase the aberrations and optical blur θ_o significantly. Thermal distortion or excessively long and wobbly structures, for example, can result in huge misalignments between elements. In these cases, Eq. (7.17) is not valid; the system will likely need to be redesigned in order to maintain alignments and avoid excess aberrations, either by adding a control system to remove low-frequency jitter or by stiffening the structure to reduce the dynamic displacements and misalignments at higher frequencies. As shown in the next section, such redesign is most efficient when using software that allows an *integrated* or *collaborative* approach to the thermal, structural, and optical designs.

7.5 Component Specifications

Few optical systems are simple enough to be analyzed using the order-of-magnitude equations in this chapter. The calculation of thermal distortion, natural frequencies, and the like can be extremely complex, and the back-of-the-envelope estimates given in this book are often insufficient for anything other than first-order sizing. The results are useful to the optical systems engineer for purposes of initial design and trend analysis, but they are seldom accurate enough to be used for developing component specifications such as tube diameter and material stiffness. Simply choosing a large diameter and stiff material is unacceptable in light of the additional constraints of system size, weight, and cost. This means that techniques are needed for the more accurate prediction of optical system performance, particularly when high-precision tolerances or high-risk systems are involved.

Accurate predictions of performance must address two issues: complex geometries and the availability of integrated design tools that enable the simultaneous trading off of optical, thermal, and structural parameters. Complex geometries can be analyzed using finite-element analysis (FEA) and analytical cases from design handbooks such as *Roark's Equations for Stress and Strain*[23] and the *Handbook of Heat Transfer*.[21] Fully integrated design software is under development and is usually described as "structural, thermal, and optical (STOP)" analysis.

A STOP analysis consists of four components: (1) developing a computer-aided design (CAD) model of the optical and mechanical systems; (2) using a finite-element model to calculate the temperature and strains of the system under thermal and structural loading; (3) converting the calculated effects into wavefront errors; and (4) assessing the consequences of these WFEs for optical performance metrics such as blur size and MTF.

Finite-element models and finite-element analysis are used for geometries that are not easily broken down into the analytical cases found in design handbooks. This occurs less often in thermal than in structural design, where the tensor nature of stress and strain requires geometries such as cones, tubes, struts, and so on to have a large (though finite) number of spatial components. An example is shown in Fig. 7.35, where the entire volume of a lens-cell structure has been divided into small elements for which temperature, stress, and strain are computed separately; in turn, these factors determine the tilt, decenter, and defocus and well as optical component distortions and thermo-optic and stress optic effects. The selection of enough elements to give accurate results within a reasonable computation time is a critical skill of the thermal and structural analysts.

After creating a CAD model using a software tool such as ProEngineer, STOP analysis via finite-element analysis can take a

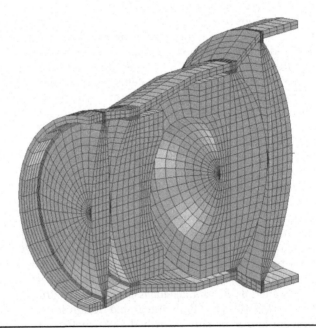

FIGURE 7.35 Finite-element analysis (FEA) of a lens requires that the lens be subdivided ("meshed") into small elements, each of which has material properties that depend on thermal and structural conditions. The thermal and structural meshes usually differ from one another. (*Credit: Sigmadyne, Inc., www.sigmadyne.com.*)

number of paths. The simplest technique for integrated design is to connect the thermal performance (Sec. 7.3) and the structural performance (Sec. 7.4) with the optical performance; as suggested by Fig. 7.36, the most accuracy is obtained by simultaneously including all the relevant thermal and structural effects and interactions. In the current state of the art, STOP analysis calculates temperature and stress distributions to determine the element displacements and wavefront error; additional structural effects (e.g., self-weight, vibrations due to motors and structural motion) can also be included.

Software packages for thermal analysis include Thermal Desktop, SINDA, and TSS. The temperature distributions are then fed into a structural analysis code such as NASTRAN, which along with the structural loads on the system determine the surface strains and element motions (tilt, decenter, and defocus).

These temperatures, stresses, deflections, and deformations determine the alignment and surface figure errors. NASTRAN and Abaqus do not calculate wavefront error from these parameters, so results from the structural models must be exported into SigFit or equivalent proprietary software. SigFit uses the estimated structural, thermostructural (CTE), thermo-optic (dn/dT), and stress-optic ($dn/d\sigma$) effects to compute the change in curvature for each optical surface, misalignments between surfaces, and change in index of refractive materials.[39]

Changes in surface curvature are represented by aspheric coefficients, and thermo-optic and stress-optic wavefront errors are calculated using polynomials originally developed by Fritz Zernike. These polynomials are used to generate an interferogram mapping of the WFE; this mapping—along with aspheric coefficients that capture the optical surface deformations—is exported to optical design codes such as Zemax or Code V, which allow the effects on image quality to be evaluated.[39] For imagers, the most important output from the optical code is WFE, from which image quality metrics (blur size, MTF, Strehl, etc.) are determined.[40]

The goal of an integrated STOP analysis is not to turn the structural designer into an optical engineer—or an optical designer into a thermal analyst. Rather, the intent is to reduce the barriers that prevent different designers from working together. Just as there are file-format and coordinate-system barriers between software packages, there are also difficulties in communication between designers with different areas of expertise. These difficulties may be greatly alleviated by optical systems engineers who have a good understanding of the relevant disciplines and can talk the same language as the structural, thermal, optical, and control system designers. Such a collaborative approach can greatly reduce the risk and cost of system development.[41]

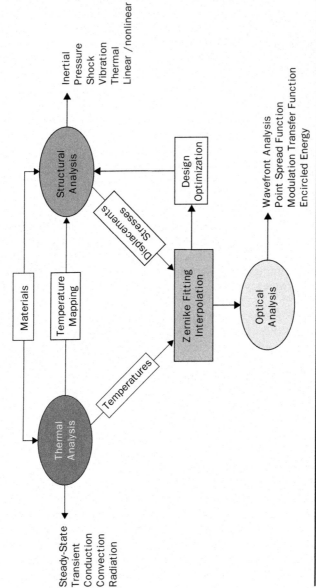

Figure 7.36 STOP analysis brings together elements of optical, thermal, and structural design to generate accurate predictions of total wavefront error. (*Credit: Sigmadyne, Inc., www.sigmadyne.com.*)

STOP analysis is usually employed on large projects and in cases (e.g., space-based instruments) where there is typically only one chance for success; for lower-budget projects, an approach mixed with experience and back-of-the-envelope calculations is the norm. In either case, an important systems engineering tool is the WFE budget that summarizes the component specifications from the design analysis (Fig. 3.43). As we have seen, excess wavefront error increases the blur size, modifies the MTF, and reduces image resolution. The budget lists the WFE associated with each component; this enables the designer to identify the largest detractors from image quality, indicating where efforts should be focused if the WFE is larger than the requirement.

The entries listed for each component in the WFE budget are those reviewed throughout this chapter: tolerances for design, fabrication (surface figure, wedge, etc.), and assembly (tilt, decenter, despace, and defocus) as well as thermal (wavefront distortion and structural deflection) and structural (wavefront distortion and structural deflection) errors. The fabrication term includes metrology errors, since the tests that verify WFE have their own tolerances. A unique aspect of space-based instruments is their inclusion of "on-orbit alignment" components, which include (among other factors) alignment changes due to escape from Earth's gravity, the thermal load transferred from the Sun and into deep space, and graphite-epoxy moisture loss (desorption). Independent errors are summed together using RSS addition to produce the total wavefront error.

Errors in the *estimation* of WFE arise because the models used to predict performance—integrated or otherwise—may not be properly (i.e., experimentally) validated. Once estimated, the allocation of WFE to the different categories is not always straightforward. Design WFE, for example, is often included as a "rolled up" entry summarizing the total WFE of the overall optical design, not the individual components. In contrast, thermal WFEs are listed separately for component distortion errors (due to CTE and dn/dT) and alignment errors between components. Thermal WFE is also often divided into (1) instrument-level thermal distortions and deflections due to internal heat loads and (2) system-level WFE due to external solar or deep-space loading; structural WFE is often treated the same way.

The allocations chosen should closely map the test procedures on the component, subsystem, and system levels so that the WFE entries can be verified by a combination of test, analysis, and assessment of similarity to other designs. At the same time, the allocations should also reveal the physical reasons for low-performance areas with large WFE. The coordination of these two goals underscores the need for "integration and test" engineers to be involved in developing the system architecture as early as possible in the design process (see Fig. 1.12).

Alignments with other major subsystems—for example, a telescope collecting photons to be fed into an instrument—must also be part of the system-level WFE budget; these include assembly misalignments and any effects (e.g., heat loads) that the instrument may have on the telescope. The wavefront error budget is perhaps one of the most visible results of the optical systems engineer's efforts, and it receives close scrutiny during preliminary and conceptual design reviews. These reviews involve more than just go–no-go checkoffs on a piece of paper, for if the actual WFE turns out to be worse than the budgeted requirement then the subsequent redesign may be time-consuming and expensive.

Given that thermal and structural misalignments cause both WFE and LOS errors, LOS budgets are also used to assess the full impact of the operating environment. Line-of-sight error budgets include separate entries for long-term, typically thermal changes (drift) as well as short-term, typically vibrational changes (jitter) that vary over the integration time. In order to properly manage risk, good systems engineering practice requires that both WFE and LOS budget allocations distribute errors more or less equally over as many elements as possible, thus reducing impacts of the failure or degradation of any one component.

Problems

7.1 For a thin lens with a tolerance of 0.5 percent on a 100-mm effective focal length, what is the tolerance on image distance for a nominal one-to-one ($2f : 2f$) imager? What is the tolerance on image distance for an object at infinity? How do the tolerances compare?

7.2 What tolerance on f/# is implied by the use of a single-digit spec such as f/2? Is it reasonable to specify f/#, or does it make more sense to specify the entrance pupil diameter and effective focal length?

7.3 In addition to field flatteners, what other types of lenses require a precision spec for bubbles and inclusions?

7.4 A mirror is fabricated with 8 fringes of power as measured at a wavelength of 633 nm. How many fringes of power are then expected at the design wavelength of 1.55 μm?

7.5 What is the allowable SFE for a mirror with a requirement of 1 fringe of irregularity at a wavelength of 0.5 μm?

7.6 Ignoring the tilt, decenter, and despace tolerances, what is the total WFE (design plus fabrication) for a lens consisting of four refractive elements if each has a WFE of $\lambda/5$? Is the system diffraction limited? If not, then what is the maximum WFE that each element can contribute?

7.7 A design team suggests that an RSS of tolerances underestimates the total. The team proposes the use of straight addition whereby 0.2 waves in one element adds directly with 0.3 waves in another to produce a total of 0.5 waves. Are there any conditions under which this approach is legitimate? Would it work in a manufacturing environment?

7.8 Why does 1 watt of optical power focused down to a 100-μm spot on a block of aluminum increase the aluminum temperature more than would an unfocused spot?

7.9 At what wavelengths would a temperature-control coating need to exhibit high emissivity for a cryogenically cooled object at 80 K?

7.10 Discuss the advantages and disadvantages of using a (visible) black paint for the interior surface of a baffle tube. How does it compare with (visible) white paint? Assume that the LWIR emissivity is the same for both paints. Does it matter if the instrument is intended for visible or LWIR use?

7.11 Is a thin window better or worse than a thick window for minimizing distortion due to a thermal gradient across the window? If defocus can be removed with a focus adjustment, is the thin window better or worse than the thick one?

7.12 Why do thin-film filters shift their peak transmission to longer wavelengths at higher temperatures? *Hint:* Assume that $dn/dT > 0$.

7.13 In selecting refractive materials for thermal design, indicate whether the following parameters should be large or small: absorption coefficient α_m, coefficient of thermal expansion α_t, thermal slope dn/dT, thermal conductivity k, and the ratio α_t/k. Do your answers depend on the application? If so, give examples of when it is better for a parameter to be large in one application yet small in another.

7.14 Given that strain is defined as $\varepsilon = \Delta L/L$, what is the thermal strain ε_t in terms of the CTE α_t?

7.15 The decenter due to deflection for the geometry shown in Fig. 7.29 is given by $\delta = FL^3/3EI$. Why isn't the decenter equal to the tilt angle multiplied by the length?

7.16 A guitar string clamped at both ends has the natural frequencies given in Fig. 7.30. Do the frequencies we hear depend on where along its length the string is plucked? Why or why not?

7.17 How does the natural frequency ω_0 given in Eq. (7.15) vary with length for a rectangular cantilever? How does it vary with the width and height of the beam? The bending moment of inertia of a rectangular beam is $I = bh^3/12$, where b is the width and h is the height.

7.18 Which is bigger and heavier: an f/2 or an f/4 imaging telescope? Assume diffraction-limited optics in both cases. Does the answer depend on how the f/# is changed (i.e., via the focal length versus the diameter)? Are normalized metrics such as f/# useful when dealing with absolute quantities such as size or weight?

Notes and References

1. Robert E. Fischer, B. Tadic-Galeb, and P. R. Yoder, *Optical System Design*, 2nd ed., McGraw-Hill (www.mcgraw-hill.com), 2008, chaps. 16–18.
2. Paul R. Yoder, Jr., *Opto-Mechanical Systems Design*, 3rd ed., CRC Press (www.crcpress.com), 2005.
3. Paul R. Yoder, Jr., "Mounting Optical Components,", in M. Bass, E. W. Van Stryland, D. R. Williams, and W. L. Wolfe (Eds.), *Handbook of Optics*, 2nd ed., vol. 1, McGraw-Hill (www.mcgraw-hill.com), 1995, chap. 37.
4. Paul R. Yoder, Jr., *Mounting Optics in Optical Instruments*, 2nd ed., SPIE Press (www.spie.org), 2008.
5. Daniel Vukobratovich, *Introduction to Optomechanical Design*, SPIE Tutorial Short Course (www.spie.org), 2009.
6. Daniel Vukobratovich, "Optomechanical Systems Design," in J. S. Acetta and D. L. Shoemaker (Eds.), *The Infrared and Electro-Optical Systems Handbook*, vol. 4, SPIE Press (www.spie.org), chap. 3.
7. R. R. Shannon, *The Art and Science of Optical Design*, Cambridge University Press (www.cambridge.org), 1997, chap. 6. See also Robert R. Shannon, "Optical Specifications" and "Tolerancing Techniques," in M. Bass, E. W. Van Stryland, D. R. Williams, and W. L. Wolfe (Eds.), *Handbook of Optics*, 2nd ed., vol. 1, McGraw-Hill (www.mcgraw-hill.com), 1995, chaps. 35, 36.
8. Warren J. Smith, *Modern Optical Engineering*, 4th ed., McGraw-Hill (www.mcgraw-hill.com), 2008, chaps. 10, 14, 16, 20.
9. R. Parks, "Optical Fabrication," in M. Bass, E. W. Van Stryland, D. R. Williams, and W. L. Wolfe (Eds.), *Handbook of Optics*, 2nd ed., vol. 1, McGraw-Hill (www.mcgraw-hill.com), 1995, chap. 40.
10. David Anderson and Jim Burge, "Optical Fabrication," in D. Malacara and Brian J. Thompson (Eds.), *Handbook of Optical Engineering*, Marcel Dekker (www.dekker.com), 2001.
11. Eric P. Goodwin and James C. Wyant, *Field Guide to Interferometric Optical Testing*, SPIE Press (www.spie.org), 2006.
12. Daniel Malacara, *Optical Shop Testing*, 3rd ed., Wiley (www.wiley.com), 2007.
13. D. Malacara, "Optical Testing," in M. Bass, E. W. Van Stryland, D. R. Williams, and W. L. Wolfe (Eds.), *Handbook of Optics*, 2nd ed., vol. 2, McGraw-Hill (www.mcgraw-hill.com), 1995, chap. 30.
14. H. H. Karow, *Fabrication Methods for Precision Optics*, Wiley (www.wiley.com), 2004.
15. A. B. Shorey et al., "Surface Finishing of Complex Optics," *Optics and Photonic News*, 18(10), 2007, pp. 14–16.
16. Ed Friedman and J. L. Miller, *Photonics Rules of Thumb*, 2nd ed., SPIE Press (www.spie.org), 2004, chap. 15.
17. Philip C. D. Hobbs, *Building Electro-Optical Systems*, Wiley (www.wiley.com), 2000, chap. 4.
18. Rudolf Kingslake, *Lens Design Fundamentals*, Academic Press, 1978.
19. Arthur Cox, *Photographic Optics*, 15th ed., Focal Press, 1974.
20. Dave S. Steinberg, *Cooling Techniques for Electronic Equipment*, 2nd ed., Wiley (www.wiley.com), 1991. Thermal resistance equations for specialized geometries can be found in this reference.

21. W. Rohsenow, J. Hartnett, and Y. Cho, *Handbook of Heat Transfer*, 3rd ed., McGraw-Hill (www.mcgraw-hill.com), 1998.
22. William P. Barnes, Jr., "Some Effects of Aerospace Thermal Environments on High-Acuity Optical Systems," *Applied Optics*, 5(5), 1966, pp. 701–711.
23. Warren C. Young and Richard G. Budynas, *Roark's Formulas for Stress and Strain*, 7th ed., McGraw-Hill (www.mcgraw-hill.com), 2001.
24. Michael H. Krim, *Athermalization of Optical Structures*, SPIE Tutorial Short Course T44, 1989.
25. P. J. Rogers and M. Roberts, "Thermal Compensation Techniques," in M. Bass, E. W. Van Stryland, D. R. Williams, and W. L. Wolfe (Eds.), *Handbook of Optics*, 2nd ed., vol. 1, McGraw-Hill (www.mcgraw-hill.com), 1995, chap. 39.
26. The dn/dT for the IR materials were calculated from the CTEs listed in Table 7.5 and the refractive index n and thermo-optical constant data given in M. Bass, E. W. Van Stryland, D. R. Williams, and W. L. Wolfe (Eds.), *Handbook of Optics*, 2nd ed., vol. 1, McGraw-Hill (www.mcgraw-hill.com), 1995, chap. 39, table 3.
27. J. E. Gordon, *Structures: Or Why Things Don't Fall Down*, DaCapo Press, 2003.
28. Roger A. Paquin, "Properties of Metals," in M. Bass, E. W. Van Stryland, D. R. Williams, and W. L. Wolfe (Eds.), *Handbook of Optics*, 2nd ed., vol. 2, McGraw-Hill (www.mcgraw-hill.com), 1995, chap. 35.
29. Marvin J. Weber, *Handbook of Optical Materials*, CRC Press (www.crcpress.com), 2003.
30. Material property data for the correct wavelength and temperature range is often difficult to obtain, so approximations must be used. Vendor catalogs are excellent sources of data, including Schott' s Web site for glass data (www.schott.com) and Crystran's publication *The Crystran Handbook of Infra-Red and Ultra-Violet Optical Materials*, Crystran Ltd. (www.crystran.co.uk), 2008.
31. As a rule of thumb, a minimum thickness of 2 mm is required for flatness during fabrication of standard-size elements, independent of aspect ratio.
32. David A. Bath and Eric A. Ness, "Applying Silicon Carbide to Optics," *Optics and Photonic News*, 19(5), 2008, pp. 10–13.
33. Virgil M. Faires, *Design of Machine Elements*, 4th ed., Macmillan, 1965.
34. J. P. Den Hartog, *Strength of Materials*, Dover (www.doverpublications.com), 1949.
35. For a load that is not centered, the tube will also twist. The amount of twist is determined in part by a material property known as the *shear modulus*, symbol G.
36. It is unfortunate that the kilogram (kg) is commonly used as a unit of weight. The actual unit of weight in the MKS system is the newton (N); it is equal to the mass multiplied by the acceleration due to gravity ($F = mg$), or kg × 9.81 m/sec^2, such that 1 pound = 0.454 kg × 9.81 m/sec^2 = 4.45 N. The use of "kilogram" for weight is fine for grocery stores but leads to serious problems in engineering.
37. J. P. Den Hartog, *Mechanical Vibrations*, Dover(www.doverpublications.com), 1985.
38. Dave S. Steinberg, *Vibration Analysis of Electronic Equipment*, 2nd ed., Wiley (www.wiley.com), 1988.

39. Keith B. Doyle, Victor L. Genberg, and Gregory J. Michels, *Integrated Optomechanical Analysis*, SPIE Press (www.spie.org), 2002.
40. Depending on the FPA frame rate, drift dominates in a thermal-based STOP analysis with relatively long time constants, whereas jitter is more important in a structural analysis that includes dynamic effects at about 1 Hz or faster. Micron-scale drift can also result from spontaneous changes in material dimensions; for details, see C. W. Marschall and R. E. Maringer, *Dimensional Instability*, Pergamon Press, 1977.
41. J. Geis, J. Lang, L. Peterson, F. Roybal, and D. Thomas, "Collaborative Design and Analysis of Electro-Optical Sensors," *Proceedings of the SPIE*, 7427, 2009.
42. Thomas D. Rossing and Neville H. Fletcher, *Principles of Vibration and Sound*, Springer-Verlag (www.springeronline.com), 1995.
43. Gerald C. Holst, *Electro-Optical Imaging System Performance*, 5th ed., SPIE Press (www.spie.org), 2008.

Index

"f" indicates material in figures. "n" indicates material in endnotes. "p" indicates material in problems. "t" indicates material in tables.

A

Abaqus, 405
Abbé diagram, 103f
Abbé factor, 100f
Abbé number. See V-number
aberrations
  about, 69–71
  astigmatism. See astigmatism
  chromatic. See chromatic aberration
  coma. See coma
  distortion. See distortion
  field curvature. See field curvature
  image quality and, 67–68, 117
  spherical. See spherical aberration
  spot diagrams for. See spot diagrams
absorption
  attenuation coefficient and, 138–139
  emissivity and, 191–194, 192f
  heat load from, 372, 374, 379
  in photodiodes, 249
  QE and, 307–308
  responsivity and, 243, 307–308
  thickness and, 137–138
  transmission and, 136–139

absorption coefficient, 238, 245, 255
absorption layer, 242, 259
Academy Awards, 21n12
acceleration, 396–400, 399f
acceleration spectral density (ASD), 398
achromats
  aberrations and, 71, 104, 105
  dispersion in, 105, 348
achromatic doublets, 36f, 71, 78f, 104–105, 104f
acid resistance, 100f
active-pixel sensor (APS), 273
adaptive optics, 129, 339n22
ADC. See analog-to-digital converter
Aeroglaze Z306 coating, 160
afocal systems
  description of, 24, 56–64
  scan mirror in, 60–61, 61f
  types of, 57f
air, refractive index of, 26
air resistance, 396
Airy pattern
  angular size of, 113, 261
  definition of, 113
  diffraction and, 113–116
  resolution and, 113, 116, 116f
aliasing, 261–269, 268f, 331–333

413

Index

alignment tilt, 364
alignment WFE, 119, 120f, 364, 365, 407–408
altitude, 33, 129, 155
aluminum
  CTE of, 376t
  elastic modulus of, 389t, 394
  reflectance of, 144, 145f
  specific stiffness of, 389t, 391
  specific thermal diffusivity of, 376t
  thermal conductivity of, 376t, 380
  thermal distortion of, 376t
  weight of, 391
aluminum gallium arsenide, 206, 221
aluminum gallium indium phosphide, 206, 206t
aluminum indium gallium nitride, 206t
amorphous silicon, 239t, 257–258
amplifier noise, 256–257, 279, 280f, 286, 292
AMTIR, 140f, 385t, 386, 386t, 389t
analog-to-digital converter (ADC)
  definition of, 279, 286
  dynamic range and, 319
  noise from, 279, 287–288, 288f, 292
anastigmat lens, 87–89, 89f, 92, 108, 176
angle of incidence (AOI)
  aberrations and, 70
angle of refraction, 41, 93, 94
Angstroms, 174–175
angular cutoff frequency, 124
angular deviation, 360
angular FOV, 29, 38, 148
angular IFOV, 154
angular magnification, 58–62
angular resolution, 115–116, 315, 331
angulon lens, 36f

anti-reflection (AR) coating
  definition of, 142
  Fresnel reflections and, 142, 172
AOI. *See* angle of incidence
APD, 244, 249–250, 250f, 256–257, 271, 339n16
aperture. *See also* entrance pupil area; lens diameter
  aberrations and, 70, 81, 109f, 111t
  atmospheric distortion and, 128–129
  blur size and, 235n20, 261, 331–332
  etendue and, 154–156, 334
  relative, 34–36, 149, 315–316, 340n37
  resolution and, 334
  sensitivity and, 334
aperture stop
  in afocal systems, 60, 61f, 63f
  chief ray and, 97t
  coma and, 71, 81–83, 83f, 90f, 108, 112
  definition of, 50
aplanat lens, 129p3.5
apochromats, 105
APS, 273
arc lamps
  about, 202–204
  applications for, 183, 202–203
  classification of, 186, 230t
  in comparison matrix table, 230t, 231
arcminute, 360
Arhenius relationship, 234n9, 256
ASAP software, 21p1.3, 224–225
ASD, 398
aspect ratio, 355–356, 390
aspheric surfaces, 76–77, 107–108, 112, 115, 345, 350
assembly tilt, 362, 365
astigmatism
  about, 84–91
  aperture and, 111t
  aperture stop and, 83, 89, 108
athermalization, 384–387

Index

atmosphere
  distortion from, 128–129, 339n22, 401
  solar excitance at top of, 232p5.1
  wavelengths absorbed by Earth's, 5t
attenuation coefficient, 138–139
avalanche photodiode (APD), 244, 249–250, 250f, 256–257, 271, 339n16
axial color, 96–98, 104–105, 111t, 112
axial ray, 44

B

back focal length (BFL), 30f, 40f, 44–49, 45f, 46t, 48t
background current
  background flux and, 282, 291
  BLIP and, 297, 340n33
  SNR and, 277, 295–300, 304
background flux
  NEP and, 301
  sensitivity and, 300–301
  well capacity/depth and, 319–320
background noise
  cold stop and, 194
  shot, 282–283, 291–293, 297–300, 310, 319, 336
  solid angle and, 282–283
background-limited infrared photodetector (BLIP), 297–304, 299f, 303f, 310, 340n33
baffles, 169–170, 170f, 171f, 194, 283
BaK1 glass, 100f, 140f
BaK2 glass, 104f, 105
bandgap energy
  cutoff wavelength and, 244
  definition of, 205–206
  photon energy and, 206, 244, 246
  responsivity and, 243

bandgap wavelength, 254
band-pass filters, 143, 297–298
bar, load on, 388
bar chart, 121, 122f
barium glass, 102–103, 102t
barrel distortion, 94–96, 94f
BDI, 288
beam area, 178
beam convergence, 144t
beam diameter, 56, 91
beam splitters
  astigmatism with, 89–91, 90f
Beer's law, 138, 139
beginning-of-life (BOL) value, 226, 378
bending moment of inertia, 394–395, 409
beryllium, 357–358, 376t, 389t, 391
best form singlets, 71, 77, 83
BFL, 30f, 40f, 44–49, 45f, 46t, 48t
bi-concave lens, 30f
bi-convex lens, 30f, 71, 72f, 360, 360f
bidirectional reflectance distribution function (BRDF), 177f
bidirectional scatter distribution function (BSDF), 175–179, 176f, 282–283, 298, 357
binoculars, 17–18, 17t, 59–64
bipolar TIAs, 286
birefringence, 100f, 140f, 181n6, 347t, 348–349, 391–392
blackbodies
  about, 187–195
  absorption by, 187, 191–193
  classification of, 186, 230t
  in comparison matrix table, 230t, 231
  definition of, 187
  spectral exitance of, 188–190, 189f, 196f
BLIP, 297–304, 299f, 303f, 310, 340n33
blur angle, 113
blur energy distribution, 119

blur size. *See also* resolution
  Airy pattern. *See* Airy pattern
  aliasing and, 261, 267, 331
  aperture and, 235n20, 261, 331–332
  detector size and, 126
  diffraction limited, 113–117, 115f
  f-number and, 331
  lens diameter and, 75, 81
  pixel pitch and, 267–269, 268f
  Q-parameter and, 267–269, 268f, 332
  spherical aberration, 73-75
Bode plot, 396, 397f
BOL, 226, 378
bolometers
  applications for, 259
  definition of, 242, 257
  for LWIR systems, 242, 257–259, 303t, 326, 328–330, 329t
  vs. photon detectors, 257, 259, 328
  vanadium oxide in, 257–258
Boltzmann constant, 189
boresight alignment, 16t
boron glass, 102, 102t
borosilicate glass
  Pyrex, 376t, 389t
boxcar averaging, 296
BRDF, 177f
Breault Research Organization, ASAP software, 21p1.3, 224–225
Brewster windows, 140f
BSDF, 175–179, 176f, 282–283, 298, 357
bubbles, 347t, 349
buffered direct injection (BDI), 288

C

C wavelength, 99, 104
CA. *See* clear aperture
CAD, 403–404
cadmium selenide, 246

cadmium sulfide, 239t, 246, 253–254
cadmium telluride, 254
calcite, 140f, 348
calcium fluoride, 140f, 387
calibration, 11, 193
cameras
  Academy Awards for designers of, 21n12
  EFL of, 15t, 33, 56
  field angle for, 33
  FOV of, 15t, 17t, 33, 56
  FPA for, 15t, 33, 260, 315
  IFOV of, 15t, 33
  MTF for, 15t, 127t
cantilevered beam, 393–396, 393f, 397f
capacitive TIAs, 280f, 285–288
carbon dioxide, 184f
carbon fiber. *See* graphite epoxy
Carlson, Chester, 237
Cassegrain telescopes, 36f, 76, 106, 107f, 107t, 126f
catadioptrics, 108, 393
CCDs. *See* charge-coupled devices
CDS, 289–290
center thickness (CT)
  focal length and, 345, 359
  illustrations of, 30f, 359f
  of thin lens, 26, 359
  tolerancing, 359, 361t
centration, 360–362, 361t
CFLs, 185, 204, 222
chalcogenides, 140f
change table, 346
charge capacity, 316. *See also* well capacity/depth
charge-coupled devices (CCDs)
  cooling of, 324
  dynamic range for, 318
  frame rate of, 272, 315
  frame transfer from, 339n22
  illustration of, 273f
  noise in, 273, 288, 292, 340n28
  orthogonal transfer from, 339n22

QE of, 309, 309f
well capacity/depth for,
    287–288, 292
charge-transfer capacitance,
    340n28
charge-transfer efficiencies
    (CTEs), 272, 273
chemical vapor deposition
    (CVD), 391
chief ray
  in afocal systems, 58, 64, 166
  aperture stop and, 97t
  distortion of, 94–96
  etendue and, 165
  in finite conjugates, 47
  tracing of, 43
chlorine, 185
chromatic aberration
  about, 96–98
  achromatic doublets for,
    104–105
  dispersion. *See* dispersion
  lateral color, 90f, 108–109, 109f,
    110f, 111t, 112
  V-number and, 104–105
CI Systems, 193
circle of least confusion, 73
clamped-clamped beam, 397f
clamped-free beam, 396, 397f
clamped-hinged beam, 397f
cleanliness
  BRDF and, 177f
  BSDF and, 176–178, 176f, 177f,
    282–283
  TIS and, 175
clear aperture (CA)
  about, 362
  in coating specification,
    144t
  in component drawing,
    346f
  in optical prescription,
    110–112, 111t
Cleartran, 141t, 181n6
climate resistance, 100f
clouds, reflectivity of, 160
cluster, dead, 323f

CMOS. *See* complementary
    metal-oxide semiconductor
    (CMOS) devices
coatings
  anti-reflection. *See* anti-
    reflection coating
  AOI of, 144t, 160
  reflectivity of, 142–144, 142f,
    144t, 169–170, 350, 378
  specification for, 144t
  surface roughness and, 193,
    358
  transmission and, 141–144, 147,
    350
Code V software, 21p1.3, 49, 127,
    394, 405
coefficient of thermal expansion
    (CTE), 100f, 140f, 356, 376t,
    380–386
coherence length, 128–129
cold stop, 194–195, 195f, 300, 311
collimation, 211–214, 219, 225
collimators, 210–214, 210f, 220
color fidelity, 70
column, dead, 323f
coma
  about, 79–84
  aperture and, 81, 111t
  aperture stop and, 71, 81–83,
    83f, 90f, 108, 112
  definition of, 77
  tilt and, 360, 367
compact fluorescent lamps
    (CFLs), 185, 204, 222
comparison matrix table, 229,
    230t
complementary error function
    (ERFC), 295
complementary metal-oxide
    semiconductor (CMOS)
    devices
  applications for, 273
  APS for, 273
  microlens arrays in, 270
  power consumption by, 324
  silicon for, 15t, 240, 273, 318
  windowing in, 273, 315

computer-aided design (CAD), 403–404
concave-convex lens, 72
concentration ratio, 254
concentricity, 368
conduction, 375–380
conductivity, 245
conic constant, 112
conical surfaces, 354f
contour fringes, 350–352, 351f
contrast
 definition of, 132n20
 image. *See* image contrast
contrast transfer function (CTF), 121, 122f
convection, 375–377
Cooke-triplet anastigmat, 88, 89f
copper, 376t, 380, 389t
correlated double sampling (CDS), 289–290
coupling lens, 214–216, 214f
Crawford, Frank, *Waves*, xi
critically damped structures, 398
crosstalk, amplifier, 265
crowns, 102–103, 102t, 140f
cryogenic liquid coolers, 324
Crystran, 411n30
CT. *See* center thickness
CTE, 100f, 140f, 356, 376t, 380–386
CTEs, 272, 273
CTF, 121, 122f
cutoff frequency, 123–125, 261–268, 266f
cutoff wavelength, 244, 310
CVD, 391
CVI Melles Griot, 20p1.1, 47, 48t, 201
cylindrical lens, 85, 354f, 378

D

Dall–Kirkham telescopes, 107t
damping, 396–399, 397f, 400f
dark current
 definition of, 256
 factors affecting, 256, 284, 312
 noise and, 247, 283–284, 292, 298, 319, 335
 SNR and, 276–277
dark current shot noise, 283–284, 292, 298, 319, 335
DAS, 31
"Death of a Spy Satellite Program" (Taubman), 4, 341
decenter
 aberrations and, 367
 about, 367–368
 definition of, 343
 from deflection, 392–394, 400
deep ultraviolet (DUV) light, 5t, 6
deflection, 387, 390–403, 393f, 397f
defocus. *See also* depth of focus
 active adjustment of, 386
 blur size and, 369–370, 369f
 definition of, 343
 temperature and, 371, 381, 383
degree to arcminute conversion, 360
degree to radian conversion, 83
density, 100f
depth of focus (DOF), 15t, 104, 369–370, 369f. *See also* defocus
design WFE, 119, 120f, 363, 363t, 407
desorption, 392, 407
despace
 CTE and, 383
 definition of, 343
 defocus. *See* defocus
detection probability, 293–295, 295f, 296f
detectivity
 definition of, 302
 specific, 302–304, 303f, 303t, 310–312, 310f, 311f
detectors
 array formats for, 272
 FPA. *See* focal plane array
 imagers. *See* imagers
 MTF of, 125, 264–266

noise and, 278–293
photoconductors. *See*
photoconductors
photodiodes. *See* photodiodes
quantum. *See* quantum
detectors
selection of, 327–336
specification for, 306–326
thermal. *See* thermal detectors
types of, 241–242
detector angular subtense
(DAS), 31. *See also*
instantaneous field of view
detector current
background. *See* background
current
dark current. *See* dark current
signal current. *See* signal
current
SNR and, 296
detector size
EE and, 126–127, 128f
field angle and, 29, 35
FOV and, 37, 55, 152–153
IFOV and, 126
solid angle for, 152–153
detector-limited (DELI) SNR,
298, 300
deuterium, 186t
dewar
cold shield/stop in, 195f
in MWIR system, 68f, 326
DI, 288, 318
dielectric mirrors, 143
diffraction
blur size, 113–117, 114f, 115f
EPD and, 113
f-number and, 113–115, 115f
focal length and, 114f
MTF and, 123–124
Rayleigh's criterion for, 116,
116f
Sparrow's criterion for, 116f
spot diagrams and, 75
diffuse surfaces, 158, 158f
diffusers, 225
diffusion, 265

digital single-lens reflex (DSLR)
cameras, 24, 133, 151
digital zoom, 273
digs, 358–359
diopters, 85
direct current (DC) systems, 328
direct injection (DI), 288, 318
dispersion
in glass, 99–104, 99f, 103f
Sellmeier formula for, 98–99,
100f
V-number for, 99–105, 103f,
347t, 348
dispersion curves, 98–99, 99f,
103f
displacement, 360, 397f
display
electroluminescent, 185
LCD, 183
LEDs for, 204, 207
dissipation, 221f, 374
distortion
about, 93–96
atmospheric, 128–129, 339n22,
401
barrel, 94–96, 94f
pincushion, 94–95, 94f
DOF, 15t, 104, 369–370, 369f. *See
also* defocus
double-Gauss lens, 36f, 110–112,
111t, 112f, 366f
doublets
achromatic, 36f, 71, 78f,
104–105, 104f
corrected, 77–78, 83–84
drift, 16t, 401, 412n40
DSLR cameras, 24, 133, 151
durability, 144t
DUV, 5t, 6
dwell time. *See* integration time
dynamic deflection, 387,
395–401, 397f
dynamic loads, 343, 395–401
dynamic range, 15t, 318–321,
329t, 334
dynamic resistance, 304–305,
305f

Index

E

Earth
  geosynchronous orbit above, 155
  radiation from (at 300K), 298
  satellite image of, 24f, 183, 184f
Eastman, George, 237
Eddington, Arthur, 66n1
edge filters, 143
edge thickness (ET), 359
edge thickness difference (ETD), 361
Edmund Optics Web site, 20p1.1
EE. *See* encircled/ensquared energy
effective focal length (EFL)
  of cameras, 15t, 33, 56
  FOV and, 55, 260
  GSD and, 155
  in multilens systems
    f-number and, 50–51
    ray tracing, 43, 45–47, 45f, 46t
  refractive power and, 39–40
effective responsivity, 269, 273
EFL. *See* effective focal length
Einstein, Albert
  The Evolution of Physics, 13
  general theory of relativity of, 66n1
elastic modulus, 388, 390, 394. *See also* Young's modulus
ELDs, 185
electric field strength, 358
electroluminescent displays (ELDs), 185
electromagnetic radiation, 185–188, 190, 377
electron, charge on, 281
electron-multiplying CCDs (EM-CCDs), 241
Electro-Optic Industries, 193
elliptical mirror, 76, 107t
EM-CCDs, 241
emissivity
  about, 190–195
  heat transfer and, 377–378
  reflectivity and, 191
  of sources, 188–197, 201, 303
encircled/ensquared energy (EE)
  Airy pattern and, 127–128, 127t, 212, 261, 269
  spot diagram of, 128f
end-of-life (EOL) value, 226, 378
energy, 162t
entrance pupil (EP)
  aberrations and, 71, 79–81, 95, 110
  about, 51–54
  etendue and, 155, 164–166, 165f
  image brightness and, 58
  power collection and, 110, 135, 148–149, 218
entrance pupil area. *See also* aperture
  detector signal current and, 277
  etendue and, 155, 164–166, 165f
  flux and, 149
  intensity and, 161
  irradiance and, 148–149
  power collection and, 148–149, 218
entrance pupil diameter (EPD). *See also* lens diameter
  Airy pattern and, 113
  diffraction and, 113
  f-number and, 16, 148
  resolution and, 129
  sensitivity and, 241
entrance window, 56
EOL, 226, 378
EP. *See* entrance pupil
EPD. *See* entrance pupil diameter
equiconvex lens, 73, 77
ER, 59, 62, 62f, 64
erector, 55f
ERFC, 295
error budget development, 18
ESA, 23
ET, 359
ETD, 361

etendue
  in afocal systems, 166
  aperture and, 154–156, 334
  basic, 167
  conservation of, 163–167
  EP and, 155, 164–166, 165f
  f-number and, 154–156, 334
  pixel size and, 284, 315–316, 334–335
  power collection and, 154–156, 154f
  sensitivity and, 334–335
  solid angle and, 154–156, 164–165, 165f, 334
European Space Agency (ESA), 23
Evolution of Physics, The (Einstein & Infeld), 13
excimer lasers, 140f
excitation frequency, 396
exit pupil (XP)
  about, 51–54
  of afocal systems, 58–60, 61f, 62–63, 62f, 63f
  cold stop at, 194–195
  etendue and, 164–166, 165f
exit pupil diameter (XPD), 62, 113
exit window, 56
exitance
  of blackbodies, 189–190, 191t
  directionality of, 163
exposure time, 35, 148–149, 315, 401
external transmission, 136–141, 138f, 141f, 181n5, 375f
extrasolar planets, 116, 133, 134f, 173
eye, human
  pupil size, maximum, 62
  sensitivity of, 197–198, 198f, 201
  spectral response of, 200t, 201
eye lens, 55f, 60f, 62–64, 62f, 108, 356
eye relief (ER), 59, 62, 62f, 64

F

F wavelength, 99, 104
f/#. *See* f-number
F2 glass, 100f, 140f
fabrication WFE
  description of, 118–119
  index inhomogeneities and, 348–349
  SFE and, 355–357
false alarm probability, 293–296, 296f
false alarm rate (FAR), 295–296
fans, 201
FAR, 295–296
FEA, 403–405, 404f
FET-based TIAs, 286
field angle
  detector size and, 29, 35
  focal length and, 29, 35, 152
  FOV and, 29, 33
  GSD and, 33
field curvature
  about, 91–93
  aperture and, 111t
  field angle and, 92–93, 111t
  Petzval, 89, 91–93, 93f, 112
field of illumination (FOI), 210f, 211–213, 220, 224
field lens
  in binoculars, 64
  field curvature and, 92
  FOV and, 59, 61
  illustrations of, 62f
  in Keplerian telescopes, 57
  surface quality of, 358
field of regard (FOR), 61
field stop
  about, 55–56
  scattering and, 176, 177f
  stray light and, 170
field of view (FOV)
  detector size and, 37, 55, 152–153
  etendue and, 154–156
  illustrations of, 32f
  instantaneous. *See* instantaneous field of view
field flattener, 93, 93f, 349, 358

422 Index

fill factor
  about, 260
  QE and, 309
  responsivity and, 309
  sensitivity and, 310
film, photographic
  etendue and, 155, 315
  exposure time for, 35, 148, 315
filters
  band-pass, 143, 297–298
  cold, 195
  edge, 143
  temperature of, 143, 195, 297, 311
  transmittance of, 350
finite conjugates
  blur size, 83, 235n20
  f-number for, 37
  magnification in, 37–38, 58
  NA for, 37f
  principal planes and points in, 47, 49f
  Scheimpflug condition, 365–367
finite-element analysis (FEA), 403–405, 404f
finite-element model, 403–404
first-order optics, 24
fish-eye lens, 7, 8f, 36f
fixed-pattern noise (FPN), 289, 292
Flagstaff, streetlights in, 183
flat fielding, 271
flicks, 163
flints, 102–103, 102t, 140f
floating ship, 397f
flowdown process, 16
fluorescence, 140f
fluorine glass, 102, 102t
flux
  background. See background flux
  definition of, 162
  signal, 160, 280, 291
  spectral, 277–278
f-number (f/#)
  aperture and, 334–335, 344

  cutoff frequency and, 123–124
  definition of, 16
  defocus and, 369–370, 387
  diffraction and, 113–115, 115f
  etendue and, 154–156, 334
  illustrations of, 78f
  image brightness and, 34–35, 134–135
  image quality and, 331–332
  image radiance and, 157–158
  integration time and, 331
  NA and, 36f, 340n37
  NEP and, 304
  optimum, 115
  SNR and, 331
  spatial frequency and, 123
  t-number and, 179–180p4.7
focal length
  back, 30f, 40f, 44–49, 45f, 46t, 48t
  definition of, 25
  effective. See effective focal length
  equation for, 27
  etendue and, 334
  f-number and, 34–35, 148, 334–335
  FOV and, 29–33
  illustrations of, 9f, 31f, 34f
  radius of curvature and, 27, 106, 352
  refractive index and, 26–27, 345
  refractive power and, 27, 350
  relative aperture and, 34
  temperature and, 382, 384–387
focal plane array (FPA)
  about, 259–260
  frame rate. See frame rate
  GSD and, 33
  of HST, 11
  for James Webb Space Telescope, 238–239
  of Kepler Space Telescope, 6, 7f, 11f
  NUC of, 271
  operability of, 322–323, 323f

power incident on
  etendue and, 154–156, 154f
  irradiance and, 149–151, 161
  radiance and, 157, 160
  uniformity of output from, 271, 320–322
  well capacity/depth of. *See* well capacity/depth
  windowing of, 272
FOI, 210f, 211–213, 220, 224
FOR, 61
force transmissibility, 398
FOV. *See* field of view
FPA. *See* focal plane array
FPN, 289, 292
fracture strength, 391
frame, 271, 272, 289
frame rate
  about, 271–272
  bandwidth and, 313–314
  windowed, 315
frame time, 313–314
FRED Web site, 21p1.3
free-free beam, 397f
frequency
  cutoff, 123–125, 261–268, 266f
  fundamental, 16t, 395–396, 397f, 399, 401
  natural, 387, 395–396, 397f, 399–400
  Nyquist, 262–263, 266, 268–269, 332, 403
  resonant, 397f, 398–401
  spatial. *See* spatial frequency
Fresnel reflections
  coatings and, 142, 172
  equation for, 137
  QE and, 308–309
  responsivity and, 242
  transmission and, 138–139, 141, 181n5
friction, 396
Fried parameter, 128–129
full width at half-maximum (FWHM), 230t
fundamental frequency, 16t, 395–396, 397f, 399, 401

fused quartz, 140f, 140t, 181n6
fused silica
  for SWIR systems, 139, 141t
  UV-grade form of, 139, 140f, 141t
  for visible light systems, 141t
  water-free grade of, 139
  wavelength range of, 141t
FWHM, 230t

G

Galilean telescope, 57, 57f
Galileo Galilei, 23, 57
gallium arsenide, 140f, 254
gallium arsenide phosphide, 206
gallium arsenide selenium, 140f
gallium nitride, 206, 206t, 208f, 239t
gallium phosphide, 206t
gases, greenhouse, 7, 185
gate averaging, 296
Gaussian beam blur size, 235n20
General Electric Research Labs, 186
generation-recombination noise, 246
geometrical optics, 9, 23, 25–36
geosynchronous orbit, 155, 290–291
germanium
  bandgap wavelength of, 254
  for LWIR systems, 103, 139, 141t, 385t, 386–387, 386t
  mount for lens of, 386–387, 386t
  refractive index of, 76, 140t, 384f, 386, 386t
germanium arsenic selenide, 140f
ghosting, 171–172, 172f
glare stop, 168f, 170, 176, 177f, 194
Glass, Philip, 275
glass constant, 385, 385t
glass map, 103, 103f
gold, 143, 144, 145f
Google Earth, 4, 23, 33, 94
graphite cyanate, 392

graphite epoxy
 CTE of, 376t, 381, 383, 386
 for optical bench, 392
gravitational acceleration, 396, 398
gravity, 343, 407
graybodies, 188, 191, 196f, 200t, 378
GRD, 331
greenhouse gases, 7, 185
Gregorian telescopes, 76, 107f, 107t
ground, 128, 377
ground resolved distance (GRD), 331
ground sample distance (GSD), 33, 155, 331, 335

H

half field angle, 35, 36f, 153
half field of illumination (HFOI), 212
half field of view (HFOV), 31, 63, 168f, 170
half well, 319
halogen gas, 186t. *See also* tungsten-halogen lamps
Handbook of Optics (Bass et al), 169, 238, 241
HB-LEDs, 185, 204, 206, 215–216, 230t
HDTV, 98, 149, 150f, 272
heat load, 372–380
heat sinks, 374–377, 374f, 379f, 381f
heat transfer, 373–381, 374f, 381f
heat-transfer area, 380
helium, 99
helium-neon (HeNe) laser, 118, 219, 352
hemisphere, 153
Herschel Space Telescope, 391
HFOI, 212
HFOV, 31, 63, 168f, 170
HgCdTe. *See* mercury cadmium telluride
HHT/HT grades, 350

HID lamps, 185, 202
high-brightness LEDs (HB-LEDs), 185, 204, 206, 215–216, 230t
high-definition television (HDTV), 98, 149, 150f, 272
high-intensity discharge (HID) lamps, 185, 202
high-operating-temperature (HOT) materials, 241
high-reflectivity (HR) optics, 143, 143f
high-transmission (HT) grades, 350
hinged-free beam, 397f
hinged-hinged beam, 397f
homogeneity, 100f, 347t, 348–349
HOT materials, 241
HR optics, 143, 143f
HT grades, 350
Hubble Space Telescope (HST)
 design of, 16
 despace tolerance for, 380–381, 383
 flowdown process for, 16
 illustration of, 2f, 12f
 NASA *Failure Report* on, 3–4, 21n10
 SA on primary mirror, 1–3, 67, 76, 344–345
hybrid FPAs, 272, 273–274, 274f, 287–288
hydrogen, 99, 104
hyperbolic mirror, 76, 106, 107, 107t

I

IBCs, 247
ICD, 18–19
IDCA, 325–326, 326f
IFOV. *See* instantaneous field of view
illumination, target, 210
image
 brightness of. *See* image brightness
 contrast in. *See* image contrast

Index **425**

distance from lens to. *See* image distance
height of, 44, 87, 88f, 92, 95–96, 150f
intensity modulation of, 122
irradiance of, 149, 150f, 151, 157–158
location of, 32–33, 33f, 52, 73, 90f, 93
orientation of, 9f, 25, 31f, 59, 66n9
quality of. *See* image quality
radiance of, 157–158
size of. *See* image size
image brightness
　entrance pupil and, 58
　etendue and, 156
　f-number and, 34–35, 134–135
　image size and, 34
　and source brightness. *See* radiance
image contrast
　MTF and, 125, 261
　scattering and, 173
　stray light and, 167
image distance
　for finite conjugates, 37, 47, 49f
　in linear magnification, 37
　object distance and, 27–28
image quality
　aliasing and, 261, 266, 332–333
　vs. aperture, 331–332
　atmospheric distortion and, 128–129
　MTF. *See* modulation transfer function
　pixel spacing and, 260
　Q-parameter and, 332–333
　SNR. *See* signal-to-noise ratio
　WFE. *See* wavefront error
image size. *See also* field of view
　chief ray and, 52
　determining, 32–33, 33f
　image irradiance and, 157–158
imagers
　cameras. *See* cameras
　description of, 4–5

function of, 6
Q-parameter for, 268
imaging radiometers, 5
impurity-band conductors (IBCs), 247
incandescent bulbs, 185, 187, 196, 222, 226
incident angle of refraction, 41, 93
inclusions, 347t, 349
index inhomogeneities, 348, 349
index ratio, 23
indium antimonide (InSb)
　dark current in, 283t
　for MWIR systems, 239t, 327, 339n17
　specific detectivity of, 303f, 303t, 310, 310f
　temperature of, 283t, 310–311, 310f
　uniformity of, 322
　wavelength range of, 239t, 246
indium arsenic, 303f
indium bumps, 274, 274f
indium gallium arsenic phosphide, 206, 206t
indium gallium arsenide (InGaAs)
　dark current in, 283t
　for NIR systems, 239t
　QE of, 308
　specific detectivity of, 303f, 303t
　for SWIR systems, 239t, 242
　temperature of, 283t
　uniformity of, 271, 321–322
　wavelength range of, 242, 243f, 244, 246
indium gallium nitride, 206, 206t
indium gallium phosphide, 206, 206t, 254
Infeld, Leopold, *The Evolution of Physics*, 13
infinite-conjugate achromats, 79f

infrared (IR) radiation
  absorption of, 187
  emission of, 10, 183, 187
  from graybodies, 196f
  longwave. See longwave infrared (LWIR) radiation
  midwave. See midwave infrared (MWIR) radiation
  near. See near infrared (NIR) radiation
  shortwave. See shortwave infrared (SWIR) radiation
  systems. See IR systems
Infrasil, 140f
instantaneous field of view (IFOV)
  aliasing and, 267
  angular, 154
  blur size and, 261, 267, 331
  detector size and, 126
  GSD and, 33
  illustrations of, 32f
  periodic patterns and, 260
  pixel size and, 31, 260–261, 315–316, 334–335
  power collection and, 154–155
  Q-parameter and, 268
  resolution and, 261, 331
  sensitivity and, 241
Institute of Electronic and Electrical Engineers (IEEE), LEOS Web site, 20p1.2
integral threads, 391
integrated detector-cooler assembly (IDCA), 325–326, 326f
integrating sphere, 225
integration capacity. See well capacity/depth
integration time. See also shutter time
  bandwidth and, 278, 291, 298, 313–314
  pixel size and, 315–316, 318, 335

  sensitivity and, 241, 300, 334
  shot noise and, 289, 291–292, 297–298
  SNR and, 297–299
  well capacity/depth and, 319
intensity
  about, 161–162
  entrance pupil and, 161
  Lambert's equation for, 159
intensity modulation, 122–123, 125, 132n20
interface control document (ICD), 18–19
interference angle, 114f
interference pattern, 351f, 352
interferometers, 4, 349, 355
intermediate image
  field stop at, 55, 55f, 166, 170
  FOV and, 63
  in Galilean telescopes, 57f
  in Keplerian telescopes, 57f, 170
intermediate pupils, 52, 356
internal transmission, 137–139, 138f, 350
internal transmission coefficient, 139
International Society for Optical Engineering (SPIE) Web site, 20p1.2
intrinsic layer, 249
Invar, 376t, 386–387, 386t, 389t, 392, 392f
iodine, 196
IR radiation. See infrared (IR) radiation
IR systems
  BLIP. See background-limited infrared photodetector
  FPAs for, 260
  longwave. See LWIR systems
  midwave. See MWIR systems
  near, 102, 139, 141t, 239t, 246, 328
  shortwave. See SWIR systems
  SNR of, 297–298
iron, 358

Index

irradiance. *See also* power density
  about, 148–151
  entrance pupil and, 148–149
  focal length and, 148–149, 162
  of image, 149, 150f, 151, 157–158
  power collection and, 148–149, 178
  solar, 150–151, 160, 166–167, 233p5.4, 251–252
  spectral. *See* spectral irradiance
  uniformity of, 224–225
ISO 10110 standard, 100f, 346f

J

James Webb Space Telescope, 238–239, 367
jitter, 16t, 401–403, 402f, 412n40
Johnson noise
  about, 284–286, 292
  JOLI, 298–301, 299f, 311–312
Johnson-limited (JOLI) situations, 298–301, 299f, 311–312
Joule–Thomson gas-expansion coolers, 324

K

Kepler, Johannes, 23, 57
Kepler Space Telescope, 6, 7f, 11f, 133, 134f
Keplerian telescope, 57–58, 57f, 176
Kirchhoff, Gustav, 191
Knoop hardness, 100f
K-parameter, 75, 75t, 114–115
Krugman, Paul, ix

L

LA, 72f
Labsphere Web site, 225
LaF15 glass, 111t, 112
LaF20 glass, 111t

Lagrange invariant, 165. *See also* optical invariant
LaK8 glass, 111t
LaK12 glass, 111t
LaK21 glass, 140f
Lambertian surfaces, 158–160, 159f, 173–176, 176f, 190
Land, Ed, 237
landscape lens, 36f
lanthanum glass, 102–103, 102t, 105, 111t, 112, 140f
Large Binocular Telescope (LBT), 342f, 345, 393
lasers
  applications for, 209
  classification of, 186
  scattering of light from, 173, 178
  stability of, 223–224
Lasers and Electro-Optics Society (LEOS) Web site, 20p1.2
LaSFN9 glass, 100f, 142f
lateral color, 90f, 108–109, 109f, 110f, 111t, 112
lateral SA, 73, 75
lateral-effect cells, 241
LBT, 342f, 345, 393
LCD, 183
lead glass, 102, 103f
lead selenide, 239t, 246, 303f
lead sulfide, 239t, 246, 303f
lens
  anastigmat, 87–89, 89f, 92, 108, 176
  angulon, 36f
  aplanat, 129p3.5
  bi-concave, 30f
  bi-convex, 30f, 71, 72f, 360, 360f
  combinations of. *See* multilens systems
  component drawing for, 346f
  concave-convex, 72
  double Gauss, 36f, 110–112, 111t, 112f, 366f
  equiconvex, 73, 77
  erector, 55f

lens (cont.)
  eye lens, 55f, 60f, 62–64, 62f, 108, 356
  fabrication of, 11, 342–364
  field. See field lens
  fish-eye, 7, 8f, 36f
  meniscus, 30f, 36f
  negative. See negative lens
  objective, 36f, 57–63, 60f, 62f, 168f
  plano-concave, 30f, 47
  plano-convex. See plano-convex lens
  positive. See positive lens
  radius of. See lens radius
  relay, 59–60, 60f, 68, 77, 92
  thick. See thick lens
  thin. See thin lens
lens cell, 368, 368f
lens diameter. See also aperture
  aperture stop and, 50–51
  blur size and, 75, 81
  clear aperture and, 362, 362f
lens radius
  of curvature. See radius of curvature
  plano-convex lens, 48t
  refractive power and, 49
  shape factor and, 73
  sign conventions for, 26–27, 27f
light
  UV. See ultraviolet (UV) light
  visible. See visible light
  wavelength range of. See wavelength
light pipes, 225
light-emitting diodes (LEDs)
  about, 204–209
  color of, 205–207, 230t
  in comparison matrix table, 230t, 231
  finite-conjugate coupling of, 214f, 215–216
  Fresnel reflections in, 207–208
  high-brightness, 185, 204, 206, 215–216, 230t

  temperature of, 221, 231, 232, 372, 372f
  thermal resistance in, 376f
  TIR of, 207–208
  wall-plug efficiency of, 221, 221f, 230t
  wavelength range of, 186t, 204–207, 205f, 206t, 218, 230t
lightning, 143, 160–161, 202, 279, 282, 290–291
linear magnification, 37–38
line-of-sight (LOS), 16t, 401–403, 402f, 408
line-spread function (LSF), 121, 122f
liquid nitrogen, 256
liquid-crystal display (LCD), 183
longitudinal aberration (LA), 72f, 96
longwave infrared (LWIR) radiation
  from arc lamps, 203
  from blackbodies, 186t
  systems. See LWIR systems
  wavelength range of, 5t, 6, 186t
LOS, 16t, 401–403, 402f, 408
LSF, 121, 122f
lumens, 197, 217
luminance, 230t
luminous efficiency, 198–199, 201, 222, 230t
lux, 218
LWIR radiation. See longwave infrared (LWIR) radiation
LWIR systems
  athermal designs for, 385, 386–387
  bolometers for, 242, 257–259, 303t, 326, 328–330, 329t
  comparison matrix table for, 328–330, 329t
  photoconductors for, 247, 328–330, 329t
  photodiodes for, 256, 328–330, 329t
  SFE of, 356

SNR of, 297–298
SWaP of, 325–326
V-numbers for, 102–103
Lyot stop, 170, 194. *See also* glare stop

M

magnesium fluoride, 140f, 142, 144
magnetorheological finishing (MRF), 358
magnifying glasses, 10, 133, 148, 156, 181n3
marginal rays
 aperture stop and, 51f
 illustrations of, 45f, 49f, 80f
 semidiameter and, 111t
 tracing of, 43
Martin Black anodize coating, 160, 169, 169f, 192–193, 192f, 193f
mass, 223, 314–315, 395–396
MCT. *See* mercury cadmium telluride
mechanical loads, 343, 348, 364, 387–401, 397f
medial focus, 86
melt-to-melt mean index tolerance, 100f
MEMS, 238, 258
meniscus lens, 30f, 36f
mercury arc lamps, 186t, 202–203, 202f, 230t
mercury cadmium telluride (MCT) (HgCdTe)
 for James Webb Space Telescope, 238–239, 239f
 uniformity of, 321–322
 wavelength range of, 239t, 246
Mersenne telescope, 176, 177f, 367
Messier, Charles, 79, 131n8
metal-insulator-semiconductor (MIS) capacitor, 339n24
metal-oxide semiconductor (MOS) capacitor, 272, 339n24

metals, 143
metering tubes, 375, 378–379, 381f, 388f
microbolometers. *See* bolometers
microelectromechanical systems (MEMS), 238, 258
microlens arrays, 269–270, 270f
microscopes, 9, 23, 36f, 37
midwave infrared (MWIR) radiation
 from blackbodies, 186t
 systems. *See* MWIR systems
 wavelength range of, 5t, 6, 186t
Mikron, 193
mil, 361t
mirrors
 aspect ratio, 355–356, 390
 emissivity of, 194
 fabrication of, 11, 342–345, 353, 355
 surface figure of, 355–356
 WFE of, 355–356, 390
modulation transfer function (MTF)
 aberrations and, 122–126, 123f, 125f
 about, 121–126
 at cutoff frequency, 123–125, 261, 264–266
 detector, 125, 264–266
 f-number and, 331
 image contrast and, 125, 261
 MTF squeeze, 266
 Nyquist frequency and, 266, 332
 pixel pitch and, 264–266
 plot of. *See* MTF curve
 Q-parameter and, 268–269, 268f
 spatial frequency and, 121–126, 127t, 261, 264–266, 402
 specification of, 15t, 125–126, 127t
moiré patterns, 263, 263f
Moon, 10, 133, 159, 167, 180p4.10, 184–185

430 Index

moonlight, 10, 133, 167, 184–185
Mouroulis, Pantazis, 13
MRF, 358
MTF. *See* modulation transfer function
MTF curves
  about, 122–123
  cutoff frequency, 261, 265–266, 266f
MTF squeeze, 266
multilens systems
  about, 38–40
  afocal. *See* afocal systems
  BFL of, 45–47, 46t
  EFL of
    f-number and, 50–51
    ray tracing, 43, 45–47, 45f, 46t
    track length and, 40f
  entrance pupil, 50–56, 51f, 53t, 54f, 55f
  exit pupil, 50–54, 51f, 53t, 54f, 55f
  f-number of, 50–52, 115
  focal length in, 38f, 39–40
  FOV for, 40, 55–56
  Petzval sum for, 92–93
  rays in, 40–47, 46t, 51–54, 51f, 53t, 55f
  refractive power of, 39–40, 43, 45–47, 46t, 93
  telephoto, 40, 40f, 43–47, 46t, 49f, 69f
multiplexer, 274
MWIR radiation. *See* midwave infrared (MWIR) radiation
MWIR systems
  athermal designs for, 385
  cooling of, 256
  dark current in, 312
  noise in, 299–300

N

NA. *See* numerical aperture
NASA, 3–4
NASTRAN, 405
National Aeronautics and Space Administration (NASA), 3–4
National Image Interpretability and Rating Scale (NIIRS), 331
natural frequencies, 387, 395–396, 397f, 399–400
natural position, 83, 89
near infrared (NIR) radiation
  from arc lamps, 186t, 203
  BK7 glass and, 102
  from blackbodies, 196f
  from graybodies, 196f
  from LEDs, 186t, 205, 206
  systems. *See* NIR systems
  from tungsten-halogen lamps, 186t, 196f, 199–201
  wavelength range of, 5t, 6, 141t, 186t
near ultraviolet (NUV) light, 5t, 6
NEE, 302
negative lens
  definition of, 29
  field flattener, 93, 93f, 349, 358
  illustrations of, 30f, 33f, 38f, 40f
NEI, 302
NEP, 240, 287f, 301–302, 304, 312, 338p6.27
NETD, 298, 302, 320, 320f, 323
New York Times, "Death of a Spy Satellite Program," 4, 341
Newport, 20p1.1, 193, 201
newton, 411n36
Newton, Isaac, 23, 187
Newtonian afocal telescopes, 57f
Newtonian reflecting telescopes, 76, 107f, 107t
Newton's fringes, 354f
Newton's rings, 350
nickel, 358
NIH attitude, 3
NIIRS, 331
NIR radiation. *See* near infrared (NIR) radiation

Index   431

no lead, arsenic, or cadmium in glass (N-glass), 102t, 103f
noise
  about, 278–293
  from ADC, 279, 287–288, 288f, 292
  amplifier, 256–257, 279, 280f, 286, 292
  background. See background noise
  dark current shot, 283–284, 292, 298, 319, 335
  G-R, 246
  Johnson. See Johnson noise
  for MWIR systems, 299–300
  from photoconductors, 246, 247
  from photodiodes, 246, 250, 255–257
  quantization, 279, 286–288, 288f, 292, 320, 335
  read noise. See read noise
  shot. See shot noise
  solid angle and, 156
  spatial, 279, 288–289, 290f, 292, 321, 336
  well depth and, 287
noise spectral density (NSD), 281, 284, 286, 292
noise-equivalent electrons (NEE), 302
noise-equivalent irradiance (NEI), 302
noise-equivalent power (NEP), 240, 287f, 301–302, 304, 312, 338p6.27
noise-equivalent temperature difference (NETD), 298, 302, 320, 320f, 323
nonuniformity
  in arc lamp output, 224f
  of irradiance, 149, 224
  photo-response, 270–271, 288–289
  in pixel output, 270–271, 271f
  residual, 271
  of responsivity, 260, 288–289, 320–322
  spatial noise from, 279, 288, 292
nonuniformity correction (NUC), 271, 321–322, 321f, 322f
"Not invented here" (NIH) attitude, 3
NSD, 281, 284, 286, 292
numerical aperture (NA)
  diffraction and, 114f
  equation for, 35
  for finite conjugates, 37, 37f
  f-number and, 36f, 340n37
NUV, 5t, 6
Nyquist frequency, 262–263, 266, 268–269, 332, 403

━━━ O ━━━

object
  brightness of, 121–122, 122f, 159
  distance from lens to. See object distance
  extended, 116, 121, 162, 164
  sign conventions for, 29t
object distance
  for finite conjugates, 37, 37f, 47, 49f
  image distance and, 27–28
  in linear magnification, 37
  wavefront curvature and, 28f
objective, 36f, 57–63, 60f, 62f, 168f
obscuration MTFs, 125, 126f
OPD, 118, 352, 355, 388f
open-circuit voltage, 252–253, 336p6.3
operability, 15t
optical bench, 392, 392f
optical engineering, 7–9, 8f, 11
optical fibers, 214–216, 214f, 235n20
optical invariant, 58, 165
optical path difference (OPD), 118, 352, 355, 388f
optical prescription, 110, 111t, 342

Optical Society of America (OSA) Web site, 20p1.2
optical systems
  alignment of components in, 11, 342–343, 364–370
  design of, 11–13, 210–216
  fabrication of, 341–364
  flowchart for acquisition of, 12f
optical systems engineering, 13–20, 19f, 133–134
optical transmission. *See also* power collection; throughput
  atmospheric loss in, 148
  budget allocations, 146–148
  coatings and, 141–144
  material properties and, 135–141
  vignetting and, 144–146
outgassing, 392
out-of-band rejection, 15t
overcoats, 144
oxygen, 197f

P

paints, 158, 169–170, 192, 194, 378
Palmer, Jim, 163
parabolic mirror, 36f, 75t, 76, 106, 107, 107t
paraxial angle, 47, 181n15
paraxial rays, 76, 78, 80f, 87
pascals, 389
Pasteur, Louis, 13
peak-to-valley (PV) acceleration, 398
peak-to-valley (PV) WFE
  blur size with, 118–119
  defocus, 118, 118t, 370
  fabrication, 118, 118t, 363
  index inhomogeneities and, 348
  for multilens systems, 119
  quarter-wave, 119, 362–363
  SFE and, 354f, 355, 356
pentaprisms, 392
Perkin-Elmer, 3–4

Petzval curvature, 89, 91–93, 93f, 112
Petzval radius, 92, 131n9
Petzval sum, 92–93, 105, 131n9
Petzval surface, 36f, 87f, 88f
Philips, 185
phosphate glass, 102, 102t
phosphate resistance, 100f
photoconductors
  about, 244–247
  definition of, 237
  illustrations of, 245f, 247f
  for LWIR systems, 247, 328–330, 329t
  noise of, 246, 247, 340n34
photodiodes
  about, 247–250
  avalanche, 244, 249–250, 250f, 256–257, 271, 339n16
  dark current in, 253, 256, 279, 283–284, 328
  for LWIR systems, 256, 328–330, 329t
  noise from, 246, 250, 255–257, 278–293
  photovoltaics. *See* photovoltaics
  PIN. *See* p-i-n photodiodes
  responsivity of, 252–253, 255, 257, 307–308
  reverse-bias. *See* reverse-bias photodiodes
  SNR of, 277–306
  specification for, 306–326
photomultiplier tubes (PMTs), 241, 281
photon detectors
  about, 242–244
  vs. bolometers, 257, 259, 328
  photoconductors. *See* photoconductors
  photodiodes. *See* photodiodes
  responsivity of, 242–243, 243f, 252–253, 255, 257, 307–308
  sensitivity of, 242, 255–257, 300
photon energy, 189–190, 204, 206, 242–246

photon noise, 280, 290f
photo-response nonuniformity (PRNU), 270–271, 288–289
photovoltaics
   classification of, 244
   p-i-n photodiodes as, 339n15
   solar cells. *See* solar cells
photovoltaic effect, 248
picture elements, 6, 238. *See also* pixels
p-i-n photodiodes
   intrinsic layer in, 249, 249f
   noise from, 281–284
   as PV detectors, 339n15
   reverse-bias. *See* reverse-bias photodiodes
   schematic of, 285f
pincushion distortion, 94–95, 94f
pinhole cameras, 66n4
pixels
   bandwidth of, 329t, 330
   FOV and, 31, 33
   inoperable, 321–323, 323f
   pitch. *See* pixel pitch
   size. *See* pixel size
   uniformity of output from, 240, 260, 270–271, 271f, 288–289, 320–322, 321f
pixel number
   FAR and, 296
   FOV and, 260
   image size and, 33
pixel pitch. *See also* pixel spacing
   aliasing and, 262–264
   cutoff frequency and, 262, 265–266, 268
   detector size and, 264, 268
   fill factor and, 269, 270f
   MTF and, 264–266
   Nyquist frequency and, 262–263, 266, 268
   Q-parameter and, 267–269, 332, 333, 333f
   spatial frequency and, 262–266, 262f
pixel size
   dark current and, 284, 292, 312
   etendue and, 284, 315–316, 334–335
   fill factor and, 269, 270f
   FOV and, 316
   GSD and, 335
   IFOV and, 31, 260–261, 315–316, 334–335
   integration time and, 315–316, 318, 335
   resolution and, 239, 260, 316, 335
   saturation and, 318, 335
   sensitivity and, 300, 334–335
   SNR and, 316, 335
   well capacity/depth and, 316, 318, 320, 335
pixel spacing, 31, 33, 260, 262, 269. *See also* pixel pitch
Planck, Maxwell, 188–189
Planck constant, 189
planets
   Earth. *See* Earth
   extrasolar, 116, 133, 134f, 173
plano-concave lens, 30f, 47
plano-convex lens
   definition of, 47
   illustrations of
      focal length, 30f
      SA, 9f, 78f
   shape factor for, 73
PMTs, 241, 281
point source
   Airy pattern and, 115, 116f, 132n17, 269
   classification as, 220
   definition of, 132n17
   FOI for, 212, 220
   oversampling of, 263, 264f
   Q-parameter for, 269
   visibility factor, 277
point spread function (PSF), 75, 161, 264. *See also* blur size
point visibility factor (PVF), 277
Poisson's ratio, 100f, 389t, 390
polar scattering profile, 158–159, 159f, 173
polarimeters, 4

434 Index

polarization, 137, 144t, 181n6, 348
polarizers, 140f, 345, 349, 353, 392
position factor, 73, 81
position-sensitive detectors, 241
positive lens
   definition of, 28–29
   illustrations of, 30f, 33f, 38f, 40f
pounds per square inch (psi) to MPa, 389t
power, scattered, 175, 178
power collection
   aperture and, 135, 152, 155–156
   entrance pupil and, 110, 135, 148–149, 218
   etendue and, 154–156, 154f
   f-number and, 154–155, 154f
   irradiance and, 148–149, 178
   pixel size and, 154–155
   radiance and, 156–157
   relative aperture and, 34
   solid angle and, 157–158, 157f
   throughput and, 135
power conservation, 163–164
power consumption
   ADC and, 325
   of arc lamps, 230t
   of blackbodies, 230t
   design options for, 22n22
   of LEDs, 230t
   of LWIR systems, 329t
   specification of, 222–223, 306, 324–325
   of tungsten-halogen lamps, 217, 230t
power density, 135, 149. *See also* irradiance
power spectral density (PSD)
   detector, 281
   mechanical, 398–400, 399f
precision molding, 103f
pressure, 388
principal planes, 45, 45f, 47, 49f
principal points, 30f, 47, 49f
principal ray, 55f, 85f. *See also* chief ray

prisms
   color separation by, 96
   deviation through thin, 130p3.10
   fabrication of, 11, 345, 353
   wedge specs for, 361
PRNU, 270–271, 288–289
project engineering, 18, 22n20
projected solid angle, 153, 157–160, 175
PSD, detector, 281
PSD, mechanical, 398–400, 399f
PSF, 75, 161, 264
PV acceleration, 398
PV detectors. *See* photovoltaics
PV WFE. *See* peak-to-valley WFE
PVF, 277
Pyrex, 376t, 389t
pyroelectrics, 257

Q

QDIPs, 241
QE, 307–309, 307f, 309f
Q-parameter, 267–269, 268f, 332–333, 333f
quad cells, 241
quantization noise, 279, 286–288, 288f, 292, 320, 335
quantum detectors. *See* photon detectors
quantum efficiency (QE), 307–309, 307f, 309f
quantum noise, 280
quantum sources
   arc lamps. *See* arc lamps
   lasers. *See* lasers
   LEDs. *See* light-emitting diodes
quantum-dot infrared photodetectors (QDIPs), 241
quantum-well infrared photodetector (QWIP), 239t, 298, 322, 328–330, 329t
quarter-wave optics, 119, 126, 142
quartz, crystal, 140f, 181n6, 348

quartz, fused, 140f, 140t, 181n6
quartz bulbs, 197, 202, 202f
quasi-static vibrations, 396
QWIP, 239t, 298, 322, 328–330, 329t

R

radiance
  about, 156–161
  basic, 167
  conservation of, 163–164, 166–167, 173
  detector signal current and, 277
  of image, 157–158
  of Lambertian surfaces, 159, 173, 175–176
  power collection and, 156–157
  scattered, 175
  solid angle and, 156–160, 167
  spectral. *See* spectral radiance
  of tungsten-halogen lamps, 212–213
radiant energy, 162t
radiated power, 188, 190
radiation
  from blackbodies, 186t, 196f
  from Earth, 298
  electromagnetic, 185–188, 190, 377
  from graybodies, 196f
  heat transfer via, 375, 377
  infrared. *See* infrared (IR) radiation
radio-frequency (RF) noise, 285
radiometers
  definition of, 5
  Q-parameter for, 268
  resolution of, 331–332
radiometry
  about, 133–135
  etendue. *See* etendue
  exitance. *See* exitance
  flux. *See* flux
  intensity. *See* intensity
  irradiance. *See* irradiance

names vs. common names, 162t, 188
optical transmission. *See* optical transmission
power conservation in, 163–164
radiance. *See* radiance
stray light, 167–179
radius of curvature
  BFL and, 49
  EFL and, 49, 344
  focal length and, 27, 106, 352
  in optical prescription, 110
  refractive power and, 26, 49
  shape factor and, 73
  sign conventions for, 26, 27f
  of spherical mirrors, 106
  of test plate, 352
  TPF and, 350–352, 351f
random forces, 395
rays
  angle of. *See* ray angle
  chief. *See* chief ray
  FOV, IFOV, and, 31–32, 32f
  height. *See* ray height
  marginal. *See* marginal rays
  paraxial, 76, 78, 80f, 87
  tracing of, 40–44, 42f, 45f, 49f, 51f
  wavefronts and, 29, 31f
ray angle
  in afocal systems, 58, 63–64
  etendue and, 165
  in ray tracing, 41, 42f, 45f
  TIR angle and, 215
ray height
  in afocal systems, 58, 63
  etendue and, 165
  in ray tracing, 41, 42f, 45f
  refractive angle and, 42f
  vignetting and, 146
Rayleigh's criterion, 116, 116f, 123, 129, 370
read noise
  about, 289–290
  of CCDs, 318, 324
  charge-transfer capacitance and, 340n28

read noise (*cont.*)
  definition of, 279–280
  frame rate and, 289, 290f
  from ROIC, 279–280, 289–290
  in total noise budget, 292
readout integrated circuit (ROIC)
  CTIAs in, 286
  direct injection, 288
  gain adjustments, 321
  in microbolometer, 258f
  quantization noise and, 288, 292
  read noise from, 279–280, 289–290
  schematic of, 276f
  spatial noise and, 279, 288
reflectance
  bidirectional distribution function, 177f
  coatings and, 143f, 169, 169f, 309
  emissivity and, 194–195
  refractive index and, 142f
  spectral radiance and, 160
  wavelength and, 142f, 169f
reflectance curves, 142f, 143f, 145f, 169f
reflection
  angle of, 106
  anti-reflection coatings and, 142
  of coating, 144t, 350
  diffuse, 158f
  Fresnel. *See* Fresnel reflections
  ghosting, 171–172, 172f
  in reflective systems, 139–141, 141f
  refractive index and, 137
  specular, 159f
  surface roughness and, 158f
  total internal, 207–208, 215
  transmission and, 136–141, 141f
reflective systems
  aberrations in, 106, 107f, 108
  disadvantages of, 107–108
  emissivity of, 195
  track length in, 106
  weight of, 106
reflectivity
  absorption and, 136–137, 139, 191–193, 242, 378
  coatings and, 142–144, 142f, 144t, 169–170, 350, 378
  emissivity and, 191
  of Lambertian surfaces, 159
  of paints, 378
  radiance and, 159
  wavelength and, 144, 145f
refraction, 42f. *See also* birefringence
refractive angle, 42f, 77f, 86, 89–91
refractive index
  in Abbé diagram, 103f
  aberrations and, 70
    astigmatism, 89, 367
    chromatic, 98–99, 103, 105, 109
    coma, 81
    field curvature, 92
    SA, 75, 75t, 76, 105
  of air, 26
  angular deviation and, 360
  of atmosphere, 128
  birefringence and, 348
  dispersion and, 103, 348
  focal length and, 26–27, 345
  of fused silica, 142f
  of germanium, 76, 140t, 384f, 386, 386t
  of glass
    of BK7 glass, 76, 98–99, 99f, 140f, 142, 142f, 389t
    of crowns, 102
    of flints, 102–103
    grade and, 347t
    of LaSFN9 glass, 142f
    of SF6 glass, 98, 99f, 103
    of SF11 glass, 48t, 49, 140f, 142f, 389t
    tolerances for, 347t
    grade and, 347–348

inhomogeneities, 348, 349
of IR systems, 76, 115
K-parameter and, 75
of magnesium fluoride, 142
in ray tracing, 41
reflectance and, 142f
reflection and, 137
of reflective systems, 106
of silicon, 76, 140f, 179p4.2, 384f
solid angle and, 167
striae, 347t, 349
temperature and, 17, 101f, 128, 373, 384–385, 384f
thermal slope of, 385t, 386, 386t
tolerancing, 347–349, 347t
transmission range and, 140f
for visible light systems, 76, 96, 103
of water, 137
wavelength and, 25f, 26, 98, 137, 213–214
of zinc compounds, 140f, 386, 386t
refractive power
aberrations and, 70, 93, 99
of afocal systems, 58
EFL and, 39–40
focal length and, 27, 350
lens radius and, 49
radius of curvature and, 26, 49
in ray tracing, 41
relative aperture, 34–36, 149, 315–316, 340n37. *See also* f-number
relative edge illumination, 15t
relative intensity, 205f
relative irradiance, 179–180p4.7, 224f
relay lens, 59–60, 60f, 68, 77, 92
residual nonuniformity (RNU), 271
resolution. *See also* blur size
Airy pattern and, 113, 116, 116f
angular, 115–116, 315, 331
aperture and, 334

Rayleigh's criterion for, 116, 116f, 123, 129, 370
Sparrow's criterion for, 116f, 123–124
wavelength and, 129, 261, 334
resonance, 395–400
resonant frequency, 397f, 398–401
responsivity
absorption and, 243, 307–308
of bolometers, 258–259
of CCDs, 272
of CMOS devices, 273
definition of, 242
detector signal current and, 277, 281
effective, 269, 273
electron generation and, 340n35
factors affecting, 306
fill factor and, 309
NEP and, 301–302
nonuniformity of, 260, 288–289, 320–322
operability, 322–323
of photon detectors, 242–244, 243f, 252–253, 255, 257, 307–308
QE and, 307–308
of quantum detectors, 242–243
of silicon, 251–252, 252f, 255, 257
of solar cells, 252–253
of thermal detectors, 243, 258–259
wavelength and, 242–244, 243f, 306, 310
retro-focus lens, 36f
reverse-bias photodiodes
about, 255–256
dark current and, 284, 306, 313, 328
I-V curve and, 251f, 285
responsivity and, 255
riflescopes, 59
rigid-body motion, 392, 394
risk, 17t, 18, 408

Ritchey–Chretien telescopes, 36f, 106, 107t
RMS acceleration, 398–399
RMS roughness, 357–358, 357f
RMS WFE, 15t, 118–120, 118t, 123f, 124, 144t
RNU, 271
ROIC. *See* readout integrated circuit
root sum of squares (RSS) noise, 280, 293, 340n31
root sum of squares (RSS) WFE, 119, 355, 363–364
root-mean-square (RMS) acceleration, 398–399
root-mean-square (RMS) roughness, 357–358, 357f
root-mean-square (RMS) WFE, 15t, 118–120, 118t, 123f, 124, 144t
rotational moment of inertia, 394
RSS noise, 280, 293, 340n31
RSS WFE, 119, 355, 363–364

S

Santa Barbara Infrared, 193
sapphire, 140f, 208f, 389t, 390
saturation, 151, 240, 306, 316–319, 317f, 335
scan mirror, 60–61, 61f
scatter angle, 176f
scattered power, 175, 178
scattering
  about, 172–174
  attenuation coefficient and, 138
  background noise and, 282
  BRDF, 177f
  BSDF, 175–179, 176f, 282–283, 298, 357
  by coating, 144t
  by diffuse surfaces, 158, 158f
  etendue and, 166
  factors affecting, 160
  image contrast and, 173
  at Lambertian surfaces, 158–159, 159f, 173, 176f
  of laser light, 173, 178
  MTF and, 125, 173
  noise from, 240
  polar profile, 158–159, 159f, 173
  SNR and, 298, 346
  stray light from, 168–171
  surface roughness and, 176, 357
  total integrated, 174
Scheimpflug condition, 365–367
Schmidt telescopes, 36f, 106
Schmidt–Cassegrain telescopes, 36f, 106, 108, 108f
Schott 517642 material, 347, 348
Schott Glass Web site, 132n13
scratches, 358–359
second law of thermodynamics, 35, 163
secondary color, 105
selenium, 237
Sellmeier coefficients, 99, 100f
Sellmeier dispersion formula, 98–99, 100f
semidiameter, 111t
semi-FOV. *See* half field of view
semikinematic mounts, 391
sensitivity
  aperture and, 334
  background current and, 299
  background flux and, 300–301
  of bolometers, 259
  definition of, 240, 277, 300
  etendue and, 334–335
  factors affecting, 241, 300, 312
  fill factor and, 310
  of human eye, 197–198, 198f, 201
  IFOV and, 241
  integration time and, 241, 300, 334
  NEP and, 301
  noise and, 255, 289, 297
  nonuniformity of, 320, 320f
  of photon detectors, 242, 255–257, 300
  pixel size and, 300, 334–335
  of thermal detectors, 259

sensitivity table, 346, 363
SF1 glass, 111t
SF2 glass, 104f, 105, 140f
SF6 glass, 98, 99f, 103, 376t, 385t, 389t
SF10 glass, 140f
SF11 glass, 140f
SF15 glass, 111t
SF57 material, 348
shape factor, 73, 74f, 81–83, 82f
shear modulus, 411n35
shielding, 194–195, 195f
shock, 343
short-circuit current, 252–253, 253t
shortwave infrared (SWIR) radiation
   from arc lamps, 186t
   BK7 glass and, 102
   from blackbodies, 196f
   systems for. *See* SWIR systems
   from tungsten-halogen lamps, 186t, 196f, 199–201
   wavelength range of, 5t, 6, 141t, 186t
shot noise
   background, 282–283, 291–293, 297–300, 310, 319, 336
   dark current, 283–284, 292, 298, 319, 335
   definition of, 279
   integration time and, 289, 291–292, 297–298
   signal, 280–282, 291, 294–295, 297
shutter time, 148, 152, 315. *See also* integration time
SigFit, 405
signal current
   aperture and, 297
   to background current ratio, 293
   dark current and, 296
   detection probability and, 294–295
   electron generation and, 291
   flux and, 277, 291

   shot noise and, 280–282, 291
   in SNR, 293–297
   statistical analysis of, 279, 294f
   wavelength and, 277, 281
signal shot noise, 280–282, 291, 294–295, 297
signal-to-background ratio, 293
signal-to-noise ratio (SNR)
   about, 293–296
   background limited, 297–298
   detection probability, 293–295, 295f, 296f
   electrical vs. optical, 293–294
   false alarm probability, 293–296, 296f
   f-number and, 304, 331
   Johnson limited, 299–300, 299f, 301
   pixel size and, 316, 335
   RSS noise for, 293
   saturation and, 306
   scatter and, 298, 346
   signals and, 277–278, 294, 297
   stray light and, 358
silicon
   amorphous, 239t, 257–258
   bandgap energy of, 244, 251–252
   for bolometers, 257–258
   for CCDs, 273f
   for CMOS systems, 15t, 240, 273, 318
   CTE of, 376t, 383
   dark current in, 283, 283t
   doped, 239t, 247, 328–330, 329t
   for LWIR systems, 139, 239t, 328–330, 356
   manufacturing processes developed for, 237–238
   MEMS and, 238
   for MWIR systems, 102–103, 139, 141t, 385t
   for NIR systems, 239t
   for photon detectors, 242, 246–247, 248, 250–257
   Poisson's ratio for, 389t

silicon (*cont.*)
  refractive index of, 76, 140f, 179p4.2, 384f
  responsivity of, 251–252, 252f, 255, 257
  ROIC, 274, 274f, 383
  in solar cells, 179p4.2, 238f, 251–253, 252f
  specific detectivity of, 303t
  specific stiffness of, 389t
  specific thermal diffusivity of, 376t
  temperature of, 283t, 384f
  thermal conductivity of, 376t
  thermal distortion of, 376t
  thermal slope of, 385t
  thermo-optic coefficient of, 385t
  for UV systems, 239t
  for visible light systems, 239t, 327
  for VLWIR systems, 239t
  V-number for, 102–103
  wavelength range of, 140f, 239t, 244, 247
  Young's modulus for, 389t
silicon carbide, 239t, 376t, 389t, 391
silicon dioxide, 181n6, 274f
silver, 143
silver halide, 148
simple lens, 9f. *See also* thin lens
simply supported beam, 397f
single lens systems
  aberrations and, 9f, 67, 71, 76
  aperture stop in, 83f, 97t
  FOV of, 29–33
  imaging with, 25–29
  Petzval sum for, 92
  rays through, 29, 41, 44, 47, 48t
  relative aperture of, 34–36, 149, 315, 340n37
  sign conventions, 29t
  sign conventions for, 27f
  thick lens. *See* thick lens
  thin lens. *See* thin lens
  wavefronts in, 25f, 28f, 31f

singlets
  astigmatism and, 87
  athermal, 385, 386, 386t
  best form, 71, 77, 83
  SA and, 75
sinusoidal forces, 395
size
  of blur. *See* blur size
  of image. *See* image size
  of object, 37, 83, 95, 188
  of pixels. *See* pixel size
  of sources. *See* source size
size, weight, and power (SWaP), 22n22, 325–326
Smart, Anthony, 13
Snell's law, 23–24, 31–32, 40–41, 69–70, 70f, 93
sodium arc lamps, 230t
solar cells
  cadmium telluride for, 254
  classification of, 244
  coating for, 179p4.2
  concentrators for, 254, 254f
  description of, 250–251
  efficiency of, 251–254
  gallium arsenide for, 254
  germanium for, 254
  indium gallium arsenide for, 254
  indium gallium phosphide for, 254
  I-V curve for, 251f
  photocurrent of, 252
  power generation by, 135
  responsivity of, 252–253
  silicon for, 179p4.2, 238f, 251–254, 252f
solar concentrators, 254, 254f
solar exclusion angle, 170–171, 171f
solar rejection filter (SRF), 143, 147, 147t
solid angle
  cold stop and, 300
  detector signal current and, 277
  for detector size, 152–153

etendue and, 154–156, 164–165, 165f, 334
of FOI, 212–213
for FOV, 152–153
illustrations of, 152f
image size and, 157–158
intensity and, 161–162
noise and, 156
pixel size and, 156
projected, 153, 157–160, 175
radiance and, 156–160, 167
refractive index and, 167
sources
about, 183–186
arc lamps. *See* arc lamps
blackbodies. *See* blackbodies
classification of, 186
comparison matrix table for, 229–232, 230t
cooling of, 223, 230t
cost of, 227, 230t
coupling of, 214–216
efficiency of, 220–222
energy from, 6, 212
extended, 116, 156, 210–213, 220
graybody. *See* graybodies
lasers. *See* lasers
LEDs. *See* light-emitting diodes
lifetime of, 226–227, 230t, 234n9
natural, 10, 184–185
output spectrum of, 96, 150, 156, 185, 217–218, 230t
point. *See* point source
power consumption of, 217, 222–223
power output of, 216–226, 230t, 279–280, 334
properties of, 228t
quantum. *See* quantum sources
selection of, 227–232
size of. *See* source size
specifications for, 216–227
stability of, 223–224, 230t

of stray light, 133, 167–168, 168f, 183
thermal. *See* thermal sources
tungsten-halogen lamps. *See* tungsten-halogen lamps
types of, 186–187, 230t
uniform output by, 224–225, 224f, 230t
source size
angular
point source, 132n17, 220
power collection and, 157
resolution and, 115–116
collimation and, 211–212, 220
FOI and, 211–212, 220
heat dissipation and, 156
image size and, 212, 220
radiated power and, 156, 188
Sparrow's criterion, 116f, 123–124
spatial distribution chart, 203f, 205f
spatial frequency
aliasing and, 262–263, 266
f-number and, 123
FPA sampling, 261–263
MTF and, 121–126, 127t, 261, 264–266, 402
in MTF curves. *See* MTF curves
normalized, 124
Nyquist frequency and, 262–263, 266
pixel pitch and, 260, 262–266, 262f
unit of measurement, 123
spatial noise, 279, 288–289, 290f, 292, 321, 336
specific detectivity, 302–304, 303f, 303t, 310–312, 310f, 311f
specific stiffness, 389t, 391, 396
specific thermal diffusivity, 376t, 381
specifications, 14–17, 228–229
spectral emissivity, 194, 378
spectral emittance, 311
spectral emitters, 197

spectral exitance
  of blackbodies, 188–190, 189f, 196f
  of graybodies, 196f
  of tungsten-halogen lamps, 196f, 199
  Wien's Law on, 190
  of xenon arc lamp, 203f
spectral flux, 277–278
spectral intensity, 161
spectral irradiance
  blackbody curves for, 189–190, 189f
  Planck formula for, 188–190
  reflected, 175
  solar, 150, 160, 252
  spectral radiance and, 160, 175
spectral NEP, 301
spectral power, 218
spectral radiance
  definition of, 156–157
  emissivity and, 192
  specification of, 219
  spectral exitance and, 190
  spectral irradiance and, 160, 175
Spectralon coating, 160
spectrometers, 4
specular angle, 158–159, 174–175
sphere, 153
spherical aberration (SA)
  about, 71–78
  aperture and, 111t
  blur size with
    about, 73–75
    of best form singlets, 77
    K-parameter and, 75t
    refractive index and, 75t
  f-number and, 75, 76, 114, 115f
  on HST's primary mirror, 1–3, 67, 76, 344–345
  lateral, 73, 75
  lens orientation and, 9f
  longitudinal, 72f, 96
  marginal rays and, 79
  MTF and, 122–123, 123f
  position factor and, 73
  reducing, 76–78
  shape factor and, 73, 74f
  splitting lens and, 77f
  transverse, 72f, 75
  WFE and, 84, 84f, 117–118, 118t
spherical mirror, 75t, 106, 107t, 108
spherical surfaces, 354f
SPIE, 20pl.2
splitting lens, 36f, 68, 76, 77f
spot diagrams
  for astigmatism, 88f, 125f, 128f
  for coma, 81, 82f
  diffraction and, 75
  for SA, 73–75, 74f
spring stiffness, 395
SRF, 143, 147, 147t
SRR, 14
stain resistance, 100f
stainless steel, 376t, 389t
stars
  color of, 185
  diffraction and, 115–116
  source classification of, 162
  Sun. See Sun
static deflection, 392–401
static loads, 343, 392–395
Stefan–Boltzmann constant, 188
steradians, 153
stiffness, 395–396, 400
stiffness-to-density ratio, 391, 395
Stirling-cycle coolers, 256, 324, 325, 398
STOP analysis, 343, 386, 391, 403–407, 406f, 412n40
strain, 388–392, 404
strain optic coefficient, 391
stray light
  about, 167–179
  design options for, 178–179
  finish specifications on, 357
  ghosting, 171–172, 172f
  noise from, 240

reducing, 168–171
SNR and, 358
sources of, 133, 167–168, 168f, 183
Strehl ratio, 120, 121f
stress, 391, 404–405
stress birefrigent, 392
stress optic coefficient, 389t, 391
stress-induced index difference, 348–349
striae, 347t, 349, 349f
structural, thermal, and optical (STOP) analysis, 343, 386, 391, 403–407, 406f, 412n40
Subaru Telescope, 129
Sun
  angular size of, 151, 167
  color of, 185, 187
  exitance from, 232p5.1
  irradiance from, 150–151, 160, 166–167, 233p5.4, 251–252
  light from. *See* sunlight
  in optical system, 6f
  peak exitance wavelength for, 191t
  radiance of, 167
  solid angle of, 167
  surface temperature of, 189f, 191t
sunshades, 168–171, 168f, 170f, 171f, 283
super-achromats, 105
Suprasil, 140f
surface curvature
  aspheric, 76–77, 107–108, 112, 115, 345, 350
  mapping of, 352
  pressure and, 388
  radius of. *See* radius of curvature
  refractive power and, 350
  test plate measurement of, 351f
  tolerancing, 350–353
surface figure, 352, 353–357, 358t, 382–383, 388

surface figure error (SFE), 353–356, 354f, 387, 390–391
surface quality, 358–359, 358t
surface roughness
  background noise and, 282
  BSDF and, 176, 357
  coatings and, 193, 358
  definition of, 357
  emissivity and, 192–193
  measurement of, 357f
  radiance and, 160
  RMS, 357–358, 357f
  scattering and, 176, 357
  TIS and, 174–175, 174f
  tolerancing, 357–358, 358t
SWaP, 22n22, 325–326
SWIR radiation. *See* shortwave infrared (SWIR) radiation
SWIR systems
  materials for, 138f, 139, 141t, 239t, 246, 303t
  sources for, 186t
system requirements review (SRR), 14

T

TA, 72f, 75
Taubman, Philip, "Death of a Spy Satellite Program," 4, 341
TCR, 258
TEC, 256, 286, 324, 325f, 326, 336
technology readiness level (TRL), 17t, 18
telephoto imager, 40, 40f, 43–47, 46t, 49f, 69f
telephoto ratio, 47
telescopes. *See also specific telescopes*
  afocal, 24, 56–64, 57f
  designing of, 266–267
  EFL of, 56
  FOV of, 17t, 56
  performance metrics for, 115
  resolution of, 129
  throughput budget for, 147
  trade study on, 17–18, 17t

444 Index

temperature
  absolute, 188, 189f, 234n4
  absorption and, 375, 379
  alignment of components and, 11, 343
  background noise and, 194, 282–283
  of blackbodies, 189f, 191t, 221
  BLIP and, 299f, 300, 310, 340n33
  dark current and
    Arhenius relationship, 256
    material and, 283t
    uniformity of, 322
  decenter and, 383
  defocus and, 371, 381, 383
  design considerations, 370–373, 405
  emissivity and, 198f
  exitance and, 190
  of filters, 143, 195, 297, 311
  focal length and, 382, 384–387
  FPA and, 371
  of lens, 136, 382
  lifetime and, 234n9
  of LWIR systems, 187, 256, 257
  management of, 373–380
  NEDT, 298, 302, 320, 320f, 323
  noise and, 194, 282–283
  peak emission wavelength and, 189
  photoconductors and, 246–247, 256
  photodiodes and, 256
  power and, 156, 163, 188, 374f
  QE and, 307f
  read noise and, 289
  refractive index and, 17, 101f, 128, 373, 384–385, 384f
  responsivity and, 310
  ROIC and, 371
  sensitivity and, 194, 310–312
  stability of, 223–224
  of thermal sources, 191t
  transmission and, 139, 371
  warm-up time to, 223
  WFE and, 119, 120f, 364
  Wien's Law on, 190
temperature coefficient of resistance (TCR), 258
Tessar lens, 36f
test, 11–13, 12f
test plan, 19
test plates, 350–356, 351f, 354f
test-plate fit (TPF), 350–356
thermal conductivity, 375, 376t, 380, 381
thermal contraction, 388f
thermal detectors
  about, 241–242, 257–259
  bolometers. *See* bolometers
  manufacturing of, 238
  response time of, 315
  responsivity of, 243, 258–259
  sensitivity of, 259
  wavelength range of, 187
thermal diffusivity, 376t, 381
thermal distortion, 376t, 381, 403
thermal energy, 189–190, 247, 284
thermal expansion, 17, 343, 380. *See also* coefficient of thermal expansion
thermal fracture, 197, 214, 349–350, 373
thermal management, 373–380
thermal power, 163, 220, 284, 324
thermal resistance, 374–380, 376f, 410n20
thermal slope, 385t, 386, 386t, 411n26
thermal sources
  about, 187
  arc lamps. *See* arc lamps
  blackbodies. *See* blackbodies
  graybodies, 188, 191, 196f, 200t, 378
  temperature of, 191t
  tungsten-halogen lamps. *See* tungsten-halogen lamps
  warm-up time for, 223
  wavelength range of, 186, 188, 218, 219

thermoelectric cooler (TEC), 256, 286, 324, 325f, 326, 336
thermoelectric detectors, 257
thermo-optic coefficient, 385, 385t
thick lens, 42f, 44–56, 46t, 49f, 137–138, 359
thickness
　absorption and, 137–138
　center. *See* center thickness
　coating, 142, 144t
　edge, 359
　EFL and, 344
　lens
　　absorption and, 137–138
　　edge, 359
　　EFL and, 344
　　tolerance for, 346f
　　transmission and, 137–139, 350
　photon detector layers, 245, 249, 255, 270–271, 322
　polishing and, 355
　PV WFE and, 348
　tolerancing, 346f, 363t, 364
　transmission and, 137–139, 350
thickness-to-diameter ratio. *See* aspect ratio
thin lens
　center thickness of, 26, 359
　EFL and, 65p2.8
　finite conjugates. *See* finite conjugates
　focal length of, 26–27, 384–385
　FOV of, 29–33
　imaging with, 25–29
　rays through, 44, 47
thin-film filters, 371, 383
Thorlabs Web site, 20p1.1
three-mirror anastigmat (TMA), 108, 176
threshold current, 294–295, 295f
throughput. *See also* transmission
　about, 135
　budget allocations, 146–148
　coatings and, 141–144

material properties and, 135–141
relative irradiance and, 179–180p4.7
signal current and, 297
specification for, 148t
TIA, 280f, 285–288, 285f
tilt
　about, 359–362, 365–367
　adjustments for, 367
　alignment, 364
　angle of. *See* tilt angle
　AOI and, 365, 367
　assembly, 362, 365
　astigmatism and, 89–91, 360, 365, 367
　blur size and, 367, 403
　coma and, 360, 367
　from deflection, 392–393, 400
　of flat surfaces, 360–361
　from jitter, 401
　line of sight and, 401
　mechanical loads and, 387, 392–393
　of mirrors, 60, 367
　Newton's fringes and, 354f
　ray deviation by, 361f
　temperature and, 383
　tolerances for, 361t, 363
tilt angle
　alignment and, 343
　of beam splitters, 91
　mechanical loads and, 394
　tolerances on, 361, 365
　weight and, 394
tip/tilt, 365
TIS, 174
titanium, 102, 102t, 376t, 387, 389t
TMA, 108, 176
t-number (t/#), 179–180p4.7
tolerances
　alignment, 364–369, 373, 380–383, 386–387
　fabrication, 345–363, 358t, 363, 363t, 388
　mechanical loads and, 387–388, 390

total integrated scatter (TIS), 174
total internal reflection (TIR), 207–208, 215
total reflected power, 174
total refractive power, 39–40
TPF, 350–356
track length, 15t, 40f, 45f, 47, 62–63, 106
trade study, 17–18
trade table, 17–18, 17t
transimpedance amplifier (TIA), 280f, 285–288, 285f
transmissibility, 16t, 398–400, 400f
transmission
  budget allocations, 146–148
  coatings and, 141–144, 144t
  diffuse, 158f
  external, 136–141, 138f, 141f, 181n5, 375f
  heat load and, 374–375
  internal, 137–139, 138f, 350
  material properties and, 135–141
  specular, 159f
  vignetting and, 144–146
transmission coefficient, 139
transmission curve, 138f
transmittance, 350
transverse aberration (TA), 72f, 75
transverse magnification, 58–60
triplets, 36f
TRL, 17t, 18
truss, 393, 393f, 395
tubes
  deflection of, 393–395, 393f
  fundamental frequency of, 395–396
  metering, 375, 378–379, 381f, 388f
  natural frequencies of, 395–396
  at resonance, 395–396
  stiffness of, 395–396
  twisting of, 411n35

Tucson, streetlights in, 183
tungsten-halogen lamps
  about, 195–201
  chemical process in, 197f
  in comparison matrix table, 230t, 231
  cooling of, 230t
  emission spectra of, 196f
  emissivity of, 196–197, 198f, 201
  etendue of, 213
  exitance of, 199–201
  graybody model for, 201
  irradiance of, 213
  lifetime of, 226, 227f, 230t
  luminance of, 230t
  luminous efficiency of, 199–201, 222, 230t
  luminous output of, 230t
  operating voltage of, 226, 227f
  output power from, 217
  peak exitance wavelength for, 191t
  power consumption of, 217, 230t
  power output of, 199, 202, 219, 230t, 279–280
  radiance of, 212–213, 219
  stability of, 223, 230t
  temperature of, 191t, 196, 201, 221
  wall-plug efficiency of, 221, 230t
  wavelength range of, 186t, 204, 218, 230t
T-value, 385, 385t
Type-II superlattice detectors, 241

U

ultra-low expansion (ULE) glass, 358, 376t, 382, 389t
ultraviolet (UV) light
  from arc lamps, 186t, 203
  from deuterium, 186t
  systems for. *See* UV systems

from tungsten-halogen lamps, 196f
wavelength range of, 5–6, 5t, 132n12, 141t, 186t
undercorrected zonal SA, 78
unity conjugates, 38
U.S. Federal Energy bill, 185
UV light. *See* ultraviolet (UV) light
UV systems
  coatings for, 169
  materials for, 139, 140f, 141t, 239t
  scatter and, 175

V

V/#, 99–105, 103f, 347t, 348
vacuum ultraviolet (VUV) light, 5t, 6
vanadium oxide, 239t, 257–258, 303t, 328, 329t
vanes, 169–170, 171f
velocity, 398
very longwave infrared (VLWIR) radiation, 5t, 6, 186t
very longwave infrared (VLWIR) systems, 239t, 256, 298, 328
VGA, 272
vibration
  alignment of components and, 11, 343
  damping and, 396–398
  frequencies of, 395–398
  jitter, 16t, 401–403, 402f, 412n40
  PSD plot for, 398, 399f
viewgraph engineering, 20, 20f
vignetting
  about, 144–146
  FOV and, 146
  image brightness and, 63
  ray height and, 146, 146f
  throughput and, 144–146
visible light
  absorption of, 187, 194
  from arc lamps, 186t, 203

from blackbodies, 196f
emission of, 187
from graybodies, 196f
from LEDs, 186t, 205
systems for. *See* visible light systems
from tungsten-halogen lamps, 186t, 196f
wavelength range of, 5, 5t, 141t, 186t
visible light systems
  blur size and, 113
  coatings for, 169
  materials for, 138f, 141t, 239t, 246, 303t
  photoconductors for, 328, 330
  SFE of, 356
  SWaP of, 325
  throughput budget for, 148t
  TPF test wavelength for, 352
  V-number for, 99, 102, 104
VLWIR radiation, 5t, 6, 186t
VLWIR systems, 239t, 256, 298, 328
V-number (V/#), 99–105, 103f, 347t, 348
volume, 15t
VUV, 5t, 6

W

wall-plug efficiency, 204–205, 220–222, 228t, 230t, 231–232, 235n22
warm-up time, 223
water, refractive index of, 137
wavefronts
  aberrated, 84
  adaptive optics and, 129
  atmospheric distortion of, 128–129
  collimated, 28, 57, 210-211
  distance between. *See* wavelength
  in imaging, 25–28, 25f, 28f
  rays and, 29, 31f

wavefront error (WFE)
  about, 117–119
  alignment, 119, 120f, 364, 365, 407–408
  astigmatism, 117, 118t
  atmospheric, 128–129
  budget, 119, 120f, 343–344, 407–408
  calculation of, 404
  collimators and, 213–214
  coma, 117, 118t
  definition of, 69, 84
  defocus and, 118, 118t, 370
  diffraction and, 119
  distortion, 390, 407
  fabrication. See fabrication WFE
  heat load and, 373
  index inhomogeneities and, 348, 349
  mechanical loads and, 343, 364
  of mirrors, 355–356, 390
  MTF and, 120, 123–124, 123f
  OPD and, 118, 388f
  PV. See peak-to-valley (PV) WFE
  PV to RMS ratio, 118t
  RMS, 15t, 118–120, 118t, 123f, 124, 144t
  RSS, 119, 355, 363–364
  SFE and, 353–356, 354f
  spherical, 84, 84f, 117–118, 118t
  Strehl ratio and, 120, 121f
  striae and, 349, 349f
  surface figure and, 352–355, 354f
  temperature and, 119, 120f, 364, 379–380, 379f
  wavelength and, 356
  Zernike polynomials for, 405
wavelength
  aberrations and, 110
  absorption and, 242
  Airy pattern and, 113
  bandgap, 254
  cutoff, 244, 310
  definition of, 5
  defocus and, 370
  diffraction and, 69, 113
  emissivity and, 190–191, 193
  focal length and, 28, 96-105, 214
  of IR radiation, 5–6, 5t, 132n12, 186t
  photon energy and, 189–190, 242–244
  QE and, 307f
  reflectance and, 142f
  reflectivity and, 144, 145f
  refractive index and, 25f, 26, 98
  resolution and, 129, 261, 334
  responsivity and, 242–244, 243f, 306, 310
  specific detectivity and, 302, 303f
  surface spatial, 357f
  throughput and, 148t
  TIS and, 174, 174f
  TPF test, 352
  of UV light, 5–6, 5t, 132n12, 186t
  of visible light, 5, 5t, 141t, 186t
  WFE and, 118, 356
waveplates, 140f
Waves (Crawford), xi
W-curve, 322, 322f
wedge, 360–361, 361t, 367
wedge angle, 360–361, 361t
weight
  deflection from, 391–395
  design options for, 22n22, 391
  of optical bench, 392f
  of reflective systems, 106
  tilt angle and, 394
well capacity/depth
  background flux and, 319–320
  of CCDs, 287–288, 292
  definition of, 316
  direct injection and, 288
  half, 319
  integration time and, 319
  of LWIR systems, 319, 329t

pixel size and, 316, 318, 320, 335
saturation and, 317f, 319
voltage and, 318
Wertheimer, Max, ix
WFE. *See* wavefront error
Wien's Law, 190
windowing, 272, 273, 274, 315
windows
  aspect ratio, 355–356, 390
  deflection of, 390
  defocus from, 388
  entrance, 56
  exit, 56
  fabrication of, 11, 345, 353
  materials for, 140f
  transmission and, 139–141

X

xenon arc lamps
  about, 202–203
  in comparison matrix table, 230t
  illustration of, 202f
  luminance of, 230t
  luminous efficiency of, 222
  photograph of, 226f
  spatial distribution chart, 203f
  spectral exitance of, 203f
  wavelength range of, 186t, 203f
XP. *See* exit pupil
XPD, 62, 113
xy alignment, 367. *See also* decenter

Y

Yerkes Observatory refractor, 106
Young's modulus, 100f, 388–389, 389t. *See also* elastic modulus

Z

Zemax software, 21p1.3, 49, 127, 394, 405
Zernike, Fritz, 405
Zerodur, 140f, 356, 376t, 382, 389t, 390

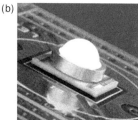

FIGURE 1.9 Optical sources such as (a) tungsten lamps and (b) high-brightness light-emitting diodes add light to the system. (*Photo credits: (a) Wikipedia author Stefan Wernli; (b) Cree, Inc., used by permission.*)

FIGURE 2.1 Composite satellite image showing cloud patterns across the planet. This image was collected using the MODIS instrument, an imaging spectroradiometer. (*Credit: NASA, www.nasa.gov.*)

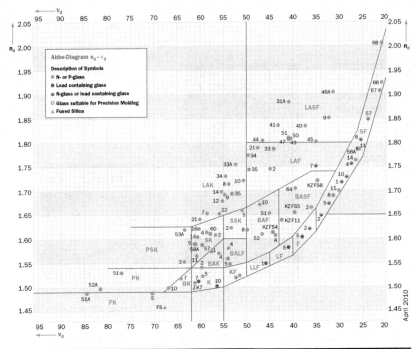

Figure 3.32 Abbe diagram showing refractive index n_d as a function of dispersion v_d for a variety of optical materials. (*Credit: SCHOTT North America, Inc.*)

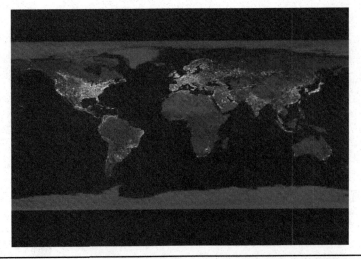

Figure 5.1 Composite image showing the planetary distribution of nighttime lights emitting visible and near-IR wavelengths. (*Credit: NASA, www.nasa.gov.*)

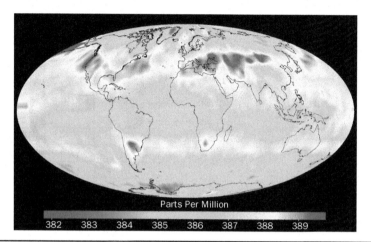

Figure 5.2 Composite image showing the planetary distribution of carbon dioxide as inferred from the atmospheric emission of MWIR and LWIR wavelengths. (*Credit: NASA Jet Propulsion Lab, http://airs.jpl.nasa.gov.*)

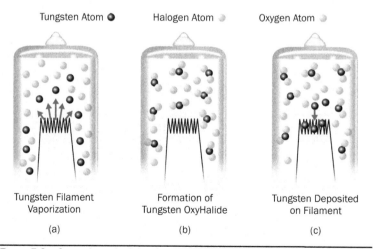

Figure 5.9 Operating principle of the tungsten halogen lamp: evaporated tungsten is redeposited on to the filament by the halogen gas. (*Credit: Carl Zeiss, Inc.*)

Figure 6.48
Operability is reduced in this FPA by hot pixels, dead pixels, dead clusters, and dead columns; dead rows are also possible. The operability of this array is only 83 percent; a value exceeding 99 percent is more common.

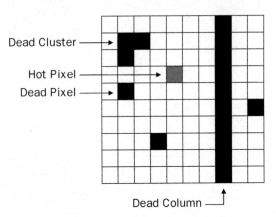

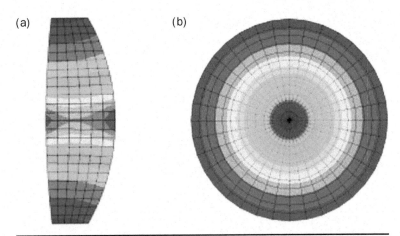

Figure 7.22 Temperature distribution (a) and average WFE (b) of a lens like the one shown in Fig. 7.19; the center is hotter owing to heat transfer out toward the heat sink at the edges. (*Credit: Sigmadyne, Inc., www.sigmadyne.com.*)

CPSIA information can be obtained
at www.ICGtesting.com
Printed in the USA
LVOW01*0829261016
509774LV00001B/1/P

9 780071 754408